P9-AGE-055

Statistics for Social Data Analysis

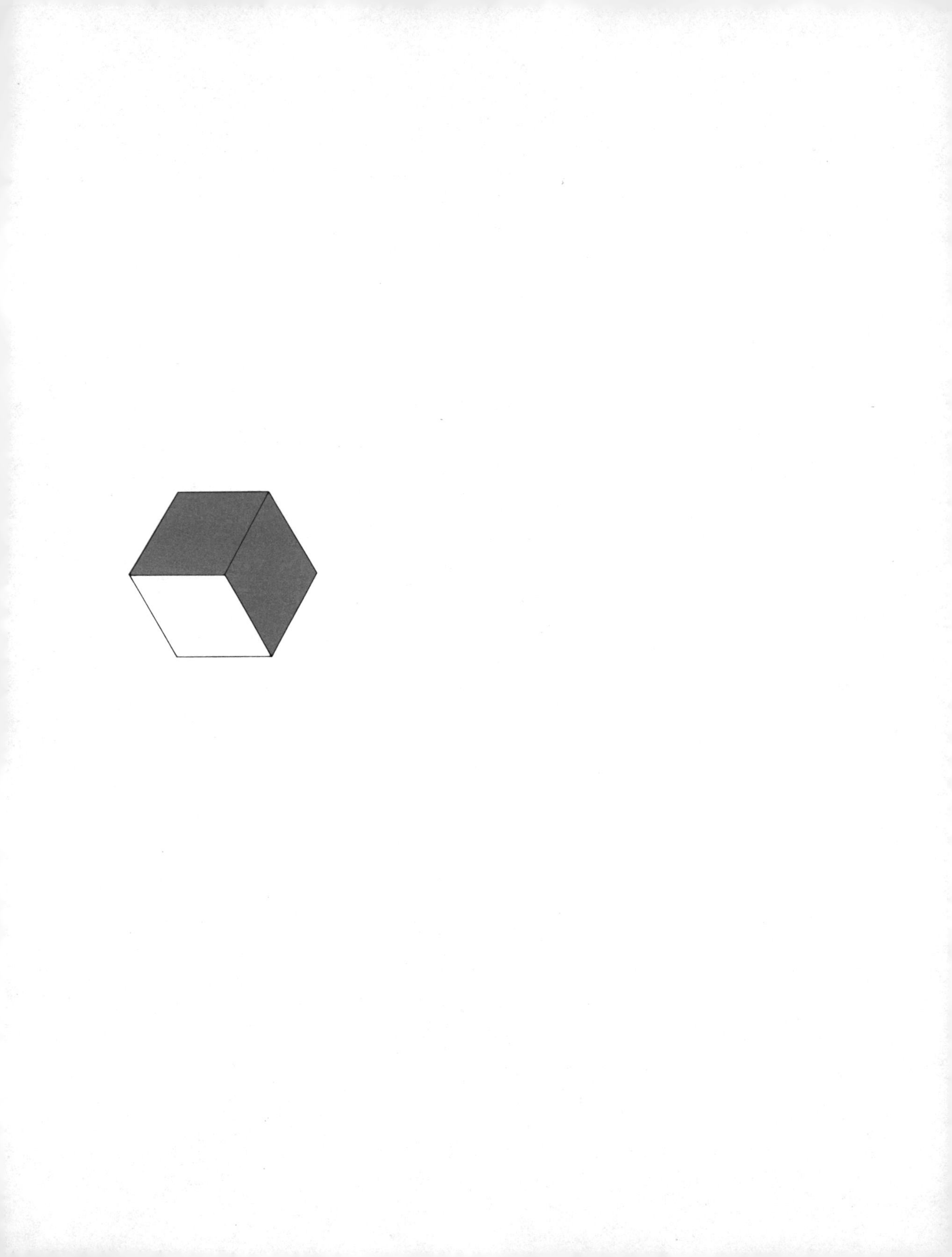

STATISTICS FOR SOCIAL DATA ANALYSIS

George W. Bohrnstedt / Indiana University

David Knoke / Indiana University

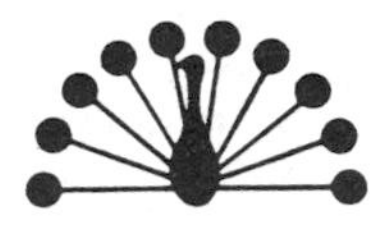

F.E. PEACOCK PUBLISHERS, INC. ITASCA, ILLINOIS 60143

For our teachers,
Edgar F. Borgatta and
David R. Segal

Cover: TUPA-F by Victor Vasarely, courtesy of the Vasarely Center, N.Y.

Copyright © 1982
F.E. Peacock Publishers, Inc.
All rights reserved
Library of Congress
Catalog Card No. 81-82889
ISBN 0-87581-275-9
Printed in the U.S.A.
Third Printing, 1984

Abbreviated Contents

Table of Contents

Preface

Every year several new textbooks on statistics appear, and most of them sound a variation on a familiar theme. This text is a sharp departure from the typical format, one we feel is more in keeping with the ways social scientists use statistics as tools for solving research problems. Although our approach is statistically correct, it does not burden the student with excessive proofs and theorems. We have not included the obligatory chapters on elementary probability; instead we have brought in this underlying mathematical basis of statistical inference whenever it supplies an essential insight into the application of statistical reasoning to social scientific problems.

The emphasis throughout this book is on relationships among variables, because covariation among measured constructs forms the theoretical and empirical heart of any social science. Students are introduced to the concept of crosstabulation early (Chapter 4), and statistical inference from samples to populations is emphasized from the beginning. A special appendix explaining summation rules (Appendix A) is included.

Instead of following the usual format of presenting statistics for nominal, ordinal, interval, and ratio levels of measurement, we have presented statistics for either continuous or discrete variables. We have taken this approach because we believe it is consistent with current practice as exemplified in the major professional journals. If social science students are to be able to read and understand the contemporary research literature, it is essential that they have a firm grasp of the dominant perspective in statistical methodology.

The most important departure from tradition in this book is its requirement that students constantly think in terms of real research questions. A hands-on approach to statistics through actual data analysis is the only way really to acquire a feel for what social research is all about. To that end, we have constructed this book, with few exceptions, around two real data sets that are accessible to students. The General Social Survey (GSS) was widely diffused during the past decade and can be acquired by any college or university at small cost, relative to the enormous benefits students and teachers can gain through classroom use. For small-sample analysis, and for those to whom access to the GSS is not possible, the 63-cities study data set is included as an appendix, and it is available in punch-card form from the publisher. Using either or both of these data sets can give students a first-hand appreciation for how theory and data work in tandem to illuminate the social world.

This text was written primarily with an undergraduate audience in mind, but the orientation and introduction to advanced topics (multivariate regression, path analysis) should lead directly into graduate statistics courses as well. Graduate students who have not had undergraduate training in statistics will find this book a particularly useful introduction.

We are indebted to A. Hald for reprint of Area Under the Normal Curve and E. S. Pearson and H. O. Hartley for reprint of the F Distribution Table. We are also grateful to the Literary Executor of the late Sir Ronald A. Fisher, F.R.S., to Dr. Frank Yates, F.R.S., and to Longman Group Ltd. London, for permission to reprint Tables III and IV from their book *Statistical Tables for Biological, Agricultural and Medical Research* (6th edition, 1974).

In addition, we acknowledge permission from SPSS Inc. to use their software package SPSS Batch System* throughout this book to illustrate the analysis of survey data with the computer. We also acknowledge the cooperation of National Opinion Research Center and the Roper Opinion Research Center for making available the General Social Surveys, and the Inter-University Consortium for Political and Social Research for providing the 63-cities study, both of which are used extensively in the book.

As with any large enterprise, many people assisted us in various capacities. We thank them here, realizing that public acknowledgment is a poor token for the great debt we owe them.

For their patient and enduring help in putting the manuscript through its many versions, we thank Susan Cohen, Rosalee Harris, Stephanie Huddle, Lois Kelly, and Katrina Wallace. We particularly appreciate the hard work and continual flow of encouragement by our publishers, particularly F. Edward Peacock, Gloria Reardon, and Joyce Usher.

The following people gave very useful comments on portions of the manuscript: Richard A. Berk, Jennifer Chaney Bohrnstedt, John Gaito, Maureen Hallinan, Craig B. Little, Rae Newton, Ira Reiss, and Robert Somers. The encouragement of Richard T. Campbell and Gerald Marwell is also appreciated. Finally, but not least, we are especially grateful to Edgar F. Borgatta for constantly prodding us to break from the pack and dare to write a statistics book that reflects the realities of contemporary social science research.

Bloomington, Indiana
December 1981

George W. Bohrnstedt
David Knoke

---

*SPSS is a trademark of SPSS Inc. of Chicago, Illinois, for its proprietary computer software. No materials describing such software may be produced or distributed without the written permission of SPSS Inc.

Statistics for Social Data Analysis

1

The Social Research Process

The social research process often begins with questions about the world: What kind of people vote for Democrats? Do lower-income people have more children than middle-income people do? What incomes can be earned in various occupations? Why do Protestants have higher suicide rates than Catholics or Jews? Do blacks achieve less education than whites because they have lower IQs or because of other differences? Each of these questions is phrased in terms of the *relationship* between two or more observable characteristics of people or groups, such as income and occupation. We will have much to say about various relationships throughout this text, since they represent the central concept in social science.

If social research is to answer questions like these, it naturally must ask where the questions come from. Personal experience, hunch, intuition, friends' suggestions, or a variety of stimuli such as newspaper and TV accounts clearly provide relevant clues. But social scientists also have an inheritance from the past from which ideas for social research can be drawn. This inheritance is a steady accumulation of social knowledge which has been painstakingly assembled by several generations of psychologists, political scientists, sociologists, anthropologists, and economists, as well as applied researchers in education, business, and law. Together, their writings contain many theories and empirical findings about social phenomena. A student's training in the social sciences begins with an introduction to the theoretical ideas of the great masters. The thoughts of Aristotle, Emile Durkheim, Karl Marx, Max Weber, John M. Keynes, Alfred Marshall, Charles Merriam, Arthur Bentley, Bronislaw Malinowski, Sir James Frazer, and other founding fathers of social science are a source of continuing inspiration for researchers. The importance of more contemporary if less renowned social scientists in providing ideas to be tested also must be recognized.

At its best, social science is firmly grounded in the real world. It seeks to explain social behavior, but it is distinct from related fields like social philosophy and theology which deal with idealizations that have few empirical referents. The more comprehensive social theories present distinctive views of reality which are sometimes labeled *paradigms,* or examples or patterns.[1] Another term which is used frequently is *model.* Paradigms are usually seen as broader and more encompassing than models, but both are abstractions and simplifications of the complex real world. Partitions of the seamless totality are essential if a theory is to be of any use in guiding social inquiry. No theory can seek to account meaningfully for all the significant aspects of social life. Instead, selective attention must be given to a few aspects of the phenomena to be explained. As a result, theory deals with only a part of the world and takes the rest for granted or, at least, assumes it to be sufficiently unobtrusive so it can be safely ignored while concentrating on the topic of interest.

Examples of such theoretical abstractions abound. One of the most popular in psychology is the stimulus-response, or S-R, paradigm. In B. F. Skinner's operant psychology theory, behavior is seen as purely reactive to external stimuli.[2] It posits that all behavior is a response to external stimuli, and there is no need to consider the mediating mental processes. In contrast, psychoanalytic explanations of social behavior rely on extensive, elaborate mechanisms of internal processes like Freud's trinity of id, ego, and superego. Many of these processes are unconscious, and their existence is inferred by observing patients' behaviors in dreams, slips of the tongue, and neurotic compulsions. Though the S-R and the psychoanalytic theories of behavior have markedly divergent elements, both concentrate their explanations on a few key aspects of reality and leave other features aside.

---

1. Thomas Kuhn, *The Structure of Scientific Revolutions* (Chicago: University of Chicago Press, 1962).

2. B. F. Skinner, *Science and Human Behavior* (New York: Macmillan Co., 1953).

Such analytic abstractions, we must stress, are true of all social theories. Even grandiose constructions such as the action theory of Talcott Parsons, with its elaborate exposition, necessarily focus attention on only a limited part of the near infinity of possible phenomena which could be investigated. This restriction may appear to prevent a faithful portrayal of social behavior, but it makes the research process manageable. It also is one reason science is so different from art, which attempts more complete representations of reality. If we had to take everything into account before we could say something about anything, we would never get started, and the results would be indigestible. Theoretical simplifications are a blessing which allow social scientists to state succinctly what they are going to study and what they expect to find, even though they are aware that their theses and models are incomplete representations of the real world.

1.1. What, Exactly, Is a Theory?

Up to this point, we have used the terms *theory, paradigm,* and *model* without being especially precise. You may be wondering exactly what a theory is, though no doubt you have heard casual remarks such as "I've got a theory about that," or "My theory about why they broke off their engagement is that they couldn't stand the same type of music." Such common uses of the term *theory*—as general ideas which explain events or conditions—capture only part of what social scientists mean by a theory. To the scientist, theory is a more precise explanation described in fairly abstract terms.

A **social theory** is a set of two or more propositions in which concepts referring to certain social phenomena are assumed to be causally related. While all social scientists do not accept this definition, the sentiments it expresses are fairly widely shared, and we will use it in this text.

social theory—a set of two or more propositions in which concepts referring to certain social phenomena are assumed to be causally related

proposition—a statement about the relationship between abstract concepts

concept—person, object, relationship, or event which is the referent of a social theory

relationship—a connection between two concepts or variables, of either a covariational or causal nature

The definition has several components. Any theory involves two or more statements, which may be written in a common language such as English or in the more elegant shorthand languages of symbolic logic or mathematics. These statements, which are called **propositions**, are comprised of two essential elements: **concepts** and their **relationship.** In this book propositions are designated by the initial *P* and a number, as in the following example:

A Politicoeconomic Mini Theory

P1: Economic instability generates disaffection with the national political regime.

P2: Disaffection with the national regime strengthens the opposition political forces.

By our definition, these two propositions qualify as a theory, in this case a theory explaining national political cycles in terms of economic factors. Notice that the theory says nothing about the effects of other factors, such as religious revivalism or demographic changes. Additional propositions could be added to this simple theory to take these factors into account if we wished to be more comprehensive about the alleged causes of the political cycle. Propositions 1 and 2 together constitute a *minimal* theory, however. All other factors are assumed to be inoperative, *ceteris paribus* ("everything else being equal"—a phrase social scientists seem to scatter about like birdseed).

The politicoeconomic mini theory contains three concepts: economic instability, regime dissatisfaction, and opposition strength. These concepts are what the theory is all about. They are objects or activities capable of existing in more than one state or condition (that is, they are not constant). An economy can exhibit varying degrees of stability or instability, and people's dissatisfaction or opposition can cause various changes in the levels or rates. Presumably the theorist's definitions of these concepts are known to those who are reading or researching the theory. Often theoretical concepts have been extensively discussed and argued by specialists in some area, so there is some shared understanding of the terms being used. But social science concepts often have meanings that are not the same as those used in everyday language. For example, status, power, culture, and response mean quite different things to sociologists, political scientists, anthropologists, and psychologists than they do to the person in the street. To be useful in theoretical discussion, therefore, a concept must be carefully and rigorously defined. Otherwise two scientists who presume they are studying the same social phenomenon may in fact be looking at two quite different things. Obviously such a situation does not promote an orderly accumulation of facts about the social world.

But a theory is more than just a list of concepts. Our definition says that a theory describes the relationship between component concepts, and our two propositions depict the relationships in an implicit, *causal* fashion. First, economic turmoil increases the level of dissatisfaction with national officeholders. As alienation grows, the political groups in opposition gain more support (presumably from the disaffected populace). Note that P1 and P2 have one concept in common—dissatisfaction—which links the propositions together and allows us to make a simple **deduction**. This is:

deduction—process of deriving a conclusion about relationships among concepts by logical reasoning about their connections to common concepts

P3: Economic instability increases the strength of the political opposition.

This ability to trace the connections between concepts makes theoretical propositions more than a haphazard assemblage of sentences. While every concept cannot be linked deductively with all others, every proposition within a theory should be connected to at least one other, so they form an interconnected complex.

A theory is often intended to apply only to certain conditions or objects. A theory's **scope conditions** include the time and place within which the phenomena supposedly occur. Marx's class struggle explanation is most relevant to industrial capitalism, for example. Cultural diffusion models may be limited to Southwestern pre-Columbian societies, but the microeconomic theory of the business corporation can be adapted to nonprofit organizations. A theory's **units of analysis** may be societies, nations, complex organizations, communities, statistical aggregates, small groups, families, individuals, or even personality traits. Propositions intended to hold at one level of analysis may not apply at a different level. For example, among individuals high education levels generally increase the rate of voting in elections. But in comparing communities, inner-city precincts, where the people are generally less well educated, may have higher voting rates than the suburbs, where the people are better educated, perhaps because the city political machine organizes registration and voter turnout. Theories are unfairly tested if research is conducted at the incorrect level of analysis or in an unintended time or place. All may be fair in love and war but not in social research.

scope conditions—the times, places, or units of analysis for which a social theory's propositions are expected to be valid

unit of analysis—the general level of social phenomena that is the object of observation (e.g., individual, nation)

Some social science theories are quite formal systems consisting of axioms, theorems, and logical deductive propositions.[3] Such systems are relatively rare in the social sciences, however,

3. Peter M. Blau, *Inequality and Heterogeneity: A Primitive Theory of Social Structure* (New York: Free Press, 1977).

compared to their proliferation in the natural sciences, where the core of theory often consists of extensive sets of logical, mathematical equations. Frequently social science theory is embedded in discursive essays written in everyday language. Superficially, these presentations appear to be directly accessible to ordinary readers of the English language, but on closer inspection such theories often are meaningless jargon to anyone who has not become acquainted with the background assumptions shared by social science professionals. Despite the jargon, a social theory must be capable of being put into proposition form if it is to qualify under our definition of an acceptable social theory. Any writings that do not attempt to order social science concepts about some phenomena into a coherent set of relationships cannot be considered a theory. Whether a theory is relevant to research depends on further considerations, however.

1.2. Theoretical Propositions and Operational Hypotheses

operational hypothesis —a proposition restated with observable, concrete referents or terms replacing abstract concepts

A social theory does not necessarily contain any information about how to verify its propositions. As the idealized depiction of the research cycle in Figure 1.1 shows, the next step is to generate **operational hypotheses** or working hypotheses from theory. In the social science literature the distinction between propositions and hypotheses is not always definite, and sometimes the two terms are used interchangeably. In this text the term *proposition* is used when the concepts referred to are *abstract,* and the terms *operational hypothesis, working hypothesis, research hypothesis,* or, simply, *hypothesis* are used when the concepts referred to are *concrete.* In the politicoeconomic mini theory introduced above, the three concepts are fairly abstract ideas which would permit a great deal of leeway in deciding how to detect the amount of economic instability, regime dissatisfaction, and political opposition present in a social system. Hypothesis generation restates the theoretical propositions in a form that is more relevant to actual observation in the particular social setting under study.

For example, in the United States we might decide that a good indicator of economic instability is the annual rate of inflation. Similarly, regime dissatisfaction might be indicated by the electorate's expressed attitudes about the president's economic policies, and opposition strength might mean electoral support for the party not in power in the executive branch. Of course, alternative indicators of the key concepts could be found (e.g., unemployment

FIGURE 1.1
The Research Cycle

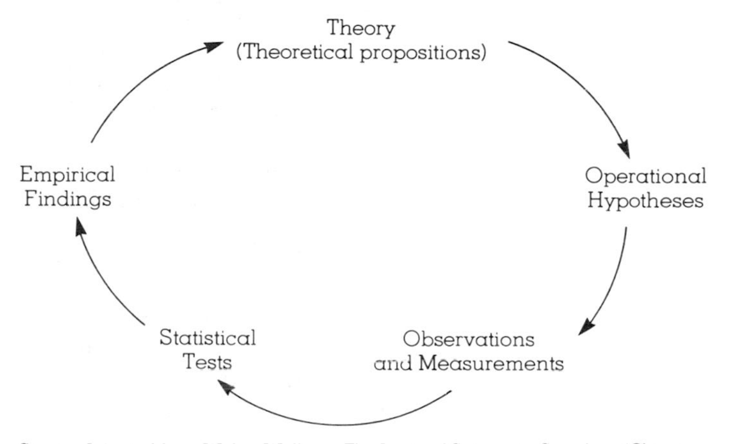

Source: Adapted from Walter Wallace, *The Logic of Science in Sociology* (Chicago: Aldine Publishing Co., 1969).

instead of inflation). And, in a different social system such as an autocratic regime, the three concepts might have different empirical referents (e.g., terrorist bombings could be used as the indicator of opposition strength). Clearly, there are many ways to turn theoretical propositions with unobservable concepts into hypotheses with empirical referents. But the process is by no means straightforward and mechanical. Indeed, much of the argumentation among social scientists flares up over the best way to formulate hypotheses for a given theory. However, *if there are several indicators of a single concept, and one is presumably as good as another, then hypotheses verified (or falsified) by one set of indicators should also be verified (or falsified) by other sets as well.* It should be clear to you why the status of a hypothesis is strengthened or weakened by using multiple indicators of its concepts.

The theoretical propositions of our theory can be restated as operational hypotheses, using the indicators designated above. Using the initial *H* instead of *P*, these hypotheses would be stated as follows:

H1: Higher rates of inflation generate negative attitudes toward the president's economic policies.

H2: Greater dissatisfaction with the president's economic policies increases the electoral support for the party not in the White House.

These hypotheses now give us something concrete to examine in the research. We are told not only what specific things to look at but how these indicators relate to each other. To be useful in promoting research on social behavior, a theory's *propositions must be translatable into testable hypotheses,* that is, statements about the relationships between *observable* phenomena. A theory whose concepts are so vague that meaningful indicators of them cannot be found would be useless for research purposes. For this reason operant psychology has generated more testable and researchable hypotheses than the psychoanalytic approach, to return to an example used above.

We would not wish to dignify some writings with the designation *theory.* Medieval scholasticism was basically a system of thought which was *atheoretical,* or without theory, and dealt with the relationships between nonobservable phenomena, such as the famous question, "How many angels can dance on the head of a pin?" While social science has become increasingly adept at spotting and weeding out such nontheories, you will often encounter such writings in your work. When you do, you should ask yourself whether the author presents a useful theory: Are the concepts related to each other, and do observable indicators exist for the more general, abstract concepts?

variable—a characteristic or attribute of persons, objects, or events that differs in value across such persons, objects, or events

constant—a value that does not change

A term which is frequently applied to the indicators in a hypothesis is *variable.* A **variable** is a characteristic of persons, objects, or events that differs in value across persons, objects, or events. Anxiety is a variable: People differ in their degrees of anxiousness. Income is a variable: People beg, borrow, steal, or earn the money they need. *Anything that does not vary cannot be a variable;* instead it is a **constant**, or a single category of a variable. Female is not a variable but a constant. But sex (gender) is a variable because there are two states into which observations can be classified.

1.3. Independent and Dependent Variables

The research scientist is interested in establishing regularities, in particular, regularities that are due to causal relationships between and among variables. These relationships are based on the assumption that change in one variable results in some predictable change in another variable. We can think of one variable as being an antecedent and another as its consequent. The antecedent variable (or variables) is called the **independent variable**, and the consequent variable is called the **dependent variable**.

independent variable— a variable that has an antecedent, or causal, role in relation to a dependent variable

dependent variable— a variable that has a consequent, or affected, role in relation to an independent variable

In its everyday meaning, the term *independent variable* is more applicable to experiments where antecedent variables can be manipulated freely by the researcher. An agricultural experimenter, for example, can freely vary the amount of fertilizer applied to various plots (independent variables) and observe the growth consequences of plants (dependent variable) for various amounts of fertilizer applied. In contrast, most of the outcomes in the social sciences can be assumed to be fixed for a given observation. A person who is selected for a study sample has a fixed religious identification, sex, and so on. But even though many variables thought to be antecedent cannot be manipulated in the social sciences, the term *independent variable* is still employed.

The terms *independent* and *dependent* are widespread in the research literature of the social sciences, and you should be certain to learn the distinction between them. Some examples may help to reinforce the distinction. If it is assumed that people's yearly earnings vary by educational attainment, then earnings is the dependent variable (consequent) and educational attainment the independent variable (antecedent). Similarly, religiosity varies by categories of religious identification; the independent variable is religious identification and the dependent variable is religiosity. For such independent variables, a statement that "A change in X creates a change in Y" makes no sense.

Variables that cannot be manipulated are sometimes called **status variables**. A person's race simply cannot be changed, for example. Nevertheless, different positions on a status variable (e.g., race) can often be related to different outcomes on another variable (e.g., yearly earnings). Thus even though race cannot be manipulated, it still makes sense to think of it as an independent variable. A better approach to status variables, instead of thinking in terms of changes in one variable producing changes in another, is to think in terms of various outcomes of the dependent variable as a function of the various conditions of the independent variable.

status variable— a variable whose outcomes cannot be manipulated

The difference between an independent variable and a dependent variable is specific to a given hypothesis. Thus, in our example on the relationships among political and economic variables, attitude toward the president's policies is the dependent variable in H1, since it is being explained. But attitude is the independent variable in H2, since it explains, or is antecedent to, electoral support.

1.4. Rejecting Hypotheses

The reason for stressing the production of operational hypotheses from theoretical propositions is simple but fundamental: *Unless it is possible to reject erroneous statements about the real world, it will not be possible to advance our theoretical knowledge of social behavior.* The idea of rejecting or falsifying propositions is an important one, since, strictly speaking, *it can never directly be proven that a proposition is true.* To try to do so would imply that the results of a single research study would hold across all time, all persons, and all cultures—and this simply cannot be known from a single study. Therefore "truth" is a goal which can be approached only incrementally. A proposition that has failed to be falsified in many studies across time and in several different cultures has more truth value than one which has been supported by only a single study. When a proposition has been falsified in a study, the implication is that the proposition is not true in general, although it *might* be true for certain groups at certain times. While social scientists hope that in the social world there are propositions that hold across time and space, they work in such an imprecise science that they cannot yet be certain that universality is the case. The important point is that falsification is easier than verification in science, and for this reason truth is approached only indirectly.

As testable statements about relationships between observable variables, hypotheses provide a capability to reject propositions. If the variables in a hypothesis are acceptable referents for the concepts of theoretical propositions, the results of research on relationships between the variables will tell something about the truth value of the theoretical propositions from which the hypotheses originally came. If in empirical analysis the hypotheses are found to be incorrect, the truth of the theory will be in doubt, at least insofar as the situation investigated is concerned. If, on the other hand, the findings are consistent with the hypothetical expectations, our belief will be strengthened that the theory helps explain something about the social world.

Imagine a theory about the effect of a certain type of teaching method on pupils' learning abilities. Suppose a well-designed experiment finds that students taught under the new method do better on a standard test of knowledge than students taught with several other methods. These results might be obtained across many replications of the experiment. Still the researchers could not conclude that the learning theory from which these hypotheses were generated has been conclusively proven. A single instance in which the new instruction method failed to yield the hypothesized outcome would cause its unqualified validity to be rejected. And, since even a very extensive experimental sequence cannot examine all possible combinations of students, subject matter, teacher experience, and so on, the possibility remains—no matter how remote and improbable—that the hypotheses one day might not be supported.

Later in this text we will show how the probabilities of error are taken into account in statistical tests. Here we are concerned with the implied strategy of hypothesis testing. Research cannot be conducted under all the relevant conditions for all the relevant populations (it is not possible to do research on the past or the future, for example), and, increasingly, the funds available for social research are being limited. Scientists would be foolish to adopt as a criterion the *proof* of a testable theory from evidence used to test hypotheses. It is better to adopt a criterion that can produce hypotheses capable of *rejection* under clearly specified conditions.

This strategy of building research foundations on hypothesis refutation was promoted by a philosopher of science, Karl Popper; you can read his arguments in the original.[4] Popper's ideal theory is one that clearly spells out in advance exactly what sort of data and results it would be necessary to find in order to reject its claims. In the natural sciences such conditions are fairly easy to find, and there is a lengthy history of theoretical progress through *critical experiments.* In these tests, the hypotheses produced by competing theories can be lined up in direct contradiction to each other, the experimental or observational data relevant to the test can be collected, and the results will refute all but one of the hypotheses. The findings can strike a devastating blow against entire theories from which the erroneous hypotheses were deduced. For example, the solar eclipse of 1911 provided evidence against Newton's theory but was consistent with Einstein's theory of the universe.

---

4. Karl Popper, *The Logic of Scientific Discovery* (New York: Basic Books, 1959).

Despite the appealing simplicity of this falsification approach to scientific rationalism, its application to social science has been open to question. The reasons are many, including imprecise concepts, propositions that use concepts without empirical referents, theories of uncertain scope or coverage, focuses on different levels of explanation, and the often debatable quality of measurement of the critical variables.

Despite these problems, we hold to the idea that the cumulation of reliable, useful knowledge in the social sciences depends on hypotheses for which refutation is sought. In sum, hypotheses should be stated so that they are capable of rejection by empirical evidence.

1.5. Operationalization

After deciding on a theory's concepts, the researcher must *operationalize* them into a set of observable variables. An **operation** is a method for observing and recording those aspects of persons, objects, or events that are relevant to testing a hypothesis.

operation—any method for observing and recording those aspects of persons, objects, or events that are relevant to testing a hypothesis

Researchers have invented and developed a wide variety of operations to meet different research needs. An entire volume would be required to detail the various approaches—historical records and archival documents; pottery shards and artifacts; participant observations; survey and questionnaire responses; content analyses of verbal exchanges; unobtrusive observations; videotapes; automatic experimental tabulations; brain-wave and galvanic skin responses. The list is potentially endless, limited only by social scientific imagination and inventiveness. These operations are all routinized, widely shared sets of activities. The forms in which observations are recorded are quite diverse, ranging from the rambling narratives of raw field notes in participant-observer studies to the highly specialized numerical coding schemes of small-group interactions in laboratory settings.

Establishing linkages between concepts and their empirical referents is a critically important step in research. Researchers want to be certain that the operations have validity, that is, they result in *valid measures* of a theory's concepts. **Validity** refers to the degree to which an operation results in a measure that accurately reflects the concept it is intended to measure.

validity—the degree to which an operation results in a measure that accurately reflects the concept it is intended to measure

> A synonym for validity is *accuracy.* To the degree that an operation results in observable measures that are accurate representations of a theory's concepts, the resulting measures are said to be *valid.*

In the politicoeconomic mini theory introduced in Section 1.1, the first proposition is:

P1: Economic instability generates disaffection with the national political regime.

The two concepts in this proposition are *economic instability* and *disaffection with the national regime.* The operational hypothesis later suggested as a test of P1 is:

H1: Higher rates of inflation generate negative attitudes toward the president's economic policies.

The concept *economic instability* is operationally measured in H1 by *rates of inflation* and the concept *disaffection with the national political regime* is measured by *negative attitudes toward the president's economic policies.* The concept rates of inflation is valid to the degree it accurately measures economic instability. And the concept negative attitudes toward the president's economic policies is valid to the degree it accurately measures disaffection with the national regime.

The topic of validity is extremely important in research, but a discussion of the methods for assessing validity is beyond the scope of this book. Good discussions can be found in other texts.[5]

Researchers must have measures that are *reliable* as well as valid. **Reliability** refers to the degree to which different operations of the *same* concept yield the same results.

reliability—the degree to which different operations of the same concept yield the same results

> Reliability refers to the degree to which observations of a study are *repeatable.* A measuring instrument is said to be reliable according to the degree to which it generates *consistent* observations at two points in time. Or a measure is reliable to the degree that two different researchers using the same instrument on the same sample would generate the same observations.

5. Recommended references are Nan Lin, *Foundations of Social Research* (New York: McGraw-Hill Book Co., 1976), and Jum C. Nunnally, *Psychometric Theory* (New York: McGraw-Hill Book Co., 1967).

Another common approach is to use several similar but not identical measures of the same concept. Industrialization, for example, might be measured by using various countries' gross national products and kilowatt hours consumed. Or people's religiosity might be measured by ascertaining their belief not only in an Almighty God but also in a life after death. To the degree that countries on the one hand and people on the other are consistently rank ordered using the two different indicators, the measurement approach is said to be reliable.

> If the observations in social research are highly unreliable, the ability to make any meaningful conclusions about the validity of social theories is seriously impaired. *Indeed, a measure of a concept cannot be valid if it is unreliable. However, a measure can be reliable (or consistent) without being valid (or a true measure).*

A bathroom scale, for example, is both *reliable and valid* if on a given morning you get on it several times and it records your weight the same each time and if the weight it records is accurate. The scale is *reliable and invalid* if it records your weight the same each time you get on but it consistently indicates you weigh five pounds below your true weight. And if the scale indicates a different weight each time you get on it, it is not only *unreliable* but clearly *invalid* as well. It cannot be valid if it indicates a different weight each time you step on it, because it does not accurately measure the concept (your weight) it is intended to measure.

1.6. Measurement

Observations and measurements are distinct processes, though we show them together in the diagram of the research cycle in Figure 1.1. Once a researcher has determined how to operationalize the variables in the hypotheses under investigation and what sample of units to observe, he or she must decide how numbers are to be assigned to the information collected so that statistical interpretation of the findings is possible. The assignment of numbers to observations according to a set of rules is called **measurement**. By translating systematic observations into a numerical system, measurement provides a means whereby mathematical

measurement—the assignment of numbers to observations according to a set of rules

manipulations can be used to assess the validity of hypothesized relationships between variables. The measurement process is one more step in the reduction of complex reality to more manageable simplicity.

The types of numbers which can be assigned to observations depend on the (assumed) underlying characteristics of the phenomena to be studied. There are two broad classes of measurements: discrete and continuous.

Discrete variables classify persons, objects, or events according to the kind or quality of their attributes. The simplest type of discrete measure is the *dichotomy,* in which observations are classified only according to whether the defining attribute is present or absent. Examples are: dead or not dead; male or not male; black or not black. Other discrete attributes are *multicategorical* (or polychotomous); that is, there are more than two outcomes. Examples are: hair color classified as blond, brunette, or redhead; country of national origin; county of residence. A further distinction can be made among discrete variables according to whether or not their outcomes are orderable. **Orderable discrete variables** have outcomes that can be ranked from low to high, or vice versa. An example of an orderable discrete variable is "How well is President Reagan doing his job?" The response categories might be: "excellent, good, fair, or poor." **Nonorderable discrete variables** cannot be ranked; examples include race, ethnicity, sex, and national origin.

discrete variable—a variable that classifies persons, objects, or events according to the kind or quality of their attributes

orderable discrete variable—a discrete measure in which the categories are arranged from smallest to largest, or vice versa

nonorderable discrete variable—a discrete measure in which the sequence of categories cannot be meaningfully ordered

Continuous variables classify persons, objects, or events according to the magnitude or quantity of their attributes. The principal difference between continuous and discrete variables is that *continuous measures allow for fractional numeric values, whereas discrete measures do not.* In the natural sciences continuous variables abound: weight, height, time, velocity, and so on. In the social sciences many concepts also are assumed to be continuous; industrialization, physical attractiveness, and political liberalism are three examples. While the concepts are clearly continuous, however, their operational measures often only approach this goal. As a result, considerably more measurement error occurs in the social sciences. Social research must rely on crude questionnaires, rating scales and the like, because the available instruments nowhere approach the accuracy of the beam scales and electron microscopes used in the physical sciences. Nevertheless,

continuous variable—a variable that, in theory, can take on all possible numerical values in a given interval

percentages and proportions do serve as continuous measures. The percentage of 18-year-olds who have ever smoked marijuana, the rate of bankruptcy among small businesses, and the proportion of registered voters who actually vote in a given election are examples.

With continuous and discrete variables alike, observations must be assigned to categories that are *both* mutually exclusive and exhaustive. **Mutual exclusiveness** refers to the need to place each observation in one and only one of the variable's categories. **Exhaustiveness** means that sufficient categories must exist for all observations to be classified into some category.

mutual exclusiveness—a property of a classification system that places each observation in one and only one category of a variable

exhaustiveness—a property of a classification system that provides sufficient categories so that *all* observations can be located in some category

level of measurement—a classification of measurement scales according to the amount of information about observations recorded. This information generates four types of scales: nominal, ordinal, interval and ratio

nominal level of measurement—scale that assigns a name or number to observations in purely arbitrary sequence

ordinal level of measurement—scale that assigns numbers to observations in sequence, from lesser to greater amounts of the measured attribute

interval level of measurement—scale that assigns numbers to observations that reflect a constant unit length between categories

ratio level of measurement—scale that assigns numbers to observations to reflect the existence of a true, or absolute, zero point

For example, if in measuring religious identification the categories Protestant, Catholic, and Baptist were used, the mutual-exclusiveness criterion would be violated, since Baptists are Protestants as well. If only the Catholic and Protestant categories were used, then the exhaustiveness criterion would be violated, since there is no category for Jews. Better categories for this variable would be: "Catholic, Protestant, Jew, other, or not ascertained." In some cases it might be better still to break the Protestant category up into several additional categories.

While "other," "not determined," or "uncertain" are convenient and often necessary categories for the recording of observations, *statistical models assume that phenomena classified the same way are in fact identical or nearly identical in how they relate to other variables.* For this reason it is better to use refined, well-thought-out categories. As much as possible, the lumping together of disparate outcomes into a single "other" category should be avoided.

The distinction between discrete and continuous variables or measures will be maintained throughout our discussion of statistical methods. Many older textbooks on social statistics devoted considerable attention to the idea of four **levels of measurement: nominal, ordinal, interval, and ratio scales.** This approach to measurement was popularized decades ago by the psychophysicist S. S. Stevens,[6] and for some time data analysts were urged to use only certain types of statistics with data measured at each of the different levels. In the approach we have

6. S. S. Stevens, *Psychophysics: Introduction to Its Perceptual, Neural, and Social Prospects* (New York: John Wiley & Sons, 1975).

adopted for this text, the levels are considered important to measurement theory, but they have no relevance to the proper choice of statistical techniques to be used to analyze social data. In modern practice, data analysts ignore the levels in favor of the simpler discrete-continuous distinction. In the brief discussion of the four measurement scales in Box 1.1, they are identified as discrete or continuous (or, in the case of ordinal scales, as both). *The distinction between discrete and continuous variables is the most meaningful criterion for the selection of statistical techniques.*

1.7. Statistical Tests

After operationalizing the measures and making observations, the researcher's next step in the research cycle (as shown in Figure 1.1) is the performance of statistical tests, or calculations. We will not say much about statistical analysis in this chapter, since the rest of the text is a detailed examination of various techniques. As a brief introduction, however, three broad categories of statistical usage can be distinguished: descriptive statistics, measures of association, and inference.

1.7.1. Descriptive Statistics

Single variables may be *described* according to their numerical properties. Newspaper and TV accounts use such figures as *averages* (mean rainfall for June; average take-home pay for workers), *rates* (annual unemployment rate; monthly rate of inflation), *proportions* (percentage of blacks earning over $30,000; proportion of first marriages ending in divorce), and *frequency counts* (populations of the largest states; number of intercontinental missiles stockpiled by major powers). Such **descriptive statistics** form the bedrock for more advanced techniques.

descriptive statistics—numbers that describe features of a set of observations; examples are percentages, modes, variances, and correlations

1.7.2. Measures of Association

The core of our approach to presenting statistics as tools for data analysis is the analysis of the *relationships* between two or more variables. The relationships to be analyzed are drawn from the research hypotheses generated by a social theory. An entire branch of statistics is concerned with **measures of association** or covariation among variables. These techniques make it possible to determine the empirical support for hypotheses such as

measures of association—statistics that show the direction and/or magnitude of a relationship between variables

BOX 1.1 Levels of Measurement

S. S. Steven's levels of measurement (which he called *scales*) are, in sequence from lowest to highest: (1) nominal, (2) ordinal, (3) interval, and (4) ratio. In general, the higher the level of the scale, the greater the information about the differences among variables measured by that scale.

A *nominal scale* assigns names or numbers to classes of outcomes in a purely arbitrary sequence. For example, the home states of students in your class can be presented in alphabetical or geographical sequence, with no effect on the number of students found in each category. The outcomes of a nominal scale are not inherently orderable.

The categories in an *ordinal scale* are arranged in a sequence from lesser to greater amounts of the measured attribute. No assumption about a constant difference between adjacent categories is made. Many social science attitude scales are ordinal measures which ask respondents to indicate the extent of their agreement with some statement.

An *interval scale* implies a constant unit length between categories. Any numbers assigned to the categories on this scale must reflect this distance information. The National Opinion Research Center (NORC) occupational prestige scale is a good example: The difference between two occupations rated at 80 and 60 in prestige is presumably the same difference as that between occupations rated 55 and 35, that is, 20 prestige points.

The *ratio scale* is defined by the presence of a true, or absolute, zero point: 0 on a ratio scale means no quantity of the attribute is present. People's ages or annual incomes in dollars are measured at the ratio level.

In terms of basic discrete-continuous measurement, the interval and ratio scales are continuous measures, and their numerical values can be analyzed by all statistics suitable to continuous variables. Nominal scales are clearly discrete variables, since their categories are arranged in an arbitrary sequence. Ordinal scales can be treated as discrete measures, particularly when only a handful of categories are involved. However, *since most ordinal scales are assumed to reflect an underlying concept which is continuous, their measures will often be treated as if they were continuous as well.*

"The more years of formal schooling, the more income people earn," or "The greater the test anxiety, the lower the test performance."

1.7.3. Inference

Much social research is performed on a **sample**, or limited number, of observations rather than observations of an entire **population**, or universe, of persons, objects, or events. Typically, the researcher collects data on some portion of all units of analysis but would like to be able to say something about the larger group from which the sample was drawn. Since selecting a sample involves a certain element of chance—who gets into or gets left out of the sample is the outcome of a probability process—it cannot be simply concluded that the results of a sample analysis hold for the population as a whole. A different sample drawn from the same population might have yielded a different set of findings. An entire statistical field addresses the problem of **inference**, or making generalizations from descriptive statistics based on sample data to the population parameters. These techniques are based on the laws of probability.

sample—a subset of cases or elements selected from a population

population—a set of persons, objects, or events having at least one common attribute to which the researcher wishes to generalize on the basis of a representative sample of observations

inference—the process of making generalizations or drawing conclusions about the attributes of a population from evidence contained in a sample

Suppose a researcher wanted to know how well you could spell, or how many of the 500,000 words in the English vocabulary you know. Obviously, you could not be asked to attempt to spell every word or give an acceptable definition of each one. The population of words is too large. But it would be possible to estimate the proportion of words in your vocabulary or the extent of your spelling skills by testing you with a **random sample** of words drawn from the dictionary. *If the sample is random and sufficiently large, the proportion of correct answers on the sample will allow an inference of how you might perform if the impossible test could have been conducted on the entire word population.*

random sample—a sample whose cases or elements are selected at random from a population

Along with descriptive statistics and measures of association, this text will expose you to the variety of appropriate inferential statistics. These statistics help determine whether the findings of the sample tested can be generalized to the population from which the sample was drawn.

1.8. Completing the Research Cycle

Statistical tests are an indication of whether or not the assertions of the research hypotheses generated by a social theory should be rejected, in line with the idea of hypothesis refutation described above. If the empirical findings do not permit the hypothesis to be rejected, there is increased confidence that the social theory is valid, at least for the population studied. If the findings cause the hypothesis to be rejected, and there is a fairly high degree of confidence that the sample results reflect the true state of affairs in the population, belief in the validity of the theory will be severely shaken. Indeed, the researcher may want to modify the theory to take into account the new findings. Such modifications, in effect, produce a new social theory, although the family resemblance to the now-discredited former theory may still be strong. This feedback of research findings into social theory completes the cycle which was opened with theoretical propositions in Figure 1.1.

Our depiction of the research cycle is a highly idealized simplification of what may be a very complex process involving many researchers working on a single social problem, with minimal coordination of their activities. We believe this idealization captures the essential spirit of the social research enterprise, but the best way for you to find out what goes on in it is to try doing some research yourself.

> There is no simple substitute for direct experience in trying to understand what any social activity is all about. We have designed this text, therefore, to provide hands-on experience in doing research through the analysis of actual data.

If you follow our examples and attempt to replicate them on topics of your choice, you will gain a greater insight into the research process than you could by merely reading about how others have done their research. Indeed, the research reports in professional journals do not convey much about the hard work, false starts, raised expectations and shattered hopes, doldrums and breakthroughs that are typically encountered on the way to discovering a few gems of social behavior. By concentrating on the statistical analysis of already collected data, this text exposes you to the heart of the research cycle—that point at which we find out whether the beliefs about social reality expressed in the hypotheses have any support. From our own experiences in learning to become social scientists (a learning experience that is never completed, as we found out while writing this book), *doing* social research can be an exhilarating or a frustrating experience—usually both! Though your experience in using this text may be at neither extreme, you are certain to gain considerable insight into an increasingly important aspect of our society.

Review of Key Concepts

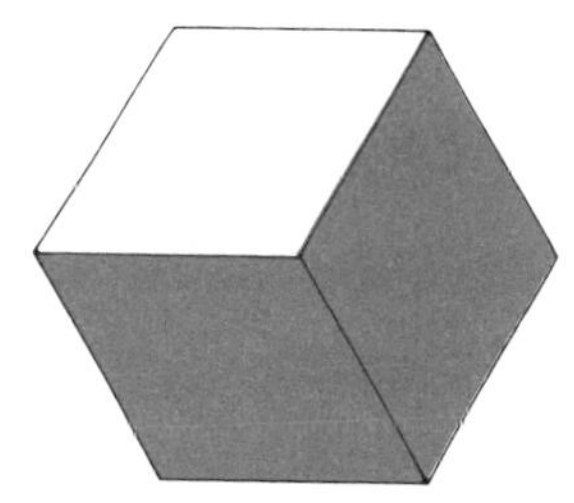

This list of the key concepts introduced in Chapter 1 is in the order of appearance in the text. Combined with the definitions in the margins, it will help you review the material and can serve as a self-test for mastery of the concepts.

social theory
proposition
concept
relationship
deduction
scope conditions
unit of analysis
operational hypothesis
variable
constant
independent variable
dependent variable
status variable
operation
validity
reliability
measurement
discrete variables
orderable discrete variables
nonorderable discrete variables
continuous variables
mutual exclusiveness
exhaustiveness
levels of measurement
nominal scale
ordinal scale
interval scale
ratio scale
descriptive statistics
measures of association
sample
population
inference
random sample

PROBLEMS

General Problems

1. Do the following three statements, taken together, constitute a social theory? Why or why not?
   a. The greater the interpersonal attraction within a group, the less the intensity of intragroup conflict.

b. Intragroup conflict is the frequency with which two or more members argue with one another in a given interval of time.
c. Task-oriented groups experience higher levels of problem solution than socioemotionally oriented groups do.

2. In each of the propositions above, identify the concepts and the nature of their relationships.

3. The following proposition goes back at least to Emile Durkheim: "An increase in group social integration causes a decrease in anomie." Restate this proposition as an operational hypothesis, using observable variables in place of the abstract concepts.

4. W. W. Rostow, in a famous paper ("The Take-Off into Self-Sustained Growth," *The Economic Journal* vol. 66, [March 1956], pp. 25–48), asserted the hypothesis that "the process of economic growth can usefully be regarded as centering on a relatively brief time interval of two or three decades when the economy and the society of which it is a part transform themselves in such ways that economic growth is, subsequently, more or less automatic. This decisive transformation is here called the take-off." What can you glean about the units of analysis and the scope conditions from these words?

5. In a classic study of the conditions of trade union democracy (S. M. Lipset, M. Trow, and J. Coleman, *Union Democracy* [Garden City, N.Y.: Anchor-Doubleday, 1956], pp. 465–67), the following two propositions appear:

   "The greater the autonomy of its member locals, the greater the chance for democracy in a trade union."

   "The greater the identification of workers with their occupations, the more likely they are to be interested and perhaps participate in the affairs of their union."

   a. What are the concepts in each proposition?
   b. What differences in units of analysis, if any, are there in these two propositions?

6. Give three operational definitions of the theoretical concept social class that could be used in a study of the social origins of American college students.

7. Alvin Gouldner ("Cosmopolitans and Locals: Toward an Analysis of Latent Social Roles," *Administrative Science Quarterly,* vol. 1 [December 1957], pp. 281–92) provided the following definitions of two latent organizational identities:

   "*Cosmopolitans:* those low on loyalty to the employing organization, high on commitment to specialized role skills, and likely to use an outer reference group orientation."

   "*Locals:* those high on loyalty to the employing organization, low on commitment to specialized role skills, and likely to use an inner reference group orientation."

   Suggest observable phenomena to be used in a study of the cosmopolitan-local orientations of college faculty members, consistent with the criteria given by Gouldner's definition.

8. Identify the following as either variables or constants:
   a. Size of community.
   b. Sexual permissiveness.
   c. Orthodox Jew.
   d. Party preference.
   e. Denver.
   f. Computer literacy.
   g. College graduate.
   h. Antiabortion demonstrator.

9. Identify the independent and dependent variables in the following hypotheses and state which are status variables:
   a. The more television a child watches, the lower her or his academic performance.
   b. Southerners have a greater tolerance for interpersonal violence than do residents of other regions.
   c. Earnings are significantly lower among black women than among black men.

10. A social scientist repeats an experiment 50 times, each time observing that men who have been exposed to sexually explicit materials (films or photographs) show higher levels of arousal than women exposed to the same material. The scientist therefore concludes that she has proven that males are always more aroused by visual stimuli than females are. Is this a reasonable conclusion? What alternative interpretation could you place on these results?

11. Briefly explain the difference between a valid measure and a reliable measure. Can a reliable measure be invalid? Explain.

12. Indicate whether the following variables are continuous, orderable discrete, or nonorderable discrete measures:
    a. Academic department size: number of full-time faculty.
    b. Region of nation: North, South, East, or West.
    c. TV watching: amount of time spent per week
    d. Crime rate: murder per 100,000 residents each year.
    e. 4–H contest results: blue, red, and yellow ribbon winners.
    f. Hometown: Indianapolis, Bloomington, Gary, other.
    g. Partisanship: Percentage voting socialist, by precinct.
    h. Self-esteem: 7-point scale from "like very much" to "hate very much."
13. *a.* What measurement criteria are violated if all the following categories are used to record respondents' occupations?
    (1). Doctor.
    (2). Teacher.
    (3). Professional, technical, and kindred.
    (4). Clerical and sales.
    (5). Craftsmen and foremen.
    b. What remedies would you suggest?
14. Using the examples in Problem 12, indicate the highest level of measurement or scale that can be attained by each variable.
15. Finish these statements:
    a. A paradigm is the shared ________________ of members of a scientific community.
    b. A concept that takes on two or more values is a ________________.
    c. The process by which an abstract scientific concept is translated into observable procedures is ________________.
    d. Evidence about the truth value of a social theory can best be gathered in an effort to ________________ the theory's null hypotheses.
    e. If an instrument satisfactorily reflects variation in the abstract concept that it was designed to measure, the instrument is ________________.
    f. To be complete, most social measurement will require at least one category into which ________________ cases can be classified.
    g. Measures of association are statistics that show the degree of ________________ between social variables.
    h. A sample is used to make ________________ about the ________________ from which it is selected.

Statistics for Social Data Analysis

Frequency Distributions 2

The usual reason a social science researcher collects data is in order to test hypothesized relationships between two or more variables. The results of these tests, which are performed with the various statistical procedures we will introduce you to throughout this text, determine whether the operational hypotheses about the aspects of social behavior under study should be rejected or supported, as we noted in Chapter 1.

2.1. Constructing a Frequency Distribution

A preliminary step in data collection is to determine the number of observations in each response category for the variables. Suppose we want to learn the proportion of heads of households who identify themselves with each major political party. We choose a random sample of 125 heads of households in Smalltown. Then we make a **tally** of the number of persons aligned with each party and use the results to construct a **frequency distribution,** which is a table of the **outcomes,** or response categories, of a variable and the number of times each outcome is observed in the sample. Table 2.1, for example, shows there are 37 Republicans, 47 Democrats, 23 independents, and 18 others in the sample. On this basis, we conclude that the town has more Democrats than members of any other political party.

tally—a count of the frequency of outcomes observed for a variable or the frequency of joint outcomes of several variables

frequency distribution—a table of the outcomes of a variable and the number of times each outcome is observed in a sample

outcome—a response category of a variable

While this is a perfectly reasonable conclusion, scientists are more often interested in how the frequencies of a distribution are

TABLE 2.1
Frequency Distribution: Political Party Identification Tally for Smalltown

Party Identification	*Tally*	*Frequency*
Republican	卌 卌 卌 卌 卌 卌 卌 II	37
Democrat	卌 卌 卌 卌 卌 卌 卌 卌 卌 II	47
Independent	卌 卌 卌 卌 III	23
Other	卌 卌 卌 III	18
Total		125

related, since they want to be able to compare the results from one sample to those of other samples. For purposes of comparison, suppose we also study Bigtown by sampling 230 persons. If we again find 37 Republicans, this is proportionately a smaller number than it is in the sample of 125 we used in Smalltown. To see this clearly, we must examine the relative frequency or proportion of cases for each outcome of the variable of interest. **Relative frequencies,** or **proportions,** are formed by dividing the cases associated with each outcome by the total number of cases. In Smalltown the proportion of Republicans is $37/125 = 0.296$; Democrats, $47/125 = 0.376$; independents, $23/125 = 0.184$; and others, $18/125 = 0.144$. We can again conclude that Smalltown has a higher proportion of Democrats than followers of other political parties or independents. Now, however, comparisons with different-size samples are easier to make and more meaningful.

relative frequency (proportion)—a number formed by dividing the cases associated with an outcome of a variable by the total number of cases

2.1.1. Percentage Frequency Distributions

A more familiar interpretation of these findings can be made by creating a percentage for the proportion of each outcome. **Percentages** are created from proportions by multiplying each one by 100. By calculating percentages we *standardize* for sample size by indicating the number of observations that would fall in each outcome of a variable if the total number of cases were 100.

percentage—a number created by multiplying a proportion by 100

Of every 100 persons in the Smalltown sample, 29.6 are Republicans. Since it makes no sense to consider six-tenths of a person, we can do one of two things. First, we can round our results up or down. Convention varies on how to do this, but generally values of 0.1 to 0.4 are rounded down to the next whole number, and values of 0.5 to 0.9 are rounded up. (More rules for rounding are given in Section 2.4.2.) Thus we conclude that 30 of every 100 persons in our Smalltown sample are Republicans, 38 are Democrats, 18 are independents, and 14 represent other parties. Second, we can multiply each percentage by 10 and say that *if* we had observed 1,000 persons, we would expect 296 to be Republicans, 376 Democrats, 184 independents, and 144 others.

With either procedure, an important point is evident: *Whenever data are summarized, some distortion almost always occurs.* The trade-off in understanding and interpretation, however, usually makes this distortion worthwhile. It is easier to comprehend that 29.6% of Smalltown's residents are Republicans than to make sense of the fact that 37 out of 125 residents are Republicans. For this reason statisticians and researchers quickly get used to the idea of dealing with fractions of persons.

Percentages usually are arrayed in a **percentage frequency distribution.** The frequency data from the tally in Table 2.1, for example, are displayed in a percentage frequency table as shown in Table 2.2.

percentage frequency distribution—a distribution of relative frequencies or proportions in which each entry has been multiplied by 100

TABLE 2.2
Percentage Frequency Distribution: Political Party Identification in Smalltown

Party Identification	*Frequency (f)*	*Percent*
Republican	37	29.6%
Democrat	47	37.6
Independent	23	18.4
Other	18	14.4
Total	125	100.0%

To facilitate discussions about frequencies and relative frequencies, a shorthand notation is used. In this system, N denotes the total sample size (in the Smalltown example, $N = 125$). And f_i denotes the number of cases or the frequency associated with the ith outcome (category) of a variable. The subscript i can take values from 1 to the number of categories into which the variable is coded.

In our example, if we code Republican = 1, Democrat = 2, Independent = 3, and Other = 4, then $f_1 = 37$ (there are 37 Republicans in the sample), $f_2 = 47$, $f_3 = 23$, and $f_4 = 18$. Notice that column 2 of Table 2.2 is labeled f, and it gives the number of cases associated with each political identification. The sum of the frequencies associated with each outcome equals the total sample size:

$$f_1 + f_2 + f_3 + f_4 = N$$

In the Smalltown example:

$$37 + 47 + 23 + 18 = 125$$

In the notation system, p_i denotes the *proportion* of cases in the ith outcome of a variable. The formula for finding this proportion is:

$$p_i = \frac{f_i}{N}$$

In the Smalltown example, as we saw earlier, the proportion of Republicans is $p_1 = f_1/N = 37/125 = 0.296$. Similarly, $p_2 = 47/125 = 0.376$, $p_3 = 23/125 = 0.184$, and $p_4 = 18/125 = 0.144$.

The sum of proportions must always equal 1.00 (except for any rounding error). This is easy to show:

$$\begin{aligned} \text{Sum of proportions} &= p_1 + p_2 + p_3 + p_4 \\ &= f_1/N + f_2/N + f_3/N + f_4/N \\ &= (f_1 + f_2 + f_3 + f_4)/N \\ &= N/N = 1.0 \end{aligned}$$

Since a percentage is simply a proportion multiplied by 100, it also follows that *the sum of the percentages associated with each outcome must total 100%.* We have verified this fact by summing down column 3 in Table 2.2.

Now suppose we want to know whether there is the same distribution of political party identification in Bigtown as in Smalltown. To begin, we choose a random sample of 230 heads of households in Bigtown. The results of the survey on party identification for both communities are shown as a frequency distribution in Table 2.3. Clearly, the Bigtown sample has more Republicans than the Smalltown sample (52 vs. 37), but it is not clear from Table 2.3 whether Bigtown has *relatively* more Republicans than Smalltown. To answer this question we look at the *percentages* of Republicans in the two towns, as shown in Table 2.4. Bigtown has

TABLE 2.3
Party Identification Frequencies in Smalltown and Bigtown

Party Identification	*Smalltown*	*Bigtown*
Republican	37	52
Democrat	47	130
Independent	23	30
Other	18	18
Total	125	230

TABLE 2.4
Political Identifications in Smalltown and Bigtown (percentages)

Party Identification	*Smalltown*	*Bigtown*
Republican	29.6%	22.6%
Democrat	37.6	56.5
Independent	18.4	13.0
Other	14.4	7.8
Total	100.0%	99.9%*
N	125	230

*Does not total to 100% due to rounding.

relatively fewer Republicans (22.6% vs. 29.6%), considerably more Democrats (56.5% vs. 37.6%), fewer independents (13.0% vs. 18.4%) and fewer followers of other parties (7.8% vs. 14.4%).

Whenever one data set is compared to another, the most meaningful comparisons usually can be made by examining the relative numbers or the percentages of the various outcomes. This principle of data analysis is so basic that social scientists just take it for granted: *To compare two or more distributions, relative rather than absolute frequencies should be used.* This basic principle is underscored by the fact that scientists virtually always compare two or more distributions; they are rarely interested in a single distribution. Indeed, science can be thought of as the attempt to explain why two or more distributions differ from one another.

A sociologist or political scientist studying the results presented in Table 2.4 would try to explain why Bigtown has relatively fewer Republicans, more Democrats, and fewer Independents and Others than Smalltown. Is the difference due to Smalltown's location in a rural, politically conservative area, while Bigtown is an industrial city with many labor unions? Is it because Smalltown has an all-white population and Bigtown has many blacks? Or what? The "Or what?" is what science is about; it seeks to explain the relative variation or the differences between distributions of the same variable.

In terms of the distinction between independent and dependent variables made in Chapter 1, in this example the dependent variable is party identification (the phenomenon to be explained

—the consequence), and the independent variable is Smalltown vs. Bigtown (the antecedent). If you are confused as to why, you should carefully review the discussion on independent and dependent variables in Chapter 1.

2.2. Frequency Distributions for Discrete Measures

You should now be ready to examine frequency distributions with actual data. Relative frequency distributions for three variables are shown in Table 2.5. These responses were drawn from the 1977 General Social Survey (GSS) of the adult population (see Box 2.1). To measure ethnic ancestry, the 1,530 respondents were asked, "From what countries or part of the world did your ancestors come?" If more than one country was named, the person was asked, "Which one of these countries do you feel closer to?" Religious preference was ascertained by the question, "What is your religious preference? Is it Protestant, Catholic, Jewish, some other religion, or no religion?" Party identification was determined with this item: "Generally speaking, do you usually think of yourself as a Republican, Democrat, independent, or what?" Notice that for none of these items was the respondent explicitly offered "don't know" as a choice. That response was recorded by the interviewer only if it was volunteered by the respondent.

These are *discrete variables,* which classify observations only according to the *quality* or kind of person, event, or object (see Chapter 1). The response categories associated with ethnic ancestry, religious preference, and party identification are clearly *not* continuous, because they do not classify respondents according to magnitude or *quantity* of the response.

So few responses or outcomes are associated with the religious preference and party identification measures that it is easy to discern that in 1977 most Americans were Protestants and Democrats. The dominant category for the ethnic ancestry measure is not so obvious. Ethnic ancestry is coded in great detail in the General Social Survey, and some categories have fewer than 25 cases. Variables with large numbers of categories often prove unwieldy for analysis and interpretation. By combining some logically similar groups together and using a residual category for unclassifiable persons ("others"), we can reduce the number of categories to a more manageable number. (Rules for deciding how many categories to use when recoding are given in Section

TABLE 2.5

Percentage Distributions in General Social Survey for Ethnic Ancestry, Religious Preference, and Party Identification

Ethnic Ancestry		***Religious Preference***		***Party Identification***	
Africa	6.5%	Protestant	65.6%	Democrat	44.1%
Austria	1.0	Catholic	24.2	Independent	36.1
Canada (French)	0.8	Jewish	2.3	Republican	20.3
Canada (other)	1.0	None	6.1	Other party, Refused to say	0.3
China	0.3	Other	1.2	Don't know	0.0
Czechoslovakia	1.2	No answer	0.5	No answer	0.3
Denmark	0.8	Total	99.9%*	Total	101.1%*
England, Wales	10.8	$N = 1{,}530$		$N = 1{,}530$	
Finland	0.8				
France	1.7				
Germany	16.5				
Greece	0.2				
Hungary	0.8				
Ireland	7.6				
Italy	4.5				
Japan	0.3				
Mexico	1.5				
Netherlands	1.8				
Norway	1.4				
Philippines	0.3				
Poland	2.2				
Puerto Rico	0.6				
Russia	1.5				
Scotland	2.9				
Spain	0.6				
Sweden	1.3				
Switzerland	0.5				
West Indies	0.3				
Other	0.7				
American Indian	2.9				
India	0.1				
Portugal	0.4				
Lithuania	0.3				
Yugoslavia	0.5				
Rumania	0.1				
Belgium	0.1				
Arabic	0.1				
Other Spanish	0.3				
West Indies (non-Spanish)	0.2				
America	1.7				
No Answer	0.6				
Not Codable	22.5				
Total	100.2%*				
$N = 1{,}530$					

*Does not total to 100% due to rounding.

Source: 1977 General Social Survey.

BOX 2.1 General Social Survey Samples

The General Social Surveys, begun in 1972 under the guidance of James A. Davis, then director of the National Opinion Research Center, combine a data diffusion project and a social indicator research program. Funded originally on an annual basis and biennially since 1978 by the National Science Foundation, the GSS conducts interviews with a cross-section sample of about 1,500 adult Americans during the spring season. In interviews lasting more than an hour, questions are asked on a variety of topics, including occupational and labor force participation, family situation, social activity, religion, leisure, political behavior, attitudes on race relations, sexual behavior, violence, political institutions, life satisfaction, child rearing, women's roles, and abortion. Some permanent items appear in every survey, while others rotate, two years on and one year off.

The data have been compiled into a comprehensive file of respondents for the first eight surveys (1972–1978 and 1980). They provide a cornucopia of high-quality information which is available at cost (for computer tape and codebook) to the general public and the academic community. The GSS program has resulted in hundreds of scholarly research publications and literally thousands of student classroom projects. We have used the GSS data throughout this text to draw attention to this valuable social resource and encourage its expanded use as a teaching aid as well as a research tool.

The GSS uses sampling design and interview specifications that are too complex to discuss here. Details may be found in Appendixes A and B of the *GSS Cumulative Codebook,* which advises that "investigators who have applied statistical tests to previous General Social Survey data should continue to apply those tests." Our purpose in using the GSS is mainly illustrative, and we will of course apply various statistical tests to this data.

2.4 of this chapter, which describes frequency distributions with continuous data.)

The recoded results for ethnic ancestry appear in Table 2.6, which has only 12 categories for ethnic ancestry. Following the

TABLE 2.6
Percentage Distributions for Recoded Categories of Ethnic Ancestry

Recoded Categories	*Original Categories*	*Percent*
German	Austria, Germany	17.5%
English	England, Wales, Scotland, Non-French Canadian	14.7
Irish	Ireland	8.4
Italian	Italy	4.5
Scandinavian	Denmark, Finland, Norway, Sweden	4.3
East European	Czechoslovakia, Hungary, Lithuania, Poland, Russia, Rumania, Yugoslavia	6.6
West European	France, Netherlands, Switzerland, Portugal, Belgium	4.4
Spanish	Spain, Mexico, Puerto Rico, other Spanish	2.9
Oriental	China, Japan, Philippines	0.8
African	Africa	6.5
Other	Greece, West Indies, American Indian, India, Arabic, West Indies (non-Spanish), American, other	6.1
Not codable	Not codable, missing	23.1
Total		99.8%*
$N = 1{,}530$		

*Does not total to 100% due to rounding error.

Source: 1977 General Social Survey.

mutually exclusive and exhaustive criteria for assigning categories stated in Chapter 1, we have recoded categories on the basis of presumed similarity (e.g., Poland, Czechoslovakia, Rumania, Russia, Hungary, Lithuania and Yugoslavia were recoded into a single East European category). The results make two facts clear. First, large numbers of the Americans surveyed (23.1%) did not provide codable responses to questions of ethnic identity, probably because they have a mixed or unknown ancestry. Second, most of the respondents who did provide codable responses came from or had ancestors in West European countries such as Germany, England, and Ireland.

2.2.1. Techniques for Displaying Data: Tables and Bar Charts

The tables of frequency distribution in this chapter illustrate one way to display data and present quantitative information in a clear,

BOX 2.2 Statistical Tables

Statistical tables are a basic tool of the social scientist's trade. The arts of constructing tables and reading and interpreting them can best be learned, as are all crafts, by much practice. Some basic pointers on how to communicate findings through tables are given here.*

Tables are of two basic types: those which display "raw data" and those which present analyses. Raw-data tables contain frequencies or counts of observations classified in various ways, such as the number of burglaries reported in each of 50 neighborhoods last year, or the number of homicides classified according to familiarity between murderer and victim in each of eight regions of the United States. Analytic tables display the consequences of some manipulation of the data by a researcher which purports to provide an interpretation of the process producing the raw data. Such tables are highly varied and may range from simple percentaging of raw data all the way to systems of nonlinear simultaneous equations for complex mathematical models.

Each table begins with a heading, usually the word *Table,* an identifying number, and a brief phrase describing its central contents. Examples are:

Table 23.6. Number of Pre-Columbian Pottery Shards Found at Each of Eight Sites

Table 6. Regression of Personality Trait Scores on Children's Family Characteristics

Table 10. Crosstabulation of Income with Education among Sex-Race-Region Groups, 1950 and 1960.

Under the heading, usually below a rule, are subheadings which label the various columns in the main body of the table. These subheadings most often are either variable names and categories or summary statistics such as column marginals (*N*s). To save space, short labels are preferred; if further clarification is required, subheadings can be footnoted, with the expanded explanations appearing at the bottom of the table.

Additional information, as in a crosstabulation between two or more variables, appears in the column farthest to the left (sometimes called the *stub*). Each entry in this column

Box 2.2 (continued)

describes the content of one of the rows forming the body of the table. For example, if attitude responses are to appear in the rows, the labels or response categories in this column, from top to bottom, might be: "very strongly agree, strongly agree, agree, neither agree nor disagree, disagree, strongly disagree, and very strongly disagree."

The main body of the table consists of the intersections of the entries under column and row headings. It displays the appropriate data in either raw or analyzed form. If the table contents are percentages, the preferable way to arrange them is so that they total to 100% down each column. A percentage (%) total is usually the next to last row entry. The last row, labeled *N* at the left, gives the base frequencies on which the percentages were calculated. Examples of percentage tables like these in this chapter are Tables 2.4 and 2.5.

The amount of missing data (observations which could not be used in the main body of the table due to lack of information) may be reported directly below the body of the table. Any additional information about the data (such as its source) or about the analyses performed should be included in notes below the table.

*An extended discussion of table construction and interpretation appears in the article by James A. Davis and Ann M. Jacobs, "Tabular Presentation," in D. L. Sills (Ed.), *International Encyclopedia of the Social Sciences*, Vol. 15, pp. 497–509 (New York: Macmillan Co., 1968).

precise manner. **Statistical tables** are an important means of scientific communication, and some rules for constructing them are given in Box 2.2.

statistical table—a numerical display which either summarizes data or presents the results of a data analysis

Tables are not the only way to display frequency distributions of single variables, however. Striking visual impact can be achieved through a variety of **diagrams** or **graphs**, such as bar charts, pie charts, and histograms. For discrete variables, a **bar chart** is one of the most effective visual images of distributions, as well as one of the easiest to construct. To form a bar chart, categories of the discrete variable are arrayed along the horizontal

diagram (graph)—a visual representation of a set of data

bar chart—a type of diagram for discrete variables in which the numbers or percentages of cases in each outcome are displayed

FIGURE 2.1
Bar Chart Showing Party Choice

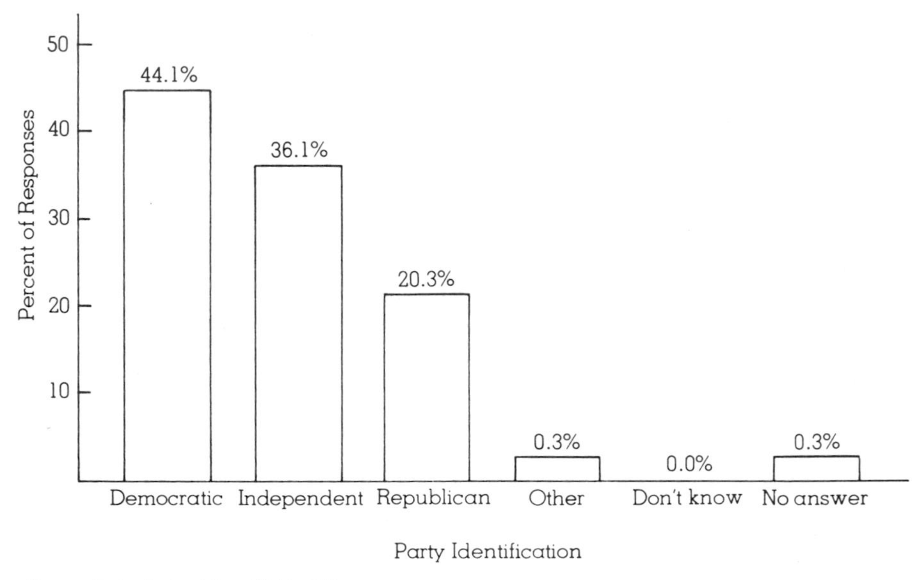

Source: Data derived from Table 2.5.

axis, and equally spaced vertical bars are erected above each category label to heights proportional to the frequency (either actual numbers or percentages) of the observations classified into each category.

Figure 2.1 is a bar chart which illustrates political party choice responses in the 1977 General Social Survey, using the data in Table 2.5. An important point to observe in constructing bar charts is that *when the discrete variable categories have no inherent order, the bars should not touch one another* but should stand alone. Only when the categories are *orderable* should the bars touch, as in the histograms described in the next section. The frequencies or the percentages are sometimes shown above each bar.

A diagram such as a bar chart does not add any information to that given in a table showing the same data. In fact, it communicates less information if the sample sizes on which categories are based are omitted and only the percentages are shown. Diagrams can be prepared only after the information has been assembled for a table.

2.3. Frequency Distributions for Orderable Discrete Variables

The frequency distributions described in the preceding section deal with *nonorderable* discrete variables, whose outcomes, as defined in Chapter 1, cannot be ranked from high to low or vice versa. The other type, *orderable* discrete variables, uses measures which classify observations into categories that can be compared by degree of *equality:* two persons or events possess either the same, greater, or lesser amounts of the measured quality. In the most extreme cases, no two observations are equal, and all the persons or events observed can be arranged in order or **ranks**, or ranked from first to last or vice versa. American states, for example, can be ranked from 1st to 50th by total land area, population size, or numerous other characteristics.

rank—the position occupied by an observation in a distribution when scores on some variable have been ordered from smallest to largest, or vice versa

More often, however, measures with ordered outcomes contain relatively few categories (perhaps half a dozen). Many observations are placed into a single category, even though they may differ somewhat in the precise amount of the variable quality they have. For example, if we construct a variable to measure how often people smoke marijuana, we might use the response categories, "regularly, often, seldom, or never." What one person means by regularly (every weekend) may not be the same as what another means by that term (twice a day), however. Even more distressing, two persons who smoke with the same actual frequency may choose different categories for their self-reports of pot smoking. Unfortunately, self-reported measures of behavioral frequency often suffer from this imprecision, although in many cases measurement could easily be improved. In this example, the researcher would do better to ask the subjects to indicate the *number* of marijuana cigarettes they smoke on average each week. While this measure is still imprecise, since it asks for an average rather than an exact number, it is clearly more exact than the one employing the broad response categories, "regularly, often, seldom, or never."

Table 2.7 displays percentage distributions for two ordered discrete variables drawn from the 1977 GSS health and health satisfaction, for males. The results clearly indicate that most American males believe they are in good or excellent health, and most

TABLE 2.7

Percentage Distributions for Health and Health Satisfaction, GSS Males

Health		*Health Satisfaction*	
Excellent	31.5%	A very great deal	29.1%
Good	41.4	A great deal	33.3
Fair	20.3	Quite a bit	12.4
Poor	6.5	A fair amount	13.4
Not ascertained	0.3	Some	4.3
Total	100.0%	A little	3.5
$N = 693$		None	3.2
		Don't know	0.3
		Not ascertained	0.4
		Total	99.9%*
		$N = 693$	

*Sums to less than 100.0% due to rounding.

Source: 1977 General Social Survey.

are satisfied with their health. Health was measured in the GSS by the item: "Would you say your health, in general, is excellent, good, fair, or poor?" Health satisfaction was also self-reported, measured earlier in the interview as part of a five-item set of life satisfaction variables: "For each area of life I am going to name, tell me the number [on the card handed to respondent] that shows how much *satisfaction* you get from that area." The area named was "Your health and physical condition," and the responses recorded were: "a very great deal, a great deal, quite a bit, a fair amount, some, a little, or none." The response categories "don't know" and "no answer," which were not on the card, were also recorded.

2.3.1. Techniques for Displaying Data: Histograms and Polygons

histogram—a type of diagram that uses bars to represent the frequency, proportion, or percentage of cases associated with each outcome or interval of outcomes of a variable

Visual displays for discrete variables were introduced in the preceding section with the bar chart. A similar diagram, called a **histogram**, can be drawn to display distributions of ordered discrete variables.

The distinguishing characteristic of a *histogram* is that the vertical bars touch one another, indicating an underlying order among categories which is absent from unordered discrete variables. Figure 2.2 shows a histogram for the GSS health satisfaction variable.

FIGURE 2.2

Histogram Showing Health Satisfaction of Males

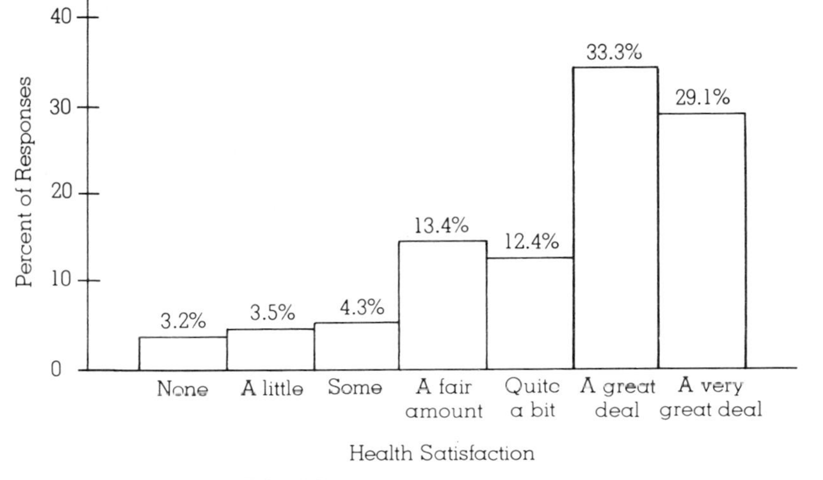

Source: Data derived from Table 2.7.

FIGURE 2.3

Polygon Showing Health Satisfaction of Males

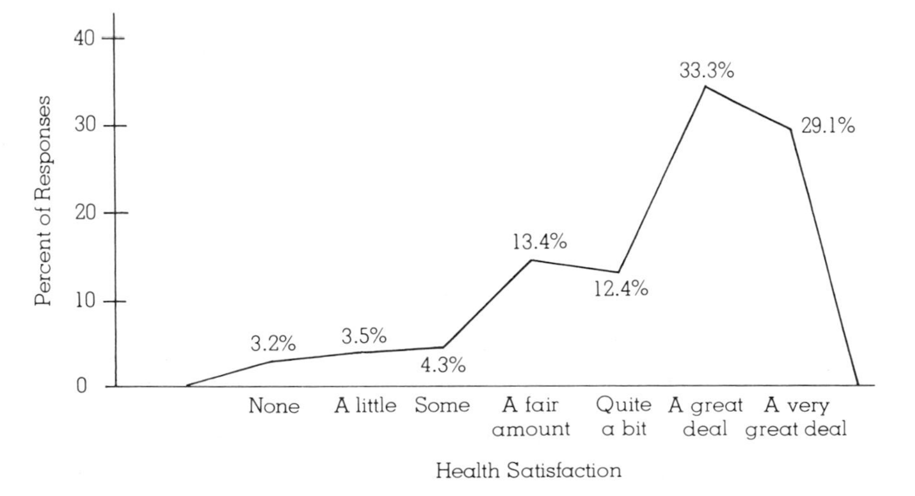

Source: 1977 General Social Survey.

polygon—a diagram constructed by connecting the midpoints of a histogram with a straight line

If instead of drawing bars, as in the bar chart and the histogram, the midpoints of each category are connected by a line, the resulting diagram is called a **polygon**. Figure 2.3 shows a polygon for the health satisfaction variable, using the data in Table 2.7. In both cases, the basis for constructing the histogram or polygon was the frequency or percentage distribution of sample observations across categories of the variable.

Another illustration of graphing or diagraming a discrete variable with ordered categories also uses the 1977 GSS data. Respondents were asked about the number of their brothers and sisters (whether living at the time of the interview or not), including stepsiblings. They were also asked about the number of children they themselves had had. The percentage distributions for these variables are shown in Table 2.8. The distributions suggest that

TABLE 2.8

Percentage Distributions for Number of Siblings and Number of Own Children, GSS Respondents

Number	*Brothers and Sisters*	*Own Children*
None	5.3%	25.0%
One	12.7	16.7
Two	16.3	23.3
Three	12.3	15.0
Four	12.0	8.5
Five	9.0	4.6
Six	8.1	2.5
Seven	6.9	1.2
Eight or more	17.4	2.3
Don't know	0.1	0.0
No answer	0.1	0.9
Total	100.2%*	100.0%
$N = 1{,}530$		

*Does not add to 100% due to rounding.

Source: 1977 General Social Survey.

the respondents' families of orientation (those in which they were born) are larger than their families of procreation (the ones in which they currently reside). However, the brothers-and-sisters variable excludes the respondent from the count, as well as families with no children. Furthermore, since most of the respondents are still in their childbearing years, we would not want to draw this conclusion without additional data and analyses.

Percentage polygons which illustrate the distribution of both the number of siblings and the number of one's own children for

FIGURE 2.4
Polygons Showing Numbers of Own Children and of Siblings

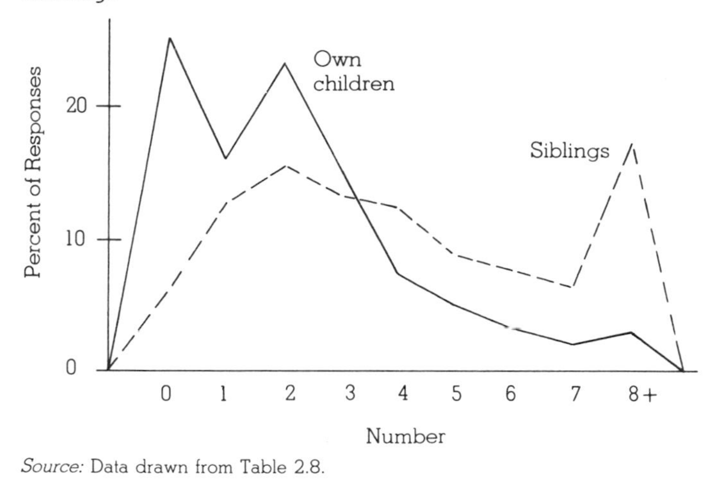

Source: Data drawn from Table 2.8.

GSS respondents are shown in Figure 2.4. These polygons make even clearer the *tentative* inference that respondents' families of orientation are larger than their families of procreation.

2.4. Frequency Distributions for Continuous Measures

Building frequency distributions is little trouble when variables are discrete. Classifying a person as male or female, for example, and then counting the numbers of each gender is easy. To count the numbers of children in families and then build a distribution of family size is also easy. When the variable of interest is *continuous,* however, constructing a frequency distribution first requires **grouped data**; that is, *the observations must be grouped in some way.*

grouped data—data that have been collapsed into a smaller number of categories

A continuous variable, as we noted in Box 1.1 in Chapter 1, can be measured at either the interval or ratio level. In theory, any two coded values on the scale can be subdivided infinitely. Take height as a example. Suppose someone's height is recorded with a mark on a tape measure, as in the illustration below:

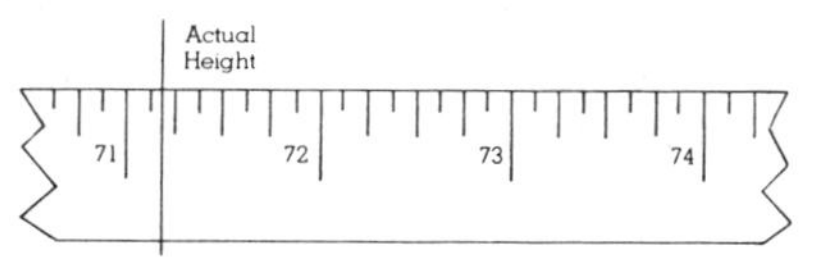

This person's height is 71.203125 inches, to be precise. But usually we do not need this much precision. In everyday conversation we would say the person is 71 inches tall; that is, we would round to the nearest whole number. For some scientific work, this value might be too imprecise. A scientist might decide, however, that measurement in 16ths of an inch is precise enough, in which case the person's height would be recorded as 71 and 3/16 of an inch, or 71.1875 inches.

> The point is that with continuous variables, measurement can be as precise as the measuring instrument will allow. At some point, however, a decision on how to group observations must be made.

2.4.1. Grouping Data in Measurement Classes

measurement class (measurement interval) —a range of scores on a variable into which observations are grouped

After deciding on the unit of measurement, observations are grouped into **measurement classes** or **measurement intervals** for ease in comprehension. For example, a report of annual income in the United States might group the data into the following categories:

$4,999 or less
$5,000–9,999
$10,000–14,999
$15,000–19,999
$20,000–24,999
$25,000 or more

How does the researcher know how wide a measurement interval to choose in working with continuous data? *The rule of thumb is to choose a measurement interval which is small enough that the original distribution of observations is not seriously distorted.* Common sense would indicate that $5,000 is a reasonable interval for the study of yearly income, but it would not do for a survey of the prices of men's suits.

For many statistical techniques presented in later chapters, *the more measurement precision, the better.* But presenting data in frequency distributions often necessitates violating this rule if the data are to be comprehended at all. On one hand, 100 income categories in $500 intervals are just too many for easy comprehension of income distribution in the United States. On the other hand, the six categories of income presented above may be too few—

too few in the sense that the categories are so gross as to seriously distort the true income distribution. Generally, *the number of intervals for presenting data in frequency distributions should be somewhere between 6 and 20,* but this is only a rule of thumb. Sometimes fewer than six intervals will not significantly distort the shape of the distribution. More than 20 categories are seldom used, however, because a large number of categories makes it difficult to interpret the distribution.

Notice that the intervals above do not permit any overlap between them. That is, when presenting the six measurement categories for income we did *not* overlap the categories' endpoints in the following way:

$5,000 or less
$5,000–10,000
$10,000–15,000
$15,000–20,000
$20,000–25,000
$25,000 or more

To have done so would have violated the principle of mutual exclusiveness discussed in Chapter 1. A person earning $5,000, $10,000, $15,000, $20,000, or $25,000 could be placed in two categories rather than in a single measurement class.

2.4.2. Rounding

The slight gap which always exists between grouped measurement categories creates another type of problem: Where could someone with a yearly income of $9,999.75, for example, be placed in the six nonoverlapping income categories? The solution is **rounding**; *that is, we would round the income to the nearest dollar—$10,000 in this case—and place that person in the $10,-000–14,999 category. The number $9,999.25 would be rounded down to $9,999 and placed in the $5,000–9,999 category. The rule is: Do not round past the original degree of measurement precision.* In the example above, income was measured in dollars, and thus we rounded to the nearest dollar. If income had been measured in *hundreds* of dollars earned, then $9,999.75 and $9,999.25 would both have been rounded to $10,000, but $9,-948 would have been rounded to $9,900. If the unit of measurement had been *thousands* of dollars earned, then all of the above amounts would have been rounded to $10,000, but $9,450

rounding—expressing digits in more convenient and interpretable units, such as tens, hundreds, or thousands, by applying an explicit rule

Box 2.3 Rules for Rounding

Generally, significant digits ending in 1 to 4 are rounded down by leaving the digit to the left unchanged, and digits ending in 6 to 9 are rounded up by increasing the digit to the left by 1. Numbers ending in 5 are allocated alternatively; the first is rounded down, the second is rounded up, the third is rounded down, and so on. Never round past the original unit of measurement or degree of measurement precision.

Several examples of rounding 999.52 and 999.37 are given below:

Unit of Measurement	*Original No.*	*Rounded No.*
Dollars	999.52	1,000
Dollars	999.37	999
Hundreds of dollars	999.52	10
Hundreds of dollars	999.37	10
Tenths of dollars	999.52	999.5
Tenths of dollars	999.37	999.4

would have been rounded to $9,000. Some rules for rounding and examples of rounding digits according to unit of measurement are given in Box 2.3.

2.4.3. True Limits and Midpoints

Grouping data in measurement classes also raises problems of determining the true limits and midpoints of the intervals. To illustrate these, we will introduce the second major set of data to be used in this text—the 63-cities data set, which is described in Box 2.4. Table 2.9 shows the raw data for city age (number of years since the census date at which a city first attains 10,000 residents), 1974 population, general expenditures in 1974, and serious crimes in 1970. Note that the index numbers for the 63-cities data start with zero, so the final city listed, Cleveland, Ohio, has an ID of 62, although $N = 63$.

true limits—the exact upper and lower limits of a measurement class or interval into which rounded values are grouped

The data for city age have been grouped into a frequency distribution with 14-year measurement intervals, as shown in the first column of Table 2.10. The interval limits are defined by whole numbers—as 31–45 years—but the **true limits** of the interval

BOX 2.4 The 63-Cities Data

In some circumstances it is necessary to illustrate relationships between variables with a much smaller data set than the typical 1,500-respondent surveys of the General Social Survey. For these purposes, the 63-cities data are ideal.

This project began at the National Opinion Research Center in the 1960s, using a sample of 51 communities selected from NORC's permanent sampling frame.* These 51 cities represent 22 different states and range in population from 50,000 to 750,000. Most are the major central city of a Standard Metropolitan Statistical Area (SMSA) or suburbs within a large SMSA (e.g., Waukegan, Illinois, in the Chicago area).

The first major research project on the 51 cities was conducted in 1967. This was a study of community decision making in which 11 informants in each city reported on four issues: urban renewal, air pollution, antipoverty efforts, and the mayoral election.† Since that time, more than 800 variables from the U.S. census and other official sources have been added to the file, including information on demographic characteristics, political activity, and governmental behavior in both the cities and their surrounding counties.

Recently the 12 largest U.S. cities were added to the data set, increasing it to the 63 cities which we analyze in this book.

---

*See Peter M. Rossi and Robert L. Crain, "The NORC Permanent Community Sample," *Public Opinion Quarterly,* vol. 32 (Summer 1968), pp. 261–72, for the background of this project.

†See Terry N. Clark, "Community Structure, Decision-Making, Budget Expenditures, and Urban Renewal in 51 American Communities," *American Sociological Review,* vol. 33 (August 1968), pp. 576–93, for a detailed analysis of this data.

TABLE 2.9
Raw Data on Four Variables from the 63-Cities Data

City ID	City Name	City Age	Population in 1974 (in thousands)	General Expenditures in 1974 (in thousands of dollars)	Serious Crimes in 1970
0	Akron, OH	86	252	268	13,252
1	Albany, NY	146	110	346	3,358
2	Amarillo, TX	46	139	131	3,991
3	Atlanta, GA	106	436	380	27,378
4	Berkeley, CA	66	110	248	6,442
5	Birmingham, AL	86	276	302	13,362
6	Bloomington, MN	16	79	179	1,792
7	Boston, MA	176	637	858	38,294
8	Buffalo, NY	126	407	648	18,284
9	Cambridge, MA	116	102	692	7,563
10	Charlotte, NC	66	281	262	12,982
11	Clifton, NJ	56	79	288	1,200
12	Duluth, MN	86	94	283	2,072
13	Euclid, OH	26	63	241	665
14	Fort Worth, TX	86	358	174	15,652
15	Fullerton, CA	16	94	156	2,832
16	Gary, IN	56	168	244	11,472
17	Hamilton, OH	76	66	163	2,067
18	Hammond, IN	66	105	189	4,437
19	Indianapolis, IN	106	715	313	25,277
20	Irvington, NJ	56	58	234	1,785
21	Jacksonville, FL	76	535	277	25,223
22	Long Beach, CA	56	336	312	15,695
23	Malden, MA	86	56	561	829
24	Manchester, NH	116	83	462	1,290
25	Memphis, TN	116	661	381	21,614
26	Milwaukee, WI	126	666	288	20,188
27	Minneapolis, MN	96	378	337	23,420
28	Newark, NJ	126	340	773	31,781
29	Palo Alto, CA	26	52	351	2,462
30	Pasadena, CA	66	108	394	7,111
31	Phoenix, AZ	56	665	255	29,483
32	Pittsburgh, PA	136	459	245	28,396
33	St. Louis, MO	126	525	371	45,915
34	St. Paul, MN	106	280	336	15,050
35	St. Petersburg, FL	46	234	150	7,886
36	Salt Lake City, UT	96	170	222	10,361
37	San Francisco, CA	116	665	763	57,136
38	Santa Ana, CA	46	177	147	5,619
39	San Jose, CA	76	556	188	14,492
40	Santa Monica, CA	46	92	191	5,588
41	Schenectady, NY	76	75	336	1,175
42	Seattle, WA	86	487	362	31,176
43	South Bend, IN	86	117	220	5,506
44	Tampa, FL	66	280	240	13,986
45	Tyler, TX	36	61	120	1,758
46	Utica, NY	116	82	281	743
47	Waco, TX	76	98	247	3,772
48	Warren, MI	16	173	219	5,257
49	Waterbury, CT	86	107	512	3,336
50	Waukegan, IL	46	65	162	2,226
51	New York, NY	186	7,482	1,382	517,716
52	Chicago, IL	126	3,099	299	128,017

TABLE 2.9 (continued)

City ID	City Name	City Age	Population in 1974 (in thousands)	General Expenditures in 1974 (in thousands of dollars)	Serious Crimes in 1970
53	Los Angeles, CA	86	2,727	256	175,719
54	Philadelphia, PA	186	1,816	454	45,734
55	Detroit, MI	126	1,335	395	127,630
56	Houston, TX	86	1,327	168	59,883
57	Baltimore, MD	146	852	897	62,150
58	Dallas, TX	86	813	266	50,391
59	San Diego, CA	66	774	201	23,232
60	San Antonio, TX	96	773	138	27,221
61	Washington, DC	146	712	1,769	59,311
62	Cleveland, OH	116	639	333	44,564

Source: NORC Permanent Community Sample (63-cities study).

are 15 years apart—such as 30.5–45.5—because of rounding, by which anything between 30.5 and 45.5 is rounded into the 31–45 interval. The second column of Table 2.10 shows all of the true limits for the measurement intervals. Actual measured observations which end in 0.5 would be alternatively allocated to first one and then the other of the two adjacent categories, according to the rounding rules spelled out above. But there are no city ages ending in 0.5 in these data.

midpoint—a number exactly halfway between the true upper and lower limits of a measurement class or interval, obtained by adding the upper to the lower limits and dividing by 2

The **midpoint** of an interval is figured by adding the true limits of each measurement category and dividing by 2. The midpoint of the interval 46–60, for example, is $(45.5 + 60.5)/2 = 53$. The

TABLE 2.10
Distribution of Age of 63 U.S. Cities

City Age	True Limits	Midpoint	f	p
16–30	15.5–30.5	23	5	.0794
31–45	30.5–45.5	38	1	.0159
46–60	45.5–60.5	53	10	.1587
61–75	60.5–75.5	68	6	.0952
76–90	75.5–90.5	83	16	.2540
91–105	90.5–105.5	98	3	.0476
106–120	105.5–120.5	113	9	.1429
121–135	120.5–135.5	128	6	.0952
136–150	135.5–150.5	143	4	.0635
151–165	150.5–165.5	158	0	.0000
166–180	165.5–180.5	173	1	.0159
181–195	180.5–195.5	188	2	.0317

Source: NORC Permanent Community Sample (63-cities data).

midpoints for all the other intervals are shown in the third column of Table 2.10. *The midpoint is the single number that best represents the entire measurement interval.*

When the measurement interval is only one unit wide, the true limits run from 0.5 below the unit to 0.5 above it. Thus we could code outcomes 0, 1, 2, 3, and so on as follows:

Outcomes	*True Limits*	*Midpoint*
0	−0.5–0.5	0
1	0.5–1.5	1
2	1.5–2.5	2
3	2.5–3.5	3
.	.	.
.	.	.
.	.	.

Notice that when the measurement interval is only one unit wide, the midpoint is simply the value of the outcome.

Once we have decided on the measurement interval, we tally the frequency *(f)*, or the number of cases, in each interval, as shown in the fourth column of Table 2.10. These frequencies then can be converted to proportions *(p)*, as in the fifth column. Note that more than one-fourth of the cities fall into the 76–90-year-old category.

> Observations are grouped into measurement intervals only when the variable of interest is a continuous one. When the outcomes are discrete, whether orderable or nonorderable, as described in the preceding sections of this chapter, the numbers of each type are merely counted.

2.5. Cumulative Distributions

Often the researcher needs to know the *relative* position of a given outcome in a distribution of scores. If a city is 126 years old, how old is it relative to other cities in the sample? Questions like this can be answered by constructing a cumulative frequency distribution or cumulative percentage distribution.

cumulative frequency—for a given score or outcome of a variable, the total number of cases in a distribution at or below that value

The **cumulative frequency** at a given score is the total number of frequencies at or below that score.

In a sample of 1,997 American adults, respondents were asked to rate how happy they were, using ten "rungs of a ladder" on which the tenth rung represented the best possible life and the first rung the worst possible life. The resulting cumulative distribution is shown in Table 2.11

In this example, the **cumulative frequency distribution** is the distribution of total scores at or below each rung. The cumulative frequencies are shown in the fourth column of Table 2.11,

cumulative frequency distribution—a distribution of scores showing the number of cases at or below each outcome of the variable being displayed in the distribution

TABLE 2.11
Cumulative Distribution of Happiness Ratings, American Adults

Rung	*f*	*%*	*cf*	*c%*
1	16	0.8%	16	0.8%
2	28	1.4	44	2.2
3	75	3.8	119	6.0
4	109	5.5	228	11.4
5	260	13.0	488	24.4
6	268	13.4	756	37.9
7	423	21.2	1,179	59.0
8	491	24.6	1,670	83.6
9	212	10.6	1,882	94.2
10	115	5.8	1,997	100.0

Note: $N = 1{,}997$.

Source: Unpublished data from a study reported in E. Berscheid, E. Walster, and G. Bohrnstedt, "The Happy American Body: A Survey Report," *Psychology Today,* December 1973, pp. 119–31.

denoted by *cf.* The cumulative frequency to rung 5 is 488, to rung 6, 756, and so on.

More interesting for interpretative purposes is the **cumulative percentage** of responses. The third column of Table 2.11 shows the percentage distribution of responses. In the fifth column (denoted by *c%*), these percentages are cumulated to form a **cumulative percentage distribution**. The table shows that only 24.4% of the sample checked the fifth rung or lower, and 83.6 percent of the sample placed themselves on the eighth rung or lower. Like the percentage distribution described at the beginning of the chapter, the cumulative percentage distribution makes clear the *relative* standing of a given observation. You can quickly see that only 11.4% of the sample chose rungs 1–4, and anyone choosing these rungs had to be *relatively* unhappy.

cumulative percentage—for a given score or outcome of a variable, the percentage of cases in a distribution at or below that value

cumulative percentage distribution—a distribution of scores showing the percentage of cases at or below each outcome of the variable being displayed in the distribution.

Cumulative percentage distributions can also be diagramed. A diagram of the ratings distribution in Table 2.11 is shown in Figure 2.5.

FIGURE 2.5
Cumulative Percentage Distribution of Happiness Ratings

Source: Data derived from Table 2.11, from a study reported in E. Berscheid, E. Walster, G. Bohrnstedt, "The Happy American Body: A Survey Report," *Psychology Today,* December 1973, pp. 119–31.

2.6. Percentiles

percentile—the outcome or score below which a given percentage of the observations in a distribution falls

A useful statistic which can be derived from cumulative distributions is the **percentile**, which is the outcome or score below which a given percentage of the observations falls. With discrete variables like the happiness ratings in Table 2.11, the percentiles are easily discernible. The 59th percentile is 7.5, since 59 percent of the ratings lie below that number, the true upper limit of the seventh rung. With continuous variables, where the data are grouped like the 63-cities sample data in Table 2.12, percentiles are computed using the following formula:

$$P_i = L_p + \left(\frac{(p_i)(N) - c_p}{f_p} \right) (W_i)$$

Where:

P_i = The score of the ith percentile.

L_p = The true lower limit of the interval containing the ith percentile.
p_i = The ith percentile written as a proportion (e.g., the 75th percentile becomes 0.75 in the formula).
N = The total number of observations.
c_p = The cumulative frequency up to but not including the interval containing P_i.
f_p = The frequency in the interval containing the ith percentile.
W_i = The width of the interval containing P_i; $W_i = U_p - L_p$

where U_p and L_p are the upper and lower true limits of the interval containing P_i.

In the cumulative frequency distribution of city age for the sample of 63 cities shown in Table 2.12, the 95th percentile is given by:

$$P_{95} = 135.5 + \left(\frac{(.95)(63) - 56}{4} \right) (15) = 149.9 \text{ years of age}$$

since $i = 95$, $L_p = 135.5$, $p_i = 0.95$, $N = 63$, $c_p = 56$, $f_p = 4$, and $W_i = 150.5 - 135.5 = 15$.

The point below which 50 percent of the cases fall, the 50th percentile, is:

$$P_{50} = 75.5 + \left(\frac{(0.50)(63) - 22}{16} \right) (15) = 84.4 \text{ years of age}$$

TABLE 2.12
Cumulative Distribution of the Age of 63 U.S. Cities

City Age	*Midpoint*	*f*	*%*	*cf*	*c%*
16–30	23	5	7.94	5	7.94
31–45	38	1	1.59	6	9.52
46–60	53	10	15.87	16	25.40
61–75	68	6	9.52	22	34.92
76–90	83	16	25.40	38	60.32
91–105	98	3	4.76	41	65.08
106–120	113	9	14.29	50	79.37
121–135	128	6	9.52	56	88.89
136–150	143	4	6.35	60	95.24
151–165	158	0	0.00	60	95.24
166–180	173	1	1.59	61	96.83
181–195	188	2	3.17	63	100.00

Source: NORC Permanent Community Sample (63-cities data).

The formula for percentiles can also be used when the scores in the frequency distribution are not grouped, as is the case in Table 2.11. For the sample of adults rating happiness on a 10-rung scale in the example above, the 50th percentile is:

$$P_{50} = 6.5 + \left(\frac{(0.50)(1{,}997) - 756}{423} \right) (1) = 7.1$$

since $i = 50$, $L_p = 6.5$, $p_i = 0.50$, $N = 1{,}997$, $c_p = 756$, $f_p = 423$, and $W_i = 7.5 - 6.5 = 1.0$.

Percentiles are useful for determining whether two or more groups have similar or different distributions on a variable of interest. Table 2.13 shows the cumulative frequency and percentage

TABLE 2.13

Cumulative Distribution of Happiness Ratings for American Females and Males

	Males (N = 995)				Females (N = 989)			
Rung	*f*	*%*	*cf*	*c%*	*f*	*%*	*cf*	*c%*
1	7	0.7	7	0.7	9	.9	9	.9
2	15	1.5	22	2.2	13	1.3	22	2.2
3	37	3.7	59	5.9	38	3.8	60	6.1
4	58	5.8	117	11.8	51	5.2	111	11.2
5	126	12.7	243	24.4	134	13.5	245	24.8
6	144	14.5	387	38.9	123	12.4	368	37.2
7	223	22.4	610	61.3	197	19.9	565	57.1
8	233	23.4	843	84.7	253	25.6	818	82.7
9	102	10.3	945	95.0	109	11.0	927	93.7
10	50	5.0	995	100.0	62	6.3	989	100.0

Source: Unpublished data from a study reported in E. Berscheid, E. Walster, and G. Bohrnstedt, "The Happy American Body: A Survey Report," *Psychology Today*, December 1973, pp. 119–31.

distributions of happiness ratings for males and females separately. Suppose we hypothesize that women are less happy than men, for whatever reason. If the hypothesis is correct, then the 50th percentile for women should be lower than for men—that is, the score below which 50 percent of the women's responses lie should be lower than that for men. Using the data from Table 2.13 for men, we can see that:

$$P_{50} = 6.5 + \frac{(0.5)(995) - 387}{223} = 7.0$$

while for women:

$$P_{50} = 6.5 + \frac{(0.5)(989) - 368}{197} = 7.1$$

The 50th percentiles are so similar for men and women that we would certainly have to reject the hypothesis that women rate themselves as less happy than men. Instead, we would conclude that both sexes appear to be equally happy—at least in this sample.

2.7. Quantiles

Percentiles are special cases of **quantiles**, which divide a set of observations into groups with known proportions in each group. Other special cases of quantiles are quartiles, quintiles, and deciles.

quantile—a division of observations into groups with known proportions in each group

Quartiles are points in a number scale which divide the observations into *four equal groups of equal size*. Q_1 is that point below which one-fourth of the observations lie, Q_2 is the point below which one-half of the observations lie, and so on. **Quintiles** divide the observations into *5 equal groups*, and **deciles**, into *10 equal groups. Percentiles* divide observations into *100 equal groups.*

quartiles—the outcomes or scores that divide a set of observations into four groups of equal size

quintiles—the outcomes or scores that divide a set of observations into five groups of equal size

deciles—the outcomes or scores that divide a set of observations into 10 groups of equal size

P_i is used to designate the *i*th percentile, D_i the *i*th decile, K_i the *i*th quintile, and Q_i the *i*th quartile. In this notation, $Q_1 = P_{25}$—the first quartile is exactly the same as the 25th percentile. Similarly, $K_1 = D_2$—the first quintile is the same as the second decile. Box 2.5 shows the relationship of the various quantiles to percentiles in tabular form.

Using the data in Table 2.12 on age of cities, we can calculate:

$$Q_1 = P_{25} = 45.5 + \left(\frac{(.25)(63) - 6}{10}\right)(15)$$
$$= 60.1 \text{ years of age}$$

$$Q_2 = P_{50} = 75.5 + \left(\frac{(.50)(63) - 22}{16}\right)(15)$$
$$= 84.4 \text{ years of age}$$

$$Q_3 = P_{75} = 105.5 + \left(\frac{(.75)(63) - 41}{9}\right)(15)$$
$$= 115.9 \text{ years of age}$$

Based on these calculations, we can conclude that in this sample of 63 cities, a city 60 years old or younger is in the lowest quartile of age, a city between 61 and 84 years old is in the

BOX 2.5 Relationships of Quartiles, Quintiles, and Deciles to Percentiles

Of all the quantiles, the percentile provides the largest number of equal-sized groups—100 of them. The relationship of quantiles, quintiles, and deciles to percentiles is therefore easily shown, as in the following table:

Quartile	*Percentile*	*Quintile*	*Percentile*	*Decile*	*Percentile*
Q_1	P_{25}	K_1	P_{20}	D_1	P_{10}
Q_2	P_{50}	K_2	P_{40}	D_2	P_{20}
Q_3	P_{75}	K_3	P_{60}	D_3	P_{30}
		K_4	P_{80}	D_4	P_{40}
				D_5	P_{50}
				D_6	P_{60}
				D_7	P_{70}
				D_8	P_{80}
				D_9	P_{90}

second quartile, a city between 84 and 116 years old is in the third quartile, and one 116 years old or older is in the top quartile.

Frequency distributions are a very important aid in summarizing a scientist's individual observations. But to account for variation in a frequency distribution, two or more variables must be examined simultaneously. This task is accomplished with cross-tabulation, a technique that is introduced in Chapter 4.

2.8. Grouping Error

grouping error—error that arises from use of the midpoint of grouped frequency distributions to represent all scores in the measurement class

In computing percentiles and other quantiles, biases in estimating can result from **grouping error.** In grouping data we assume that all cases in the group are the same or very similar. If we use only a few groups, however, this assumption cannot be realistic, and the resulting percentiles can be somewhat misleading. Therefore, whenever *un*grouped data can be used, they are preferred.

> As scientists, researchers strive for the highest degree of measurement precision possible. The grouping of observations in statistical analysis is done only to aid visual inspection of the data. Generally speaking, computing statistics with grouped data is to be avoided wherever possible.

Review of Key Concepts

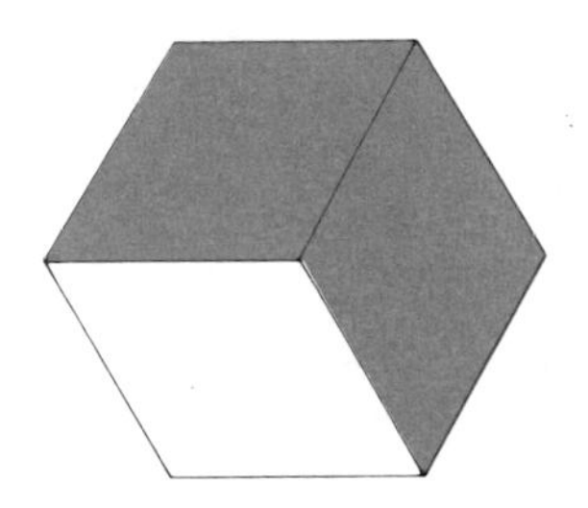

This list of the key concepts introduced in Chapter 2 is in the order of appearance in the text. Combined with the definitions in the margins, it will help you review the material and can serve as a self-test for mastery of the concepts.

tally
frequency distribution
outcome
relative frequency (proportion)
percentage
percentage frequency distribution
statistical table
diagram (graph)
bar chart
rank
histogram
polygon
grouped data
measurement class (measurement interval)
rounding
true limits
midpoint
cumulative frequency
cumulative frequency distribution
cumulative percentage
cumulative percentage distribution
percentile
quantile
quartiles
quintiles
deciles
grouping error

PROBLEMS

General Problems

1. A field researcher observed couples in a city park, chosen at random, for five minutes each. The following symbols were used to record the couples' observed behavior: H = holding hands, T = talking, K = kissing, O = other activity. Here are the field researcher's raw observations for a three-hour period:

HT	T	T	HT	TK
T	O	HT	T	T
O	TK	O	T	HK
HK	HK	HK	T	T
TK	HT	TK	HK	TK
T	T	T	TK	T

a. Construct a tally to summarize these observations.
b. Construct a table of relative frequencies for these data.
c. Transform these data into a percentage table.
d. Display the results of these data in a bar chart.

2. In the 63-cities data (see Box 2.4), 15 cities are located in the Northeast, 18 in the Midwest, 16 in the South, and 14 in the Far West. Construct a table of relative frequencies, a percentage table, and a bar chart for this variable.

3. In a sociology class there are 12 seniors, 17 juniors, 6 sophomores, and 3 special students. Construct a percentage frequency table, a table of relative frequencies, and a histogram.

4. Here is the distribution of city age in the 63-cities data set:

Y_i	f	Y_i	f
16	3	96	3
26	2	106	3
36	1	116	6
46	5	126	6
56	5	136	1
66	6	146	3
76	5	176	1
86	11	186	2

Construct a relative frequency distribution, a percentage distribution, and a histogram and a polygon.

5. The GSS asked respondents how many members over 17 years old were living in their households. The frequencies for two years are shown below:

No. of Members	1972 f	1980 f
1	195	370
2	1,007	878
3	280	145
4	96	50
5	24	14
6	7	1
7	0	3
8 or more	4	0

Superimpose two percentage polygons for these two distributions and decide whether there was any noticeable shift in adult household size over the eight years.

6. In the 1974 GSS, 0.491 of American adults thought communism was the worst form of government, 0.263 said it's bad but not the worst, 0.188 agreed that it's all right for some countries, 0.030 said it's a good form of government, and 0.028 didn't know. What were the frequencies giving each response, if the total number interviewed was 1,480?

7. Construct a histogram from the following frequencies in response to a survey item about whether a preschool child is likely to suffer if his or her mother works outside the home: strongly agree ($N = 30$); agree ($N = 70$); disagree ($N = 40$); strongly disagree ($N = 10$).

8. In the 63-cities data, the number of cities with no reform characteristics (i.e., at-large elections, nonpartisan candidates, or a city manager) is 18. Sixteen of these cities have one of these characteristics, 11 have two, and 18 have all three reform attributes. Show the cumulative percentage distribution and construct a cumulative percentage polygon.

9. The Bureau of the Census provided the following information on the number of persons with at least a high school degree, by race (white and nonwhite), for the years 1940, 1950, 1960, and 1970:

	White	*Nonwhite*
1940	22,446,000	695,000
1950	33,807,000	1,398,000
1960	45,002,000	2,688,000
1970	67,559,000	5,357,000

   a. Compute the proportion of persons receiving high school diplomas each decade who were white.
   b. Construct a histogram with decade on the horizontal axis. What do you observe?

10. From Table 2.9, recode serious crimes into categories with intervals of 5,000, with the last category at "60,001 or more." Build a percentage distribution and display it in a polygon. What distortion, if any, occurs by lumping together cities in the last category?

11. Suppose an urban planner decides to classify city populations in the following categories:

under 100,000
101,000–200,000
201,000–300,000
301,000–400,000
etc.

a. What is the width of the measurement intervals?
b. What are the true upper and lower limits to the first four categories?
c. Into which interval should the planner place a city with 100,490 residents?
d. Into which interval should a city with 100,501 residents be placed?

12. Here are data on faculty salaries compiled by a teachers' union:

Interval (thousands)	*f*
$ 6–10	10
$11–15	15
$16–20	20
$21–25	5

a. What is the width of the measurement interval?
b. Give the true limits and midpoints for each interval.

13. The number of hours worked by respondents in the labor force in the 1980 GSS is given as:

No. of Hours	*f*
00–09	17
10–19	41
20–29	65
30–39	98
40–49	436
50–59	71
60–69	53
70–79	19
80 or more	22

State the true limits of each interval, give the interval midpoints, and calculate the proportion in each interval.

14. Use Table 2.9 to construct a percentage table for 1974 general expenditures by the 63 cities, using $200 intervals (in thousands). Show the true limits and midpoint values, as well as the frequencies, cumulative frequencies, percentages, and cumulative percentage distributions.

15. Using the table you constructed in Problem 14, what is the score for the 30th percentile? For the 90th percentile?

16. A scale ranges from 1 to 32 in 16 two-point intervals measuring feminist attitudes. A sample of 50 women shows a cumulative frequency of 37 up to but not including the 14th interval. In the 14th interval are found 8 respondents. What is the score (to two decimal places) of the 80th percentile?

17. In what deciles do scores at the 12th, 47th, and 63rd percentiles fall? What quartiles? What quintiles?

18. Church attendance in the 1980 GSS showed the following distribution:

Score	*Category*	*f*
1.	Never	167
2.	Less than once a year	109
3.	About once or twice a year	230
4.	Several times a year	223
5.	About once a month	98
6.	Two or three times a month	118
7.	Nearly every week	91
8.	Every week	316
9.	Several times a week	109
Total		1,461

If these categories are scored from 1 to 9 as shown, give *(a)* the 40th percentile, *(b)* the 3rd quartile, and *(c)* the 9th decile.

19. A social psychologist reasons that "proximity breeds interaction," and so people are more likely to spend a social evening with neighbors than with friends who live outside their neighborhood. Test this proposition by comparing the 50th percentiles on these two distributions from the 1978 GSS:

Score	**Category**	***Frequency of Visits***	
		With Neighbors	**With Friends**
7.	Almost every evening	92	33
6.	Once or twice a week	359	289
5.	Several times a month	167	319
4.	About once a month	189	252
3.	Several times a year	206	324
2.	About once a year	130	130
1.	Never	379	179
Total		1,522	1,526

20. Construct a cumulative percentage graph for both distributions in Problem 19. Give a verbal interpretation of the relationship between the two lines.

Problems Requiring the 1980 General Social Survey Data

21. Some political analysts attributed Ronald Reagan's election in 1980 to public dissatisfaction with the U.S. international military posture. Use responses to the question about how much defense spending should be increased (DEFSPDR) to find where the 50th percentile lies on a seven-point scale. Display in a histogram.

22. Show how to decide whether homosexuality (HOMOSEX) or extramarital sexual relations (XMARSEX) receives greater public condemnation, by displaying the percentage distributions of these two items.

23. Historically, women have married at younger ages than men, but there may be a change over the generations. First, recode the age at first marriage (AGEWED) into decades (e.g., 10 to 19, 20 to 29, etc.). Then, using SEX and AGE, select four groups: (1) women born before 1946, (2) women born in 1946 and after, (3) men born before 1946, and (4) men born in 1946 and after. Show the four relative frequency distributions and report the 50th percentiles for each. What do you find?

24. Use the number of voluntary group memberships (MEMNUM) to construct a frequency polygon. What is the general shape of this graph?

25. Compare the relative frequency distributions for family income when the respondent was growing up (INCOM16) with present income (FINRELA). Is there any noticeable difference?

Problems Requiring the 63-Cities Data

26. Compute the frequency distribution, relative frequency distribution, and cumulative percentage distribution for the number of municipal functions performed by the sample cities (FUNCTION). Construct a histogram.

27. Display the percentage distribution of municipal birth rates in 1968 (BIRTHS). What are the 25th, 50th, and 75th percentiles?

28. Recode city population (POPULAT) in the 63-cities data into categories with 100,000 intervals. Give the true interval limits and the category midpoint values. Show the cumulative percentage table and find the 50th percentile.

29. Find the percentiles on percent black (BLACKPCT) for these communities: Detroit, Euclid, Memphis, Washington, D.C., and Palo Alto.

30. Show the frequency distribution, percentage distribution, and cumulation proportions for the percentage of the labor force that is female (FEMLABOR). What percentile of the cities has a score of 40% female?

Statistics for
Social Data
Analysis

Describing Frequency Distributions

3

By our nature, we try to simplify the world so we can understand and describe it better. The social scientist, for example, uses average income, median age of the U.S. population, and so on as convenient shorthand descriptions of entire distributions of data. This chapter introduces a number of **statistics** for summarizing distributions of discrete and continuous variables. As we noted in Chapter 1, discrete variables classify persons, objects, or events according to the *quality* of their attributes, and continuous variables classify them according to *quantity.*

statistic—a numerical characteristic of a sample, usually designated by an italic English letter

Statistics for summarizing or describing frequency distributions serve two broad purposes. We want to find a single number or label to describe the average or **central tendency** of a distribution (e.g., "The median age of persons who smoke marijuana at least once a week is . . ."; "The modal category of religious preference in France is . . ."). We also want to know how typical that number or label is of the other members of the sample. Therefore, we need a second class of statistics to assess the amount of **variation** in the distribution. For example, we may want to know whether most persons are clustered close to the mean (defined below) or the cases are dispersed widely away from it. If the cases are close to the mean, that measure describes the typical element in the distribution better than if they are dispersed.

central tendency—a value that describes the typical outcome of a distribution of scores

variation—the spread or dispersion of a set of scores around some central value

3.1. Measures of Central Tendency

Commonly used measures of central tendency include the mode, the median, and the mean. The description and use of these three statistics are presented in this section.

3.1.1. Mode

mode—The value of the category in a frequency distribution that has the largest number, or percentage, of cases

The most elementary measure of central tendency is the mode. The **mode** is the category among the various *(K)* categories in the distribution with the largest number (or percentage) of cases. Figure 2.1 in Chapter 2, for example, clearly shows that the *modal* political party preference in the United States in 1977 was "Democrat," and Table 2.6 shows that the mode of ethnic ancestry that year was "not codable." This finding suggests a trend towards the "melting pot" phenomenon, or assimilation of nationalities, since the largest category of respondents failed to provide a specific ethnic classification when asked to do so. Note, however, that the modal category does not necessarily contain a majority of cases, only *more* than any other single category.

FIGURE 3.1
Distributions of Attitudes towards Premarital Sex

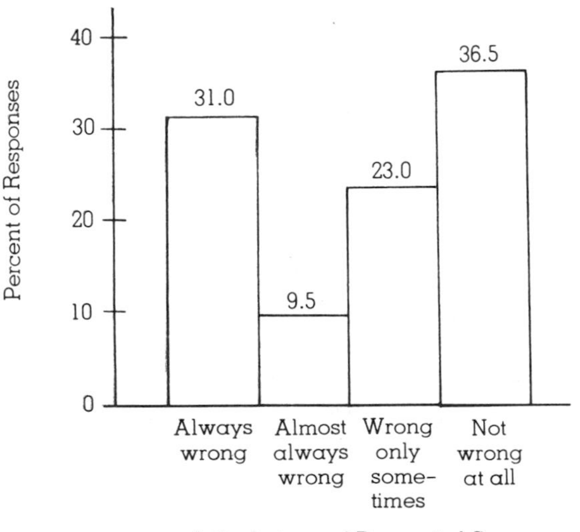

Source: Data derived from 1977 General Social Survey.

Some distributions are said to be *bimodal* (i.e., to have *two* modes). Strictly speaking, in a bimodal distribution the largest two categories each contain the same number of cases. But in practice this equality very rarely occurs, and scientists use the term to describe any distribution where two categories are roughly equal and contain the greatest frequency of cases.

An example of a distribution which is not strictly bimodal, but would be said to be, is shown in Figure 3.1. In the 1977 GSS, respondents were asked "If a man and a woman have sex relations before marriage, do you think it is always wrong, almost always wrong, wrong only sometimes, or not wrong at all?" As Figure 3.1 illustrates, respondents tended to reply either "not wrong at all" or "always wrong," with far fewer persons choosing the middle two categories.

> Any statistic that can be used to describe discrete measures can also be used to describe continuous measures. Thus the mode is a central tendency statistic for both types of variables.

3.1.2. Median

A second measure of central tendency, the median, is only suitable for describing central tendency in distributions where the categories of the variable can be ordered, as from lowest to highest. The **median** is the outcome which divides an ordered distribution into two equal-sized halves.

median—The value or score that exactly divides an ordered frequency distribution into equal halves; the outcome associated with the 50th percentile

> The median can be determined by cumulating (adding up) percentages from either end of the ranked scale until the category which contains the 50th percentile (P_{50}) has been identified. *That category is the median for that distribution.* If there is an odd number of cases, the middle case is the median. If there is an even number of observations, the median is the point half way between the two middle observations.

In distributions with only a few cases, computation of the median is easy. Consider two examples:

Distribution A: 1, 4, 6, 7, 8, 10, 22, 24, 24
Distribution B: 2, 3, 7, 9, 9, 11, 12, 13, 14, 21

In distribution A there is an odd number of cases (nine); the median is the fifth observation, or 8. In distribution B there is an even number of cases (10), so the median is halfway between the fifth and sixth cases, which is $(9 + 11)/2 = 10$.

Unfortunately, the median is not so easy to compute with most real distributions of data, given the large number of cases in each category. By using information developed in Chapter 2, however, the task can be simplified considerably. Since the median is the point above and below which the cases fall, it is the 50th percentile, or P_{50}. The formula is:

$$\text{Median} = P_{50} = L_p + \left(\frac{(0.5N - c_p)}{f_p} \right) (W_{50})$$

Where:

L_p = The real lower limit of the interval containing P_{50}.
N = The total number of observations.
c_p = The cumulative frequency up to but not including the interval containing P_{50}.
f_p = The frequency of the interval containing the P_{50} percentile.
W_{50} = The width of the interval containing P_{50}.

In 1977 the GSS respondents were asked to place themselves on a liberal-conservative scale. The results are shown in Table 3.1.

TABLE 3.1
Distribution of Political Preferences on a Liberal-Conservative Scale, GSS Respondents

Code	***Response Category***	***f***	***%***	***cf***	***c%***
1	Extremely liberal	37	2.5	37	2.5
2	Liberal	169	11.6	206	14.1
3	Slightly liberal	214	14.7	420	28.8
4	Moderate, middle-of-the-road	564	38.8	984	67.6
5	Slightly conservative	251	17.3	1,235	84.9
6	Conservative	179	12.3	1,414	97.2
7	Extremely conservative	39	2.7	1,453	99.9

Note: 77 persons who responded "don't know" or who gave no answer have been eliminated from this analysis.

Source: 1977 General Social Survey.

The mode is "moderate, middle of the road." To compute the median we coded the seven response categories from 1 to 7 and constructed the cumulative frequency and percentage distributions. The median is:

$$P_{50} = 3.5 + \left(\frac{(0.5)(1{,}453) - 420}{564} \right)(1)$$
$$= 4.04$$

Since the "moderate, middle-of-the-road" category is coded 4, the mode and the median clearly are virtually identical in this case. But they need not always be. Respondents in the 1977 GSS were also asked how much satisfaction they got from their families. As the results in Table 3.2 show, the modal response is "a very great deal" (coded 7), but the median is 5.93, which is closer to "a great deal."

TABLE 3.2

Distribution of Degree of Satisfaction Derived from the Family, GSS Respondents

Code	*Response Category*	*f*	*%*	*cf*	*c%*
1	None	26	1.7	26	1.7
2	A little	21	1.4	47	3.1
3	Some	43	2.8	90	5.9
4	A fair amount	112	7.4	202	13.3
5	Quite a bit	181	11.9	383	25.2
6	A great deal	501	32.9	884	58.1
7	A very great deal	637	41.9	1,521	100.0

Note: 9 persons who responded "don't know" or who gave no answer have been eliminated from this analysis.

Source: 1977 General Social Survey (coding was reversed from original).

When the distribution of an ordered variable is plotted, the plot may be nonsymmetric about the median. When this is the case and one end of the distribution has a long "tail" (i.e., there are many categories with few cases), the result is said to be a **skewed distribution.** The distribution of satisfaction with family shown in Figure 3.2, for example, using the data in Table 3.2, is skewed. *When distributions are skewed, the mode and the median differ, as they did in this example.* When the long tail is to the right of the median, the distribution is said to have **positive skew**; when the tail is to the left, the distribution has **negative skew.** Convince yourself that satisfaction with family has a negative skew by studying Table 3.2 and Figure 3.2.

skewed distribution—a frequency distribution that is asymmetric with regard to its dispersion

positive skew—an asymmetrical frequency distribution characteristic whereby, in a graphic display, larger frequencies are found toward the negative end and smaller frequencies toward the positive end

negative skew—a property of frequency distribution whereby larger frequencies are found toward the positive end and smaller frequencies toward the negative end

FIGURE 3.2
Distribution of Degree of Satisfaction Derived from the Family

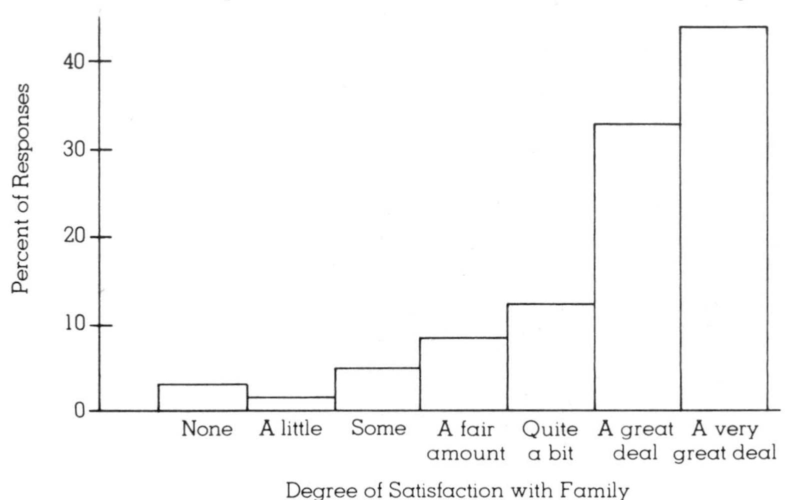

Source: Data derived from Table 3.2.

3.1.3. Mean

mean—a measure of central tendency for continuous variables calculated as the sum of all scores in a distribution, divided by the number of scores; the arithmetic average

By far the most frequently used measure of central tendency is the mean, which is used only with continuous measures. The **mean** is commonly called *the average*; that is, we simply add together the value of each observation and divide by the number of observations. If we designate the mean of a sample of N observations by $\bar{Y}$, then the statistic is given by:

$$\bar{Y} = \frac{\sum_{i=1}^{N} Y_i}{N}$$

The Greek letter *sigma* (Σ) is widely used as a symbol to signify summation. Rules on its use throughout this text are given in Appendix A, "The Use of Summations."

To illustrate the mean statistic, we can use the serious crime rates for just the 20 largest cities in the 63-cities study for 1970 (see Box 2.4 in Chapter 2). These are shown in Table 3.3. Several statistics you have already encountered can be readily calculated on these data. For example, the mode is 59.7 (two cities, Dallas and Boston, have this rate), while the median is 59.0. Because

TABLE 3.3
Serious Crimes per 1,000 Population in 20 Largest U.S. Cities

Population Rank	*City Name*	*Crime Rate*
5	Detroit	84.4
13	San Francisco	79.8
9	Washington	78.4
18	St. Louis	73.8
7	Baltimore	68.6
1	New York	65.7
3	Los Angeles	62.4
8	Dallas	59.7
16	Boston	59.7
10	Cleveland	59.3
20	Seattle	58.7
19	Phoenix	50.7
6	Houston	48.6
15	San Antonio	41.6
2	Chicago	38.0
17	Memphis	34.7
14	San Diego	34.0
11	Indianapolis	33.3
12	Milwaukee	28.2
4	Philadelphia	23.5

Source: NORC Permanent Community Sample (63-cities data), 1970.

crime rates have continuous characteristics, we can also calculate the mean, as follows:

$$\begin{aligned}\bar{Y} &= (84.4 + 79.8 + 78.4 + 73.8 + 68.6 + 65.7 \\ &\quad + 62.4 + 59.7 + 59.7 + 59.3 + 58.7 + 50.7 + 48.6 \\ &\quad + 41.6 + 38.0 + 34.7 + 34.0 + 33.3 + 28.2 + 23.5)/20 \\ &= (1{,}083.1)/20 \\ &= 54.16\text{, or, rounding, } 54.2\end{aligned}$$

As a measure of central tendency, the mean has a property no other statistic shares: *the mean is the number which has the lowest average squared distance from all the numbers in the distribution.* If we subtract the mean from each of the scores and square these differences (we square because some of the differences will be negative), the average of these squared differences will be smaller than the average we would obtain using any other number than the mean. Thus, we can say that the mean *minimizes the sum of the squared distances to all points in the data.* If you doubt this statement, try a few examples on the data in Table 3.3 using some value other than the mean. You will find that even a slightly different value for the mean (such as 54.3 instead of 54.2) will yield a higher average squared distance than the exact mean. (The sum of the squared differences from the mean of 54.2 is 6,363.26.) We will have an important use for this property shortly.

Unfortunately, the method presented above is not useful for computing the mean for data presented in the form of a frequency distribution. Since many data are presented in this way, you should learn a slightly more complex form of the formula for the mean. This is:

$$\bar{Y} = \frac{\sum_{i=1}^{K} (Y_i f_i)}{N}$$

Where:

f_i = The frequency of cases of type Y_i.

K = The number of outcomes in the distribution.

The following example shows the use of this version of the formula for computing the mean. In the 1975 and 1978 General Social Surveys, respondents were asked how many hours of television they watched on an average day. The distribution is shown in Table 3.4. The mean number of hours watched in 1975 is:

$$\begin{aligned}\bar{Y}_{1975} &= [0(57) + 1(256) + 2(395) + 3(291) + 4(217) \\ &\quad + 5(116) + 6(76) + 7(19) + 8(23) + 9(1) + 10(15) \\ &\quad + 11(1) + 12(10) + 13(1) + 14(2) + 15(1) + 16(1) \\ &\quad + 17(1) + 18(0) + 20(0) + 24(0)]/1{,}483 \\ &= 3.05 \text{ hours per day}\end{aligned}$$

Similarly, the mean number of hours watched in 1978 is:

$$\begin{aligned}\bar{Y}_{1978} &= [0(91) + 1(316) + 2(418) + 3(287) + 4(194) + 5(99) \\ &\quad + 6(55) + 7(8) + 8(30) + 9(3) + 10(12) + 11(0) \\ &\quad + 12(10) + 13(1) + 14(0) + 15(1) + 16(0) + 17(0) \\ &\quad + 18(1) + 20(1) + 24(1)]/1{,}528 \\ &= 2.79 \text{ hours per day}\end{aligned}$$

TABLE 3.4
Distribution of Hours Spent Watching Television, GSS Respondents

No. of Hours	1975		1978	
	f	%	*f*	%
0	57	3.8	91	6.0
1	256	17.3	316	20.7
2	395	26.6	418	27.4
3	291	19.6	287	18.8
4	217	14.6	194	12.7
5	116	7.8	99	6.5
6	76	5.1	55	3.6
7	19	1.3	8	0.5
8	23	1.6	30	2.0
9	1	0.1	3	0.2
10	15	1.0	12	0.8
11	1	0.1	0	0.0
12	10	0.7	10	0.7
13	1	0.1	1	0.1
14	2	0.1	0	0.0
15	1	0.1	1	0.1
16	1	0.1	0	0.0
17	1	0.1	0	0.0
18	0	0.0	1	0.1
20	0	0.0	1	0.1
24	0	0.0	1	0.1
Totals	1,483	100.0	1,528	100.0

Source: 1975 and 1978 General Social Surveys.

These results suggest that, compared to 1975, in 1978 Americans were watching about 3.05 − 2.79 = 0.26 hours, or 0.26(60 minutes) = 15.6 minutes less television per day. An important question, however, is whether or not the observed difference is due to sampling chance. We will address questions of this sort in the next chapter.

> When a distribution of scores is a relatively symmetric one, the mean, median, and mode will be very close to each other in value. However, when the distribution is skewed, the three can differ rather sharply, as the next example shows.

In the 1977 GSS, respondents were asked to rate their liking of a series of countries on a scale from −5 (dislike very much) to +5 (like very much). The distributions of responses for two countries, Egypt and Canada, are shown in Table 3.5, and these distributions are graphed in Figure 3.3. Since the like-dislike score did not contain a zero, the responses were recoded from 0 to 9.

TABLE 3.5
Distribution of Liking Scores for Egypt and Canada, GSS Respondents

		Egypt				Canada			
Score	*Code*	*f*	*%*	*cf*	*c%*	*f*	*%*	*cf*	*c%*
−5	0	82	6.1	82	6.1	14	1.0	14	1.0
−4	1	55	4.1	137	10.2	3	0.2	17	1.2
−3	2	75	5.6	212	15.8	11	0.8	28	1.9
−2	3	92	6.8	304	22.6	12	0.8	40	2.8
−1	4	254	18.9	558	41.5	27	1.9	67	4.7
+1	5	335	24.9	893	66.4	98	6.8	165	11.5
+2	6	168	12.5	1,061	78.9	119	8.3	284	19.8
+3	7	148	11.0	1,209	90.0	269	18.7	553	38.5
+4	8	63	4.7	1,272	94.6	331	23.1	884	61.6
+5	9	72	5.4	1,344	100.0	552	38.4	1,436	100.0

Source: 1977 General Social Survey.

The distribution of responses for Egypt is roughly symmetric, and the mean, median, and mode are 4.74, 4.84, and 5.00, respectively. By contrast, the mean, median, and mode for Canada are 7.57, 8.00, and 9.00, reflecting the obvious negative skew to U.S. citizens' evaluations of Canada (see Section 3.1.2).

In general, for distributions with a positive skew the mean is the largest, the median second largest, and the mode the smallest statistic. The opposite ordering holds for distributions with a negative skew. *The mean and median differ because the mean is a weighted average—extreme values affect it—whereas the median*

FIGURE 3.3
Distributions of Scores Measuring Degree of Liking by U.S. Citizens for Egypt and Canada

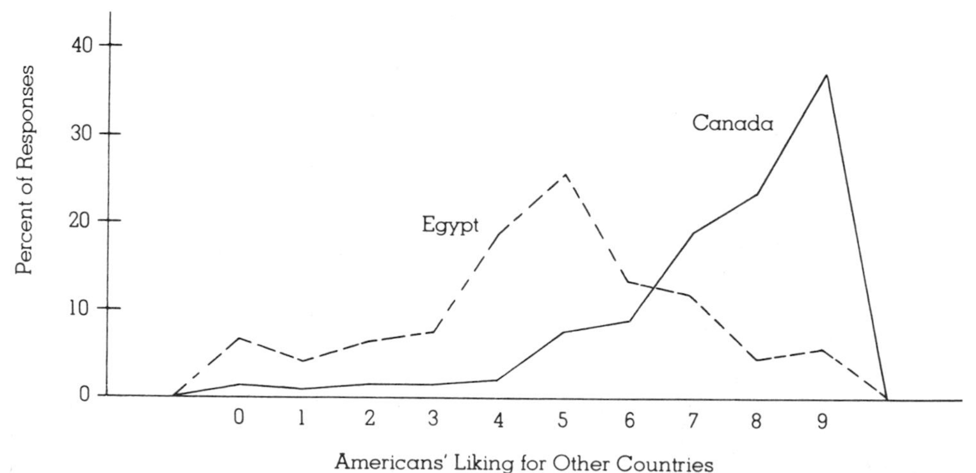

Source: Data derived from Table 3.5.

is not. In the last example, if we had recoded a 9 to a 20, the medians would be unaffected, but the means would change rather dramatically. To convince yourself of this, change the code 9 in Table 3.5 to a 20 and recompute the medians and means.

Because the mean is a weighted average, some scientists prefer to use the median as a measure of central tendency in distributions which are highly skewed, such as distributions of income.

3.2. Measures of Variation

3.2.1 Index of Diversity

Variation in discrete and continuous variables can be measured in several ways. Some of these measures, however, are better suited for discrete variables. One of the most common measures is called the **index of diversity.** This index is symbolized *D,* and it is given by:

index of diversity— a measure of variation for a discrete variable that indicates the likelihood two observations drawn at random from a population are from different categories of the variable

$$D = 1 - \sum_{i=1}^{K} p_i^2$$

Where:

p_i = The proportion of cases in the *i*th category.

This index estimates the likelihood that two observations drawn randomly from a population (see Section 1.7.3) are from different categories. To calculate the value of *D,* we simply square the proportion of observations in each category, add up the squares, and subtract them from 1.

The minimum possible value for D is zero; this occurs when all members of a population fall into a single category. The index of diversity achieves its maximum value when the population is spread evenly over all *K* categories of the variable. In general, *the maximum value occurs when the proportion of cases in each category is* $1/K$, and $D = 1 - 1/K = (K - 1)/K$. But this maximum value varies as a function of the number of categories. For a 4-category discrete variable, the maximum value of *D* is 0.75; for a 10-category variable, the maximum is 0.90. The more categories a discrete variable has, the bigger is the maximum value that D can assume. Thus, *comparing values of D for discrete variables with different numbers of categories cannot be done directly.*

A useful further calculation might be to report the maximum value of D for a discrete variable, along with the observed value of the actual distribution. For example, the full 41-category ethnic ancestry variable in Table 2.5 (Chapter 2) has an observed value of 0.902 with a maximum of 0.976, while the collapsed 12-category ethnic ancestry variable in Table 2.6 has observed and maximum values of 0.868 and 0.917, respectively. Thus, the full-category version has a more unequal concentration across its categories, relative to its maximum D, than the recoded version of the variable does.

3.2.2. Index of Qualitative Variation

index of qualitative variation—a measure of variation for discrete variables; a standardized version of the index of diversity

A second, more useful measure of variation for discrete variables is the **index of qualitative variation** (IQV), which is a standardized version of the diversity index, D. It is given by this formula:

$$\text{IQV} = \frac{(1 - \sum_{i=1}^{K} p_i^2)}{(K - 1)/K} = \frac{K}{K-1}(D)$$

This measure is very similar to D, except *its maximum value is 1.0, which occurs when the cases are spread evenly over all K categories of the variable,* that is, when $p_i = 1/K$ for all p_i. Like D, IQV has a minimum value of zero, which occurs when all cases fall into a single category. IQVs for the original and recoded distributions of ethnic ancestry in Tables 2.5 and 2.6 are 0.925 for the 41-category version and 0.947 for the 12-category variable. Thus, the conclusions differ about the amount of dispersion in the full versus recoded versions of this variable, using D and IQV. When we take the number of categories into account more variation occurs in the 12-category version, although the difference is very small. We would expect this difference because we have simply recoded the same variables, and IQV takes this recoding into account.

Using the data in Table 2.5, IQV for religion is 0.607, and for party identification it is 0.761. Therefore, we can conclude that religion is more unequally distributed in the American population than is party identification, as can readily be seen by the greater concentration of Protestants than of Democrats. Examples for computing the index of diversity and the index of qualitative variation for the distribution of religious preferences shown in Table 2.5 are given in Box 3.1 on the next page.

The 1977 GSS question about attitudes toward premarital sex which was described in Section 3.1.1 was also asked in the 1972 and 1978 surveys. Figure 3.1 (above) illustrates a distribution for the 1977 data which would be described as bimodal. In Table 3.6, which gives the data for 1972 and 1978, both distributions

TABLE 3.6
Percentage Distribution of Attitudes about Premarital Sex, GSS Respondents

	Year	
Response Categories	***1972***	***1978***
Always wrong	34.9%	28.5%
Almost always wrong	11.3	11.4
Wrong only sometimes	23.1	19.8
Not wrong at all	26.0	37.7
Don't know/No answer	4.7	2.5
Total	100.0%	99.9%*

*Does not total to 100% due to rounding.
Source: 1972 and 1978 General Social Surveys.

can be thought of as bimodal in shape, but the true mode has shifted from "always wrong" in 1972 to "not wrong at all" in 1978. This shift suggests a significant change in attitudes toward premarital sex in the six-year period.

Along with this modal shift to a more liberal view about premarital sex, we might expect to see an increase in variation due to the increased polarization of attitudes. To examine this possibility we computed the IQVs for the two distributions. The IQV for 1972 is 0.928, and for 1978 it is 0.905. While the IQVs suggest a decrease in the variability of responses to the premarital sex question between 1972 and 1978, the decrease in variation is very small. The hypothesis that there has been an increase in variation due to increased polarization is clearly *not* supported.

BOX 3.1 Calculating Variation for Discrete Variables

To illustrate more clearly how to calculate variation for discrete variables, the following example uses the data on religious preference from Table 2.5 (Chapter 2). Here again, p_i is the proportion of cases in the ith category.

Categories	P_i	P_i^2
Protestant	0.656	0.430
Catholic	0.242	0.059
Jewish	0.023	0.001
None	0.061	0.004
Other	0.012	0.000
No answer	0.005	0.000

$$D = 1 - \sum_{i=1}^{6} p_i^2 = 1 - (0.430 + 0.059 + 0.001 + 0.004 + 0.000 + 0.000)$$

$$= 1 - 0.494 = 0.506$$

$$\text{IQV} = (\text{K} / \text{K} - 1)(\text{D}) = (6/5)(0.506) = 0.607$$

3.2.3 Range

range—a measure of dispersion based on the difference between the largest and smallest outcomes in a distribution

Continuous variables also have special statistics to describe their variation or degree of spread across the scale. The simplest of these is the **range**, which indicates the two extreme scores between which all other responses are found in the sample at hand. The range of the outcomes of a distribution is defined as the difference between the largest and smallest outcomes. The range for the liberal-conservative political preference variable (see Table 3.1) is from 1 to 7, or $7 - 1 = 6$.

The range communicates relatively little information about the sample, since it is based on only the two extreme scores. It does not tell us how much the other persons or events are spread out or bunched up on the scale values between the highest and lowest scores.

interquartile range—a measure of dispersion indicating the difference in scores between the 25th and 75th percentiles

Another related measure of variation is the **interquartile range** (IQR). Like the range, it uses information on only two scores. In the IQR, however, these scores are not the extremes but the 25th and 75th percentiles—P_{25} and P_{75}. The interquartile range is given by:

$$\text{IQR} = P_{75} - P_{25}$$

The IQR gives the upper and lower scores between which the middle half of the sample may be found. For the political preference variables, these are:

$$P_{25} = 2.5 + \left(\frac{(0.25)\,(1{,}453) - 206}{214} \right) = 3.24$$

$$P_{75} = 4.5 + \left(\frac{(0.75)\,(1{,}453) - 984}{251} \right) = 4.92$$

Hence the interquartile range is 4.92 − 3.24 = 1.68. These results confirm that persons saying they are extremely liberal, liberal, conservative or extremely conservative are somewhat rare. The majority are moderate middle-of-the roaders, with only slight leanings in either direction.

3.2.4. Average Deviation

The central tendency property of the mean is highly useful in developing a measure of variability for continuous data—the spread or dispersion of scores away from the mean. The range, as we have noted, is too sensitive to the two extreme scores to tell us much about variability. A good measure of variation should take into account the location of all scores in relationship to the mean.

Note first that the distance of a score, Y_i, from the mean, $\bar{Y}$, is commonly calculated as:

$$d_i = Y_i - \bar{Y}$$

This results in some distances being positive and some being negative, depending on whether the particular observation is above or below the mean. If we were to attempt to calculate the average distance of all N scores from the mean, we should always arrive at 0. This is because another property of the mean is that it equalizes the distances in both directions. (Try this exercise on Table 3.3, if you need to convince yourself.) Thus, the average distance is useless as a statistical measure of variation for continuous variables, since all distributions will give the same value: zero.

A second approach to measuring variation which would eliminate the problem of positive and negative distance is to take the absolute values of the distances before averaging them. This cal-

average deviation—the mean of the absolute values of the difference between a set of continuous measures and their mean

culation is called the **average deviation** (AD) about the mean. The formula is:

$$AD = \frac{\Sigma |d_i|}{N}$$

where the vertical bars indicate the absolute values of the d_is.

As useful as this change is, it lacks a second criterion which is desirable for a measure of variation: Variation about the mean should be *less* than for a comparable measure calculated around any other value. That is, the average deviation about the mean should be smaller than the average deviation about another score. This is not the case with the crime data in Table 3.3, however, as can be seen by taking the average deviation about the mean and the average deviation about the median (the median in Table 3.3 is 59.0). The average deviation about the mean, 15.48, is larger than the average deviation about the median, 15.03, which is always the case. Thus, even though the mean minimizes the average squared distance to each score, the average deviation about the mean is not less than for some other values.

Although the average deviation is a meaningful statistic, it has certain properties that are less desirable as a variability measure than the variance and standard deviation measures introduced in the following section. The average deviation, therefore, rarely appears in most analyses of social data.

3.2.5. Variance and Standard Deviation

variance—a measure of dispersion for continuous variables indicating an average of squared deviations of scores about the mean

As defined in Section 3.1.3, the mean is the number which minimizes the average squared distance from all the numbers in a distribution. We can use this property of the mean to build a highly desirable measure of variability for a distribution—the **variance** of a set of continuous scores. This is the average[1] of the squared deviations of the scores about the mean. That is, the variance of a variable Y (denoted s_Y^2) is the *mean squared deviation*. This formula can be written several ways, two of which are:

1. The variance is an average even though Σd_i^2 is divided by $N - 1$ instead of N. We will explain why in Chapter 5.

$$s_Y^2 = \frac{\sum_{i=1}^{N} d_i^2}{N - 1}$$

$$s_Y^2 = \frac{\sum_{i=1}^{N} (Y_i - \bar{Y})^2}{N - 1}$$

Replacing the mean, $\bar{Y}$, in the formulas with any other number will always result in a larger value for s_Y^2. Table 3.7 shows the step-by-step calculation of the variance for crime rates in 20 cities

TABLE 3.7
Calculation of Variance for City Crime Rates Listed in Table 3.3

$(Y_i - \bar{Y})$		d_i	d_i^2
84.4 − 54.2	=	30.2	912.04
79.8 − 54.2	=	25.6	655.36
78.4 − 54.2	=	24.2	585.64
73.8 − 54.2	=	19.6	384.16
68.6 − 54.2	=	14.4	207.36
65.7 − 54.2	=	11.5	132.25
62.4 − 54.2	=	8.2	67.24
59.7 − 54.2	=	5.5	30.25
59.7 − 54.2	=	5.5	30.25
59.3 − 54.2	=	5.1	26.01
58.7 − 54.2	=	4.5	20.25
50.7 − 54.2	=	−3.5	12.25
48.6 − 54.2	=	−5.6	31.36
41.6 − 54.2	=	−12.6	158.76
38.0 − 54.2	=	−16.2	262.44
34.7 − 54.2	=	−19.5	380.25
34.0 − 54.2	=	−20.2	408.04
33.3 − 54.2	=	−20.9	436.81
28.2 − 54.2	=	−26.0	676.00
23.5 − 54.2	=	−30.7	942.49

$$\sum d_i^2 = 6{,}359.21$$

$$s_Y^2 = \frac{\sum d_i^2}{N - 1} = \frac{6{,}359.21}{19} = 334.70$$

(shown in Table 3.3), after rounding the mean to 54.2 to simplify the arithmetic. *The variance is always a nonnegative number, due to the squaring.* Note that the variance, being an average of squared distances from the mean, is a large number relative to the original scores ($s_Y^2 = 334.70$ with a scale whose highest reported score is only 84.4).

standard deviation—the square root of the variance

To restore the measure of variance to the original units of measurement, we take its positive square root, which is called the **standard deviation**. This is given by:

$$s_Y = \sqrt{s_Y^2}$$

For the 20-city crime data, the standard deviation is 18.3 crimes per thousand population, since $\sqrt{334.70} = 18.3$. Some useful properties of the standard deviation will be shown in Chapter 5, which discusses inferential statistics.

When data are grouped in a frequency distribution, as they are in Tables 3.4 and 3.5, a different formula for the variance is needed. This is:

$$s_Y^2 = \sum_{i=1}^{K} \frac{d_i^2 f_i}{N-1} = \sum_{i=1}^{K} \frac{(Y_i - \bar{Y})^2 f_i}{N-1}$$

To use this formula we first form each deviation. Then we square it, multiply it by the frequency of cases of type Y_i, sum the weighted squared deviations for the K outcomes of Y, and divide the sum by $N - 1$. This is done in Table 3.8 for the data on Egypt from Table 3.5. As the computations indicate, the variance is 4.93, and the standard deviation is 2.22.

Convince yourself that the variance and standard deviation for the data on Canada in Table 3.5 are 2.84 and 1.69. These figures, when compared to those for Egypt, make it clear that there is considerably more variability in liking for Egypt than there is in liking for Canada. Nearly all Americans in 1977 liked Canada, but opinions on Egypt were much more variable. Many liked the Middle Eastern nation, but many others did not.

TABLE 3.8
Calculation of Variance for Data in Table 3.5 on Liking for Egypt

$(Y_i - \bar{Y})$	d_i	d_i^2	f_i	$d_i^2 f_i$
(0 − 4.74)	−4.74	22.47	82	1,842.54
(1 − 4.74)	−3.74	13.99	55	769.45
(2 − 4.74)	−2.74	7.51	75	563.25
(3 − 4.74)	−1.74	3.03	92	278.76
(4 − 4.74)	−0.74	0.55	254	139.70
(5 − 4.74)	0.26	0.07	335	23.45
(6 − 4.74)	1.26	1.59	168	267.12
(7 − 4.74)	2.26	5.11	148	756.28
(8 − 4.74)	3.26	10.63	63	669.69
(9 − 4.74)	4.26	18.15	72	1,306.80

$$\sum_{i=1}^{10} d_i^2 f_i = 6{,}617.04$$

$$s_Y^2 = \frac{\sum_{i=1}^{N} d_i^2 f_i}{N-1} = \frac{6{,}617.04}{1{,}344 - 1} = 4.93$$

$$s_Y = \sqrt{s_Y^2} = 2.22$$

3.3. *Z* Scores

The researcher is often interested in comparing scores across distributions with different means and standard deviations. But when two distributions have different means and standard deviations, the same score can mean something quite different in each case. For example, in Table 3.5 a recoded score of 6 means something quite different in the data on liking for Egypt than it does in the data on liking for Canada. A 6 in the data on Canada means the respondent likes Canada somewhat less than the average American ($\bar{Y} = 7.57$). But a 6 in the data on Egypt means that the person likes Egypt somewhat *better* than the average American, since for these data $\bar{Y} = 4.74$.

To make comparisons across distributions, taking into account varying means and standard deviations, we compute *standard*

Z scores (standard scores)—a transformation of the scores of a continuous frequency distribution by subtracting the mean from each outcome and dividing by the standard deviation

scores, or, as they are more commonly called ***Z* scores**. The formula is:

$$Z_i = \frac{d_i}{s_Y} = \frac{(Y_i - \bar{Y})}{s_Y}$$

Table 3.9 shows the Z scores corresponding to the data for liking of Egypt and Canada presented in Table 3.5. The Z scores

TABLE 3.9

Standard Scores (Z Scores) for Data in Table 3.5 on Liking for Egypt and Canada

Liking for Egypt			*Liking for Canada*		
Original Recoded Scores	$(Y_i - \bar{Y})/s_Y$	Z_i	*Original Recoded Scores*	$(Y_i - \bar{Y})/s_Y$	Z_i
0	(0 − 4.74)/2.22	−2.14	0	(0 − 7.57)/1.69	−4.48
1	(1 − 4.74)/2.22	−1.68	1	(1 − 7.57)/1.69	−3.89
2	(2 − 4.74)/2.22	−1.23	2	(2 − 7.57)/1.69	−3.30
3	(3 − 4.74)/2.22	−0.78	3	(3 − 7.57)/1.69	−2.70
4	(4 − 4.74)/2.22	−0.33	4	(4 − 7.57)/1.69	−2.11
5	(5 − 4.74)/2.22	0.12	5	(5 − 7.57)/1.69	−1.52
6	(6 − 4.74)/2.22	0.57	6	(6 − 7.57)/1.69	−0.93
7	(7 − 4.74)/2.22	1.02	7	(7 − 7.57)/1.69	−0.34
8	(8 − 4.74)/2.22	1.47	8	(8 − 7.57)/1.69	0.25
9	(9 − 4.74)/2.22	1.92	9	(9 − 7.57)/1.69	0.85

indicate that a recoded score of 5 for the data on Egypt is roughly comparable to an 8 in the data on Canada, relative to what the average American feels for these two countries.

Since the data are standardized when using Z scores, you should not be surprised that a distribution of Z scores has the *same* mean and standard deviation. In particular, when Z scores are computed on *any* distribution, *the mean of the Z-score distribution will always be zero, and the standard deviation will always be 1.* This is difficult to see from the data in Table 3.9, since the observations are grouped. However, you should convince yourself that the mean of Z scores is zero, with a standard deviation of 1 for the data on crime rates in Table 3.3.

Since Z scores are used *very* often in this text, you should be certain you understand this section clearly.

3.4. The Coefficient of Relative Variation

Another descriptive statistic which is used only for continuous measures combines both the mean and the standard deviation. This is the **coefficient of relative variation** (CRV), which is useful for comparing two distributions of scores that have substantially different standard deviations and whose means lie at different points on the scale. The formula is:

coefficient of relative variation—a measure of relative dispersion obtained by dividing the standard deviation by the mean

$$\text{CRV} = \frac{s_Y}{\bar{Y}}$$

The CRV adjusts for the tendency of groups with high scores (and hence high means) to have a greater dispersion of scores. The CRV is an approximate standardization of the standard deviation by the mean. In the data on crime rates from Table 3.3, the CRV is 0.33, indicating that the standard deviation is about one-third the size of the mean. If we had data for crime rates on another set of communities with a higher or lower mean crime rate, we could determine whether the variances of the two distributions relative to their means were similar or different. Many social scientists do not report CRVs for their data; instead they routinely report means and standard deviations. This allows you to compute the CRV values for yourself, assuming the data are measured at the continuous level.

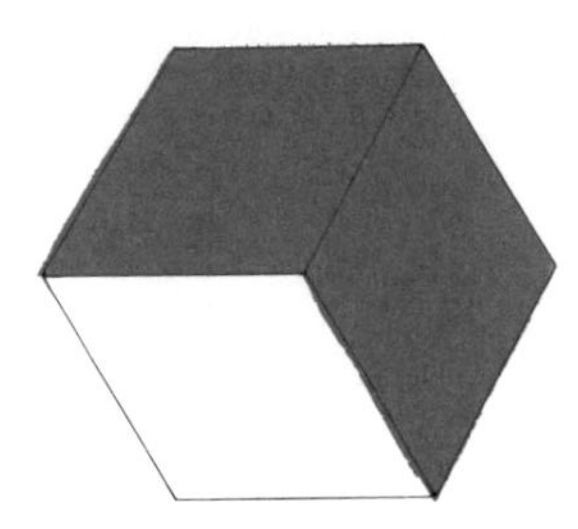

Review of Key Concepts

This list of the key concepts introduced in Chapter 3 is in the order of appearance in the text. Combined with the definitions in the margins, it will help you review the material and can serve as a self-test for mastery of the concepts.

statistic
central tendency
variation
mode
median
skewed distribution
positive skew
negative skew
mean
index of diversity
index of qualitative variation (IQV)
range
interquartile range (IQR)
average deviation
variance
standard deviation
Z scores (standard scores)
coefficient of relative variation (CRV)

PROBLEMS

General Problems

1. Here are 1970 data from 16 noncommunist European nations:

Nation	*Gross National Product (GNP) (billions of $)*	*Percent in Agriculture*	*Crude Birth Rate per 1,000*
Austria	9	18	18
Belgium	17	7	16
Denmark	10	23	18
Finland	8	38	17
France	94	25	18
Great Britain	99	5	18
Greece	6	48	18
Ireland	3	42	22
Italy	57	24	19
Netherlands	19	13	20
Norway	7	24	18
Portugal	4	48	23
Spain	18	36	21
Sweden	20	18	16
Switzerland	14	15	19
West Germany	112	8	18

Find the mode for: *(a)* the crude birth rate, *(b)* the percent employed in agriculture, and *(c)* GNP.

2. Using the data in Problem 1, find the median for *(a)* GNP, *(b)* the percent in agriculture, and *(c)* the crude birth rate.

3. In a sample of adults, respondents gave these responses to a question about the frequency of their newspaper reading: every day, 87; a few times a week, 31; once a week, 15; less than weekly, 11; never 8. Assign values from 4 to 0, respectively, to these categories, and calculate the mode and the median.

4. What is the skew, if any, for *(a)* crude birth rate in Problem 1, and *(b)* newspaper reading in Problem 3?

5. For the data in Problem 1, calculate the means of *(a)* GNP, *(b)* the percent in agriculture, and *(c)* the crude birth rate.

6. In 1980, the General Social Survey found the following data on household size; by number of persons in the household: one = 288; two = 493; three = 259; four = 248; five = 107; six = 38; seven = 22; eight = 6; nine = 1; and ten = 6. What is the mean of household size?

7. What skew, if any, does household size have in Problem 6?

8. In a word recognition test, the following number of correct recognitions were given by 14 subjects: 0, 6, 7, 2, 5, 6, 9, 4, 6, 7, 5, 8, 3, 4. Find the *(a)* mode, *(b)* median, *(c)* mean, and *(d)* skew.

9. Among a sample of veterans of the United States armed forces, the following distribution by service branch was found: Air Force, 56; Navy, 70; Army, 166; National Guard, 11; Marine Corps, 14. Calculate *(a)* the index of diversity and *(b)* the IQV for this distribution.

10. For the data in Problem 3, find *(a)* the index of diversity, and *(b)* the IQV.

11. For the data in Problem 1, find *(a)* the range of GNP, *(b)* the range of percent in agriculture, and *(c)* the range of the crude birth rate.

12. In a percentile distribution of faculty salaries at Big State University, the following values are observed: P_{15} = \$8,500; P_{25} = \$12,000; P_{50} = \$23,400; P_{75} = \$36,300; P_{85} = \$41,250. What is the interquantile range of this distribution?

13. What is the IQR for the word recognition scores in Problem 8?

14. Find the average deviation for the following scores on a mid-term statistics exam: 52, 66, 75, 78, 83, 83, 86, 89, 94.

15. For the data in Problem 14, what are *(a)* the variance, and *(b)* the standard deviation?

16. Find the standard deviations using the data in Problem 1 for *(a)* GNP, *(b)* percent in agriculture, and *(c)* crude birth rate.

17. Find the Z- scores for the following:

	$\bar{Y}$	s_y^2	Y_i
a.	50	400	85
b.	50	400	35
c.	4	9	6
d.	8.50	2.25	6.5
e.	8.50	2.25	10.0
f.	100	25	96

18. For a sample with $\mu_y = 12$ and $\sigma_y^2 = 4$, give the observed scores *(Y)* corresponding to the following Z scores:
    a. $Z = +2.00$.
    b. $Z = -1.33$.
    c. $Z = +4.50$.
    d. $Z = -0.75$.

19. At Big State University, the average grade point average (GPA) is 2.50, with variance 0.36. Who is further from the mean: Charlie Goodtime, with a GPA of 1.65, or Lester Gradegrind, whose GPA Z- score is $+1.50$?

20. Using the data in Problem 1, find the Z- scores on GNP for the following nations: *(a)* Austria, *(b)* Great Britain, *(c)* Italy, *(d)* Portugal, and *(e)* Sweden.

21. Using the data in Problem 1, which nations lie more than 1.5 standard deviations above or below the mean crude birth rate?

22. Here is a grouped frequency distribution of ideal family sizes (the number of children desired) reported by respondents in the 1980 General Social Survey:

Ideal No.	f_i
0	16
1	26
2	752
3	353
4	196
5	34
6	14
7	8

What are *(a)* the mode, *(b)* the median, *(c)* the range, *(d)* the IQR, *(e)* the mean, *(f)* the variance, *(g)* the standard deviation, and *(h)* the Z score for a family size of just one child?

23. In the South, the mean years of formal schooling is 11.5 and standard deviation is 2.25. In the North the mean is 12.75, and the standard deviation is 2.50. Which distribution has greater relative variation?

24. In Smallville, the following incomes were observed (in thousands of dollars): \$3, \$4, \$5, \$5, \$5, \$6, \$6, \$8, \$30. What different features of this distribution are revealed by the three central tendency statistics?

25. Junior faculty members of the chemistry department published the following numbers of articles last year: 3, 3, 4, 7, 10, 11, 11. What differences can you note in comparison with the publication rate last year of the junior faculty in the biology department: 5, 6, 6, 6, 6, 13.

26. The GSS responses on ideal family size in 1972 were: none ($n_0 = 27$); one ($n_1 = 19$); two ($n_2 = 638$); three ($n_3 = 375$); four ($n_4 = 286$); five ($n_5 = 49$); six ($n_6 = 43$); and seven ($n_7 = 18$). Compare the data to those in Problem 22. What conclusions can you draw about the change in attitudes over the intervening eight years?

27. Every year the General Social Survey asks respondents about their marital status. The following distributions of responses were given in 1972, 1975, and 1978;

	1972	*1975*	*1978*
Married	1,160	1,002	960
Widowed	139	144	153
Divorced	65	84	134
Separated	39	49	47
Never married	210	211	237

   a. Compute relative frequency distributions for 1972, 1975, and 1978 and compare the three. Do the data suggest any trends across time? If so, what?

   b. Compute the index of diversity and IQV for all three distributions. Does the amount of variation across time suggest a trend?

28. The GSS asks whether respondents believe abortion should be available if a woman has become pregnant because of a rape. The responses for 1972, 1975, and 1978 were:

	1972	*1975*	*1978*
Yes = 1	1,196	1,190	1,232
No = 0	316	232	249

a. Compute the mean and standard deviation for all three distributions. Does there appear to be a trend across time?
b. Compute the proportion "yes" for all three years and compare to the mean computed above. What have you discovered?
c. Compute the index of diversity and the index of qualitative variation for all three years and compare to the standard deviation. Do the coefficients yield roughly the same conclusions about the relative amount of variation across time as the standard deviation computed in 28*a* did?

29. The following distributions of Ph.D.s earned in 1952 and 1972 is from the Office of Education's National Center for Educational Statistics:

Discipline	*1952*	*1972*
Social sciences	1,538	6,150
Humanities	883	4,400
Natural sciences	3,794	15,230
Education and other	1,468	8,820
N	7,683	34,600

a. Compute the two percentage distributions. Did the distribution of Ph.D.s by area of degree change from 1952 to 1972? If so, describe the change.
b. Using the index of diversity, decide whether there is more variation in the 1952 or the 1972 distribution.

30. Consider the following Public Health Service data on the four leading causes of death in 1969 for persons between 45 and 64 years of age, by sex:

Cause of Death	*Males*	*Females*
Heart disease	130,113	47,161
Cancer	63,761	54,081
Stroke	17,523	13,893
Accidents	17,043	6,669
Total	228,440	121,804

a. What is the modal cause of death in ages 45 to 64 for men, compared to women?
b. Is there more variation in the leading causes of death for men or women of this age?

Problems Requiring the 1980 General Social Survey Data

31. *a.* Describe the distribution of respondents' occupational prestige scores (PRESTIGE), using all the central tendency and dispersion statistics discussed in this chapter. *b.* Turn to the prestige scores for specific occupational titles in Appendix F of the GSS Codebook. Give the *Z*- scores for these occupations: (1) doctor, (2) lawyer, (3) truck driver, (4) mailman (letter carrier), and (5) dishwasher.

32. For the number of voluntary associations to which respondents belong (MEMNUM), give the *(a)* mode, *(b)* median, *(c)* range, *(d)* mean, *(e)* variance, *(f)* standard deviation, *(g)* CRV, and *(h) Z*- score for belonging to five associations.

33. Compare respondents' confidence in the medical profession (CONMED) with their confidence in Congress (CONLEGIS), using whatever statistics you think appropriate.

34. Compare black and white working men and women as to relative variation in annual incomes (RINCOME). (Recode respondents' income to the category midpoints, and use $30,000 as the value for the open-ended outcome.)

35. Which is higher—satisfaction with one's family (SATFAM) or with one's friends (SATFRND)? Which shows the greater variability? Which shows the greater relative variation?

Problems Requiring the 63-Cities Data

36. For the percent employed in manufacturing, (MANUFPCT), find the *(a)* range, *(b)* mode, *(c)* median, *(d)* mean, *(e)* variance, and *(f)* standard deviation.

37. Calculate *Z* scores for the number of municipal functions performed (FUNCTION).

38. Group the distribution of percentage black (BLACKPCT) into five or six appropriate intervals and display as a percentage table.

39. Compare the relative variation in per capital income (INCOMEPC) across the four census regions (REGION). Order these regions by increasing levels of intercity variability.

40. Show the means and standard deviations of the number of municipal functions performed (FUNCTION) by cities with each type of government (REFORMGV). Is there any noticeable difference in relative variability?

Crosstabulation

4

Frequency distributions are useful displays of the quantitative attributes of continuous variables or the qualitative attributes of discrete variables. But social scientists are usually more interested in explaining *variation* in distributions. Why do some young people seek college educations, while others do not? Why do some Americans call themselves Republicans, while others consider themselves to be Democrats or independents? Why do some college students smoke marijuana, while others regard it only as a weed?

On the basis of theory, past research, or just plain intuition, we usually form hypotheses to explain variation. When we posit that X causes Y, we are suggesting that a *relationship* exists between X and Y.

4.1. Bivariate Crosstabulation

One of the most widespread and useful tools the researcher has is the **crosstabulation** or *joint contingency* table, which indicates the *joint outcomes* of two variables. Such a table provides a way to determine whether the two variables are in fact related as hypothesized; that is, whether a bivariate relationship exists.

crosstabulation (joint contingency table)—a tabular display of the joint frequency distribution of two discrete variables which has r rows and c columns

As a simple example of crosstabulation, suppose we had chosen a sample of 20 high school seniors and asked them whether they planned to go on to college. In addition, we ascertained each respondent's sex (gender), in order to test the hypothesis that the percentage of males planning to go on to college is greater than the percentage of females with college plans. The data on these responses (which are for illustrative purposes only and are not meant to be representative) are shown in Table 4.1. For each person we now have two pieces of information—sex and college plans.

TABLE 4.1
Raw Data on College Plans for a Supposed Sample of High School Seniors

Person	*Sex*	*College Plans?*
1	M	Yes
2	M	No
3	F	Yes
4	F	No
5	F	No
6	M	No
7	F	No
8	M	No
9	F	No
10	F	Yes
11	M	Yes
12	M	No
13	M	Yes
14	F	No
15	M	Yes
16	M	No
17	F	No
18	F	No
19	M	No
20	M	No

It is not easy to tell from Table 4.1, however, whether the percentage of males going on to college is greater than the percentage of females. To indicate this we build a crosstabulation in the following way. First we note there are four kinds of joint outcomes for the two variables—male/college, male/no-college, female/college, and female/no-college. Then we count how many observations there are of each of these four joint outcomes. For example, from Table 4.1 we see that person 1 falls in the male/college category, as do persons 11, 13 and 15. That is, four persons fall into the male/college category. In the same way, you can easily demonstrate for yourself the following results:

Joint Outcomes	*f*
Male/college	4
Male/no-college	7
Female/college	2
Female/no-college	7

TABLE 4.2
Crosstabulation of College Plans by Sex, Based on Data in Table 4.1

A. RAW DATA

		Sex		
		Male	Female	Total
College Plans?	Yes	4	2	6
	No	7	7	14
	Total	11	9	20

B. PERCENTAGED DATA

		Sex		
		Male	Female	Total
College Plans?	Yes	36.4%	22.2%	30.0%
	No	63.6	77.8	70.0
	Total	100.0%	100.0%	100.0%

The raw data are crosstabulated in Table 4.2, Panel A. A crosstabulation of two variables, each of which has only two categories, is called a *2 × 2 table*. The **cells** which comprise the body of any table show the joint outcomes of the two variables. The **marginal distributions** (or as they are more simply called, *the marginals*) are the row totals **(row marginals)** shown on the right and the column totals **(column marginals)** shown at the bottom of the table.

cell—the intersection of a row and a column in a crosstabulation of two or more variables. Numerical values contained within cells may be cell frequencies, cell proportions, or cell percentages

marginal distributions—the frequency distributions of each of two crosstabulated variables

row marginals—the frequency distribution of the variable shown across the rows of a crosstabulation

column marginals—the frequency distribution of the variable shown across the columns of a crosstabulation

The marginal distributions are the frequency distributions of each of the two crosstabulated variables. But a frequency crosstabulation (such as Panel A) is not very useful for deciding if two variables covary and, if so, how. When a frequency distribution has unequal marginals, direct comparisons of cell frequencies are difficult to make. We need some way to *standardize* the frequency table to a common denominator so that the pattern of covariation is more apparent.

A *percentage crosstabulation* permits such a pattern to emerge. *Percentages are calculated within categories of the independent variable,* in this case within categories of respondent's sex. Panel B of Table 4.2 is a percentage crosstabulation which shows that 36.4% of the males surveyed plan to attend college, compared to only 22.2% of the females. If the variables were unrelated, the distribution of college attendance would have been the same for both sexes, which in turn would have been the same as the row marginal distribution.

4.1.1. An Example: Political Choices and Social Characteristics

Another example of crosstabulation involves a problem in political sociology. The relationship between voters' political choices and their social characteristics has a long tradition which precedes the first cross-sectional surveys. Politicians and analysts of election results know that certain sections of the nation have persisted in voting for the candidates of the same political party in many elections. Ever since the Civil War, voters in the South have favored the Democrats in presidential elections more than the rest of the country has. And rural areas in many states are consistently more Republican in their choices than the large urban centers are.

One social basis of political participation is presumed to be a person's national origin—otherwise known as *ethnicity.* America is a nation of immigrants and their descendants, and many cultural traditions from their homelands have remained strong for generations. The first immigrants from a region often settled together in the same communities, where they could maintain their languages and other traditions. Assimilation into the larger American population was slow for many groups, and discrimination by the dominant white Anglo-Saxon Protestants helped to keep alive the identities of the ethnic groups. For many of these groups, politics was an avenue of both economic advancement and status recognition. Ethnic groups such as the Irish and the Italians used the big-city political machines of the late 19th and early 20th centuries to gain practical political experience and to secure the advantages of patronage dispensed by winning officeholders. More recently, minorities like blacks and Mexican Americans have forged political alliances with the Democratic party to improve their situations.

The era of immigrant ethnic politics has peaked, and the extent to which distinctive ethnic political orientations remain is open to question. Today, however, religion is generally regarded as a better predictor of political party support than such social class variables as occupation, education, and income.[1] One study suggests that religion alone is a satisfactory predictor of political affiliation, because there is "little intrareligious variation in a party identification by national origin."[2]

---

1. David Knoke, *Change and Continuity in American Politics: The Social Bases of Political Parties* (Baltimore: Johns Hopkins University Press, 1976).

2. Steven Martin Cohen and Robert E. Kapsis, "Religion, Ethnicity, and Party Affiliation in the U.S.: Evidence from Pooled Electoral Surveys, 1968–72," *Social Forces,* Vol. 56 (December 1977), pp. 637–53.

■ *Propositions and Hypotheses.* We can explore these questions with the 1,530 cases from the 1977 General Social Survey, which was described in Box 2.1 in Chapter 2. The particular propositions we will examine are:

P1: Ethnicity is related to political orientation.

P2: Religion is related to political orientation.

Our examination of these propositions will utilize the percentage distributions for ethnic ancestry, religious preference, and party identification presented in Table 2.5 in Chapter 2.

Neither of these propositions specifies the exact meanings of ethnicity, religion, or political orientation. Nor is the form of the relationship stated, except to indicate that political orientations will not be the same for all ethnic or religious groups. Based on the research cited above and other evidence, however, we can anticipate that Jews and Catholics are more likely than Protestants to support the Democratic party. Similarly, those with origins outside of northern Europe (Italians and East Europeans, plus the Irish) should offer less support to the Republican party than descendants of immigrants from England, Germany, and Scandinavia.[3] While political orientation embraces more than party support, including attitudes and activities such as voting, campaigning for candidates, and general psychological affinity for a party, our interpretation of the term will consider only party identification.

To turn both propositions into testable hypotheses, we must link the concepts in P1 and P2 with variables measured by the General Social Survey. Ethnicity is measured by a question asking "From what countries or part of the world did your ancestors come?" A respondent who named more than one nation was asked, "Which one of these countries do you feel closer to?" (Some respondents insist that their ancestors came from America; these persons' families have usually been in the United States for many generations.) Religion is determined by response to the question, "What is your religious preference? Is it Protestant, Catholic, Jewish, some other religion, or no religion?" (For Protestants, additional information was collected about the particular denomination, but we will not use that data here.) The political orientation variable was operationalized not by the presidential vote but by a party identification item:

3. Edward O. Laumann and David R. Segal, "Status Inconsistency and Ethnoreligious Membership as Determinants of Social Participation and Political Attitudes," *American Journal of Sociology*, Vol. 77 (July 1971), pp. 36–61.

"Generally speaking, do you usually think of yourself as a Republican, Democrat, Independent, or what?" Note that all three variables are nonorderable, discrete measures, since the response categories can be rearranged in any arbitrary sequence.

The propositions restated as operational hypotheses are:

H1: National ancestry is related to party identification.

H2: Religious preference is related to party identification.

These two hypotheses fail to capture the detail of our hunches about how the variables are related, however, just as the propositions did. Particular ancestries are presumed to be more likely to be identified with one or the other of the parties than are other ancestries. More advanced statistical texts present ways to test the richness of these hypotheses, but in this one we will only test the more global hypotheses when examining nonorderable discrete variables. Even though no formal testing of the detail of the hypotheses will be done, we will be able to get some idea of the extent of support for them by simple examination of the data. Marginal distributions for the three variables were shown in Table 2.5, but because of the large number of categories of ethnic ancestry, the recoded version shown in Table 2.6 is used. In the crosstabulations the marginal frequencies show the distribution of observations on each variable after respondents for whom information is missing have been disregarded, or removed from the data.

TABLE 4.3

Crosstabulation of Ethnic Ancestry and Party Identification Frequencies, GSS Respondents

	Party Identification			
Ethnic Ancestry	*Democrat*	*Independent*	*Republican*	*Total*
German	108	90	67	265
English	76	76	69	221
Irish	57	42	26	125
Italian	32	22	13	67
Scandinavian	30	17	19	66
East European	56	31	14	101
West European	25	22	21	68
Spanish	22	18	4	44
Oriental	3	6	3	12
African	76	19	4	99
Other	41	37	16	94
Total	526	380	256	1,162

Missing data: 368 cases.

Source: 1977 General Social Survey.

■ *Crosstabulation.* A raw-data crosstabulation for the first hypothesis, which concerns ethnic ancestry and party identification, is shown in Table 4.3. Categories of the first variable are arranged in the columns, categories of the second variable in the rows. Whether the independent variable is in the column or the rows is to some extent arbitrary. When the frequencies are percentaged in Table 4.4, for example, ethnic ancestry and party identification are switched. The cells in the body of Table 4.3 contain the frequencies: 108 persons of German ancestry are Democrats, there are 18 Independents of Spanish origin, and so on. The row and column marginals are shown on the right and bottom, respectively, as usual.

When the same data are percentaged, the crosstabulation is as shown in Table 4.4. The percentages in the Total column for party identification do not agree with those shown in Table 2.5 because 368 respondents with no information on ethnic ancestry or party identification were eliminated from Table 4.4.

With the data in this form we can now examine the first hypothesis, that national ancestry is related to party identification. Comparisons across rows in Table 4.4 indicate that respondents with African, East European, and Spanish origins most often consider themselves Democrats, followed by those with Italian, Irish, and Scandinavian origins. English, West European, and Scandinavian descendants have the highest percentage of Republicans, while half the Orientals claim to be Independents. Clearly, party identification varies considerably according to the ethnic ancestry of the respondents, as anticipated in the first hypothesis. (Surprisingly, however, Italian and Irish Americans appear to be little more Democratic in orientation than the sample as a whole.)

The second research hypothesis states that religion is also systematically related to party identification. Table 4.5, which displays the percentage crosstabulation of these two variables, confirms that expectation. Jews and Catholics are the most Democratic, Protestants are the most Republican, and a majority of persons with other religions or no religion are Independents. Note that there are 346 fewer missing data in this table (N = 1,508) compared to Table 4.4 (N = 1,162), mainly because nearly everyone could be classified according to religion.

A visual inspection of the percentage crosstabulations in Tables 4.4 and 4.5 implies, therefore, that the two hypotheses relating ethnic ancestry and religion to party preference are both supported.

TABLE 4.4

Percentage Crosstabulation of Ethnic Ancestry and Party Identification Variables

Party Identification	Ethnic Ancestry											
	German	*English*	*Irish*	*Italian*	*Scandinavian*	*East European*	*West European*	*Spanish*	*Oriental*	*African*	*Other*	*Total*
Democrat	40.8%	34.4%	45.6%	47.8%	45.5%	55.4%	36.8%	50.0%	25.0%	76.8%	43.6%	45.3%
Independent	34.0	34.4	33.6	32.8	25.8	30.7	32.4	40.9	50.0	19.2	39.4	32.7
Republican	25.3	31.2	20.8	19.4	28.8	13.9	30.9	9.1	25.0	4.0	17.0	22.0
Total	100.1%*	100.0%	100.0%	100.0%	100.1%*	100.0%	100.1%*	100.0%	100.0%	100.0%	100.0%	100.0%
N	265	221	125	67	66	101	68	44	12	99	94	1,162

*Does not total to 100% due to rounding.

Source: 1977 General Social Survey.

TABLE 4.5
Percentage Crosstabulation of Religious Preference and Party Identification Variables

Party Identification	*Religious Preference*					
	Protestant	*Catholic*	*Jewish*	*None*	*Other*	*Total*
Democrat	42.5%	53.5%	61.8%	30.8%	29.4%	44.8%
Independent	31.1	32.6	29.4	54.9	64.7	33.2
Republican	26.5	13.9	8.8	14.3	5.9	22.0
Total	100.1%*	100.0%	100.0%	100.0%	100.0%	100.0%
N	998	368	34	91	17	1,508

*Does not total to 100% due to rounding.
Source: 1977 General Social Survey.

Figure 4.1 shows one way in which data from bivariate crosstabulations can be presented as a bar chart (see Section 2.2.1). Bars representing the same category on the dependent variable (party identification) are grouped together and given different shadings to stand for the various categories of the independent variable (religious preference). We have omitted the optional placement of frequencies above each bar.

4.2. Population Inference from Samples

The data displayed in Tables 4.4 and 4.5 came from a cross-sectional sample of the American adult population: the 1977 General Social Survey. Despite the large sample size, these observations represent only a tiny fraction of the millions of persons who might have been chosen for interviews, however. Literally trillions of different samples of the same size might have been selected by the National Opinion Research Center. Since only one sample was drawn, we are faced with a question of whether the conclusions based on this data set can be meaningfully generalized to the entire population. In other words, what sort of inference can we make from this sample to the U.S. population as a whole?

This chapter introduces some basic concepts necessary for understanding statistical inference. Additional material is presented in Chapter 5, but we do not propose to give an extended formal treatment of probability in this book. Instead, we will develop whatever technical apparatus is required in the course of explaining how **statistical significance tests** can be applied to sample data.

statistical significance test—a test of inference that conclusions based on a sample of observations also hold true for the population from which the sample was selected

FIGURE 4.1
Bar Chart Showing Religious Preference and Party Identification

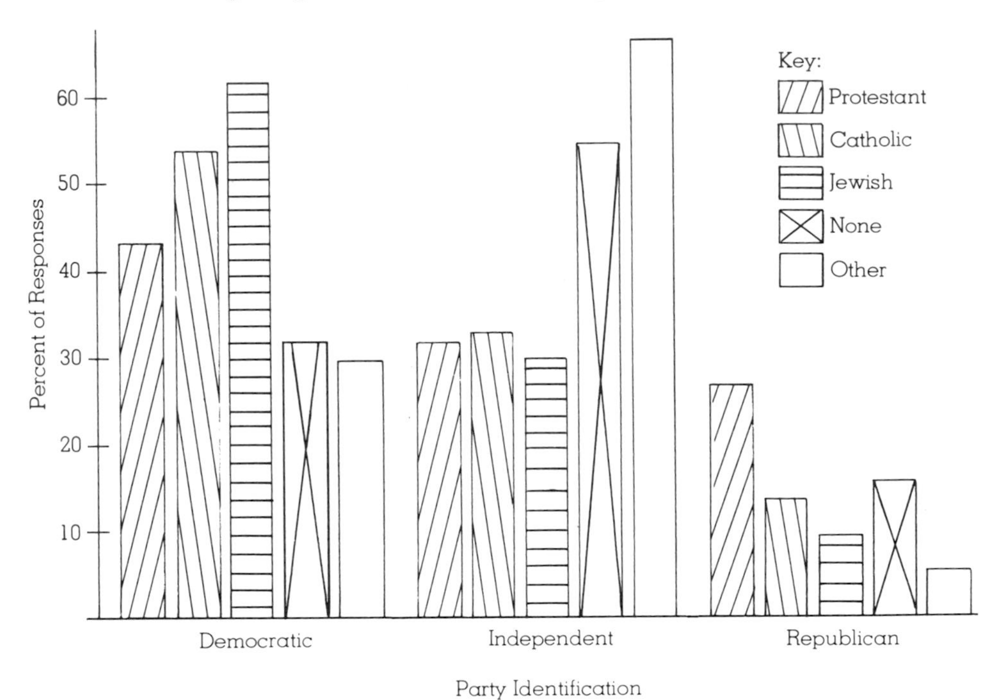

Source: 1977 General Social Survey.

The idea behind statistical significance tests is to permit a reasonable *inference* that conclusions drawn from a sample of observations would hold true of the population from which that sample was drawn. We can never be entirely certain that the sample findings accurately reflect the population situation, for we might by chance have selected an unusually deviant set of observations. But if the sampling design is reasonably representative of the population, we can draw an inference for the entire population, with a calculable chance that this inference is correct.

random sampling—a procedure for selecting a set of representative observations from a population, in which each observation has an equal chance of being selected for the sample

The typical requirement of a "reasonably representative" sample from a population is met by a procedure called **random sampling**. (The concept of a random sample was introduced in Chapter 1.) In simple random sampling, each observation (i.e., a person, an object, or an event) in the population has an equal chance of being selected for the sample. That is, if the population consists of T units, each unit has the probability of exactly $1/T$ of being chosen for the sample. If, as is typically the situation in national surveys, the adult population has a T of 150 million and a sample of 1,500 persons is to be interviewed, each American adult has one chance in a hundred thousand—or a probability of .00001—of being drawn. No wonder so few people say they have ever been interviewed for a Gallup Poll!

Unfortunately, drawing a national sample cannot be done by simple random sampling. This technique requires a complete listing of all sample candidates, a prohibitively costly task for a national or even a city population. Survey research institutes such as Harris, Gallup, and NORC select their samples according to more cost-efficient procedures. In this text we will treat the GSS data as if it had been drawn using simple random procedures. The amount of error this will introduce in inferences is small.

4.3. Probability and Null Hypotheses

For discrete data in crosstabulation form (such as Table 4.3), the basic question of inference is: *What is the probability that the relationship observed in the sample data could be obtained from a population in which there was no relationship between the two variables?* If there is a reasonably high probability that no covariation exists in the population, even though a relationship was discovered in the sample, then we would be extremely reluctant to conclude that the two variables are related. We would conclude that the hypothesis asserting a relationship between, say, ethnic ancestry and party identification should be rejected. Only if the chance were small (perhaps less than 1 out of 20) that a sample relationship could be observed when no relationship exists in the population would we decide that the research hypothesis should not be rejected.

> Tests of statistical significance are rules which aid in decisions to reject or not to reject a hypothesis. They make it possible to estimate the probability that an observed sample relationship between variables arose from a particular relationship that exists in the population.

One way to determine that two variables are related is to test

null hypothesis—a statement that no relationship exists between variables in a population

the hypothesis that they are *un*related, as we noted in Chapter 1. The way to do this is by positing a **null hypothesis**, which states that *no* relationship exists in the population between the variables. This statement is contrary to the research hypothesis, which states that there *is* an expected relationship between variables, on the basis of theory or past research.

In the example on political choices and social characteristics in Section 4.1.1, H1 and H2 can be restated as null hypotheses, applying the convention in statistics of using H_0 to designate them:

$H_0 1$: Ethnic ancestry is unrelated to party identification.

$H_0 2$: Religion is unrelated to party identification.

Our real desire is to *reject* these null hypotheses as untrue statements about the population of adult Americans in 1977. If the probability is small—again, say less than .05 (that is, less than 1 chance out of 20)—that such null relationships occur in the population, based on the sample evidence, we reject the null hypotheses. In doing so, we conditionally accept their alternatives—the research hypotheses which state that the variables do in fact covary in the population as well as in the sample. We say we conditionally accept their alternatives because *statistics can only be used to examine the truthfulness of hypotheses through the rejection of hypotheses.* In this sense, as we noted in Chapter 1, truth is not directly accessible or testable.

4.3.1. Type I and Type II Errors

probability level—the probability selected for rejection of a null hypothesis, which is the likelihood of making a Type I error

alpha—the probability level for rejecting a null hypothesis when it is true, conventionally set at .05 or lower

Whenever we deal with probabilities, we run the risk of making an incorrect decision. Otherwise we would be dealing with certainties, which is never the case. We conventionally set a **probability level** (α, or **alpha**) for rejection of a null hypothesis at .05 or lower. When we do this, we make a deliberate choice of running a particular risk of being incorrect in our inference from sample to population relationship.

In making inferences, we might make two different types of errors of judgment. First, on the basis of the significance test results, *we might reject a null hypothesis which in fact is true.* That is, we might reject the hypothesis that two variables are unrelated, based on sample results, when in the population they are in fact unrelated. This error occurs when, simply by chance, the sample is not a good representative of the population from which it was

BOX 4.1 Remembering Type I and Type II Errors

Type I and Type II errors are often confused. One way to keep them straight is to memorize this table:

In the population from which the sample is drawn, the null hypothesis is:	Based on sample results, the decision made is: Reject null hypothesis	Do not reject null hypothesis
True	Type I or false rejection error (α)	Correct decision
False	Correct decision	Type II or false acceptance error (β)

An alternative is to keep in mind this story: Once upon a time a princess was forced by her father, the king, to marry Prince Beta for reasons of state. But they couldn't stand each other and failed to consummate the marriage. Instead, the princess carried on a secret affair with Prince Alpha, the man she really loved. Thus, the princess had to pretend that a relationship with Beta existed when in fact none really did, while she pretended that no relationship with Alpha existed when in fact one really did.

drawn. This kind of error, called **Type I error**, may also be called the **false rejection error.** The chance of making this mistake is the same as the probability level, α, which we set for rejection of the null hypothesis. Hence, Type I error is sometimes called *α error,* or *alpha error.*

Type I error—a statistical decision error that occurs when a true null hypothesis is rejected; its probability is alpha

The second type of error is a **Type II error**, which may also be called a **false acceptance error.** This represents exactly the opposite case: *Although the null hypothesis is actually false, we fail to reject it on the basis of the significance test.* This type of decision-making error is also called *β error,* or *beta error.* Box 4.1 offers some help in differentiating between the two types of error.

Type II error—a statistical decision error that occurs when a false null hypothesis is not rejected; its probability is beta

The probability of making a Type II error is not simply 1.0

minus the probability of a Type I error. That is, if α is .05, β is *not* just $1.00 - .05 = .95$. Things are more complicated than that. A complete explanation of how to find the probability of Type II error would get us into an extended discussion of the "power" of statistical tests and take us away from our immediate objective, which is to develop the basis of significance tests for nominal crosstabulations.

Nevertheless, it is important to note that false rejection and false acceptance errors are related to each other, so that reducing the potential probability of making a false rejection error—by setting α at a very low level, such as .001—tends to increase the risk of making a false acceptance error. Standard procedures for offsetting false acceptance error are to increase the sample size, thereby reducing sampling error in inferring population relationships from sample data, or to replicate the study on another, independently drawn sample, in which case consistent results would strengthen our belief in the findings. The chance of making two false rejection errors in succession is the product of the α levels: $(.05)(.05) = .0025$, for example.

4.4. Chi Square: A Significance Test

chi-square test—a test of statistical significance based on a comparison of the observed cell frequencies of a joint contingency table with frequencies that would be expected under the null hypothesis of no relationship

An appropriate test for assessing the statistical significance of crosstabulated variables is the **chi-square** (or χ^2) **test**. This is based on a comparison between the observed cell frequencies of a crosstabulation or joint contingency table with the frequencies that would be expected if the null hypothesis of no relationship were in fact true. The key to the calculation of the χ^2 statistic is to obtain these expected frequencies.

statistical independence—a condition of no relationship between variables in a population

What would a contingency table look like if no relationship whatsoever existed between the two variables? That is, what would we expect to find in a situation of **statistical independence**?

> If two variables are statistically independent, within categories of the independent variable in a contingency table there are identical proportions of the dependent variable. Similarly, within each category of the dependent variable the same proportion of the independent responses occurs.

TABLE 4.6
Expected Frequencies for Ethnic Ancestry by Party Identification, Under Null Hypothesis of Independence

	Party Identification			
Ethnic Ancestry	***Democrat***	***Independent***	***Republican***	***Total***
German	120.0	86.7	58.4	265.1*
English	100.0	72.3	48.7	221
Irish	56.6	40.9	27.5	125
Italian	30.3	21.9	14.8	67
Scandinavian	29.9	21.6	14.5	66
East European	45.7	33.0	22.3	101
West European	30.8	22.2	15.0	68
Spanish	19.9	14.4	9.7	44
Oriental	5.4	3.9	2.6	11.9*
African	44.8	32.4	21.8	99
Other	42.6	30.7	20.7	94
Total	526	380	256	1,162

*Row totals do not agree with those in Table 4.3 due to rounding.

Table 4.6 displays such an independence relationship for the ethnic ancestry and party identification crosstabulation in the preceding example (Table 4.3). If you calculate the percentages within rows using the **expected frequency** for each category of ethnicity, you will find 45.3% Democrats, 32.7% Independents, and 22.0% Republicans, the same percentages for the sample as a whole. In fact, the marginal totals of the two variables in Table 4.3 provide the basis for calculating the 33 expected frequencies in the cells of Table 4.6.

expected frequency—in a chi-square test, the value that cell frequencies are expected to take, given the hypothesis under study (ordinarily, the null hypothesis)

If two variables, X and Y, are statistically independent, the formula for the expected frequency in row i and column j is:

$$\hat{f}_{ij} = \frac{(f_{i.})(f_{.j})}{N}$$

Where:

$\hat{f}_{ij}$ = The expected frequency of the cell in the ith row and jth column.

$f_{i.}$ = The total in the ith row marginal.

$f_{.j}$ = The total in the jth column marginal.

N = The grand total, or sample size for the entire table.

To illustrate, the expected frequency of Scandinavian Republicans under the null hypothesis that the two variables are unrelated is:

$$\frac{(66)\,(256)}{1{,}162} = 14.5$$

You should calculate a few more expected frequencies yourself to check on your understanding of the concept.

Tables 4.3 and 4.6, respectively, provide the observed frequencies and the expected frequencies under independence. The χ^2 statistic summarizes the discrepancy between these two tables across all 33 cells in the body of each table. (The row and column marginals in the two tables are identical, so no comparison there is helpful.) These discrepancies are noted cell by cell. If $\hat{f}_{ij}$ is the expected frequency under the null hypothesis and f_{ij} is the observed frequency for the same cell, the value of the statistic for the table is found by the formula:

$$\chi^2 = \sum_{i=1}^{R} \sum_{j=1}^{C} \frac{(\hat{f}_{ij} - f_{ij})^2}{\hat{f}_{ij}}$$

Where:

C = Number of columns.
R = Number of rows.

The difference between the observed and expected frequencies in a cell is first squared (to remove plus and minus signs) and then divided by the expected frequency for that cell. After this operation has been performed for all cells, the results are summed for all cells of the table.

The formula for finding the χ^2 statistic uses double summation signs, which we have not introduced before. The use of summation is described in Appendix A, and some rules for applying double summation in this formula are spelled out below:

1. Set $i = 1$ and $j = 1$, and evaluate the first term. In this case it is:
   $$(\hat{f}_{11} - f_{11})^2/\hat{f}_{11} = (120.0 - 108)^2/120.0 = 1.20.$$
2. The index of the second, or inside, summation oper-

ator moves from 1 to C one step at a time before the index of the first, or outside, summation operator moves to the next step.

3. At each step, as index i moves to the next higher level, index j must pass through its *entire* cycle of 1 to C.
4. The process ends when $i = R$ and $j = C$.

Table 4.7 shows the chi-square component of each cell. The larger the value, the greater the relative discrepancy between observed and expected frequencies. The value where the first row

TABLE 4.7
Chi Square Components for Ethnic Ancestry by Party Identification

	Party Identification		
Ethnic Ancestry	*Democrat*	*Independent*	*Republican*
German	1.20	0.13	1.27
English	5.76	0.19	8.46
Irish	0.00	0.03	0.08
Italian	0.10	0.00	0.22
Scandinavian	0.00	0.98	1.40
East European	2.32	0.12	3.09
West European	1.09	0.00	2.40
Spanish	0.22	0.90	3.35
Oriental	1.07	1.13	0.06
African	21.73	5.54	14.53
Others	0.06	1.29	1.07

and column intersect (1.20) was computed above. You should compute a few other entries in Table 4.7 for yourself to make certain you understand the table and the use of the formula.

The largest deviations in Table 4.7 are underscored to show where the major departures from independence occur. You must look at both Tables 4.3 and 4.6 to determine where the excesses and deficits occur, since the directions of the differences are obliterated by the squaring operation. African descendants, for example, are substantially more Democratic and less Republican than expected under the null hypothesis. But Americans of English origin are more likely to be Republicans than would be expected by chance. Only the respondents claiming African ancestry showed much lower independence than expected.

The sum of all entries in Table 4.7 is 79.7. Hence, we say that $\chi^2 = 79.7$. To evaluate the meaning of this number in making a decision whether or not to reject the null hypothesis, we must first consider how χ^2 values are distributed under the null hypothesis. The χ^2 statistic for a large N (30 or greater) follows a **chi-square distribution**, which is the topic of the next section.

chi-square distribution—a family of distributions, each of which has different degrees of freedom, on which the chi-square test statistic is based

4.5. Sampling Distributions for Chi Square

A central concept in statistical inference is the sampling distribution of a test statistic. A **test statistic** is used to evaluate a hypothesis about a population, given evidence drawn from a particular random sample (see Section 4.2). In the example in Section 4.4, the sample value of the χ^2 test statistic is 79.7. *Each test statistic has a characteristic sampling distribution that depends on the sample size and the degrees of freedom* (defined below). You will encounter different sampling distributions throughout this book as we investigate different test statistics. For now, our attention is focused on the sampling distribution of χ^2.

test statistic—a number used to evaluate a statistical hypothesis about a population, calculated from data on a sample selected from a population

In general, a **sampling distribution** is a theoretical frequency distribution of a test statistic, under the assumption that all possible random samples of a given size have been drawn from some population. For example, assume there is a population of size 50 from which we draw a sample of 45. It can be shown that there are 2,118,760 different samples of size 45 from a population with 50 elements. Suppose we collect data on two variables, X and Y, for each of these different samples. Then we compute a χ^2 test statistic for each of the X by Y crosstabulations. In theory, we could build a frequency distribution of the 2,118,760 χ^2 values.

sampling distribution—a theoretical frequency distribution of a test statistic, under the assumption that all possible random samples of a given size have been drawn from some population

This distribution would look much like a sampling distribution for χ^2. We say it would *look much like* a sampling distribution because *an actual sampling distribution of χ^2 also assumes that the population from which the samples were drawn was normally distributed.* At this point you may not know what a normal distribution is. We will discuss this concept in Chapter 5; for now, you need only know that *a normal distribution is a smooth, bell-shaped curve.* (Figure 5.3 shows two normal distributions.) Obviously, in a population composed of only 50 elements, there are too few observations for a distribution of them to form a smooth, bell-shaped curve. For that reason, our distribution of 2,118,760 χ^2 values is technically *not* a sampling distribution.

Most of the populations from which we will choose our samples are very large. Literally billions of different samples of size 100 can be chosen from a population of 100,000. Since it is not feasible to calculate test statistics for each sample, *sampling distributions are theoretical distributions.* We rely on mathematics beyond the scope of this book to derive a formula which gives us the shape of the sampling distributions for χ^2 statistics, under the assumption that the null hypothesis is true.

Notice in the preceding sentence the use of the plural, *sampling distributions.* Instead of a single distribution, there is an entire *family of χ^2 distributions,* similar to but slightly different from one another, according to the degrees of freedom involved. The best way to introduce the concept of degrees of freedom is with reference to the crosstabulation of ancestry and party identification. In a crosstabulation table, **degrees of freedom** *(df)* refers to the potential for cell entries to vary freely, given a fixed set of marginal totals.

degrees of freedom— the number of values free to vary when computing a statistic

Table 4.3, for example, includes data on a total of 265 respondents of German ancestry. We could choose any two numbers which sum to 265 or less for any pair of party identification categories. But once we knew any pair of numbers, the third value would be strictly determined by the difference of their sum from the row total. Thus we can speak of 2 degrees of freedom in the three cells of each row in Table 4.3. For example, in Row 1 we know that the row total is 265. If we know there are 67 Republicans and 90 Independents, we know there must be 265 − (90 + 67) = 108 Democrats.

Similarly, in each column of 11 entries there are 10 degrees of freedom. For the three-column table as a whole there are 20 degrees of freedom ($df = 20$). Once 20 independent cell entries are known, the remaining 13 entries are not free to vary within the constraints set by the row and column marginal sums. For *any* two-variable crosstabulation, the degrees of freedom for the chi-square statistic equal the product of the number of rows, less 1, times the number of columns, less 1. The formula is:

$$df = (R - 1)(C - 1)$$

In Table 4.3 there are $(3 - 1)(11 - 1) = 20$ *df,* as noted above. Thus, the sampling distribution appropriate to evaluating

FIGURE 4.2
Three Chi-Square Distributions

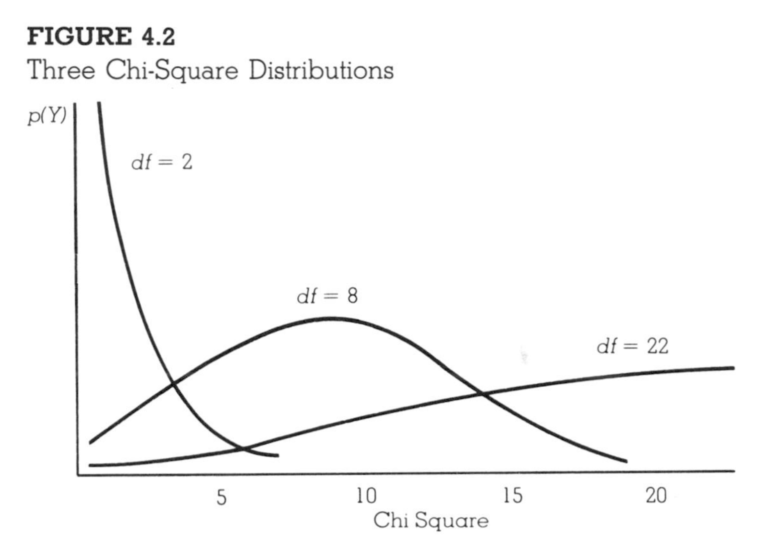

the χ^2 obtained for Table 4.3 is a sampling distribution of χ^2 with 20 degrees of freedom. Now calculate for yourself how many degrees of freedom there are in Table 4.5.

What do theoretical χ^2 sampling distributions look like, and how are they used to evaluate hypotheses? Figure 4.2 shows three different sampling distributions for χ^2 statistics with 2, 8, and 22 degrees of freedom, under the assumption that the null hypothesis is true. Three features of χ^2 distributions can be seen in this figure. First, notice that *χ^2 values are always nonnegative,* i.e., they vary in value from zero to plus infinity ($+\infty$). (This is not true of all the sampling distributions we will discuss in this text, however.)

Second, *the χ^2 value which has the greatest likelihood of being observed is in the neighborhood of its degrees of freedom.* This means that the size of a given χ^2 only makes sense relative to the degrees of freedom involved. In a 2×2 crosstabulation or joint contingency table, which has 1 degree of freedom (why?), an observed χ^2 of 8 or more would be an extremely rare event. It would occur fewer than five times in a thousand chi-square distributions with 1 degree of freedom. A χ^2 of 8 in a 3×5 table (how many *df?*) would be a relatively common outcome, however. Indeed, we would expect to observe a χ^2 of 8 or greater in nearly half of all chi-square distributions with 8 *df.*

Third, *as the number of degrees of freedom increases, the shape of χ^2 distributions becomes more bell-like.* This means that *the larger the degrees of freedom, the lower the probability of observing a chi-square value in either the extreme left or right tail of the distribution.* In a sampling distribution with only a few degrees of freedom, in contrast, rare events occur only in the extreme right tail of the distribution. For example, with 22 *df* an observed χ^2 value of either less than 1 or greater than 40 would be almost equally rare. But with 2 *df* an observed value of 3 or greater is much less probable than an observed value of 1 or less.

Finding the probabilities for specific χ^2 values in a sampling distribution with a given *df* requires the use of the chi-square table in Appendix B. To use this table, follow these steps:

1. Choose a probability level for rejection of the null hypothesis (α level).
2. Calculate the degrees of freedom in the table.
3. Enter the chi-square table at the corresponding row and column and observe the entry. This value for χ^2 tells you how large the χ^2 calculated from the data table must be in order to reject the null hypothesis, while running a chance of α of making a Type I or false rejection error (i.e., rejecting a null hypothesis which really is true—no relationship exists in the population).

The value for χ^2 determined in step 3 is called the **critical value** (c.v.). In the example of the relation between ethnic ancestry and party identification, we found in Section 4.4 a χ^2 value of 79.7 for Table 4.7. For $df = 20$ and $\alpha = .05$, we needed to have a χ^2 only equal to or larger than 31.4 in order to reject the null hypothesis. Even setting $\alpha = .001$, we are well above the critical value necessary to reject the hypothesis that no relationship between ethnic ancestry and party identification exists in the population.

critical value—the minimum value of a test statistic that is necessary to reject the null hypothesis at a given probability level

Similarly, in considering the relationship between religion and party identification shown in the percentage crosstabulation in Table 4.5, we find that a χ^2 equal to or greater than 15.5 is required to reject the null hypothesis with $\alpha = .05$, given that $df = 8$. The calculated χ^2 for Table 4.5 (which we leave for you to verify) is 61.4, more than enough to reject the null hypothesis. Thus we conclude that not only are ethnicity and religion both related to party identification in the case of the GSS samples (as we established in Section 4.1.1), but these relationships are *statistically significant, that is, these variables most probably are also related in the population from which the samples are drawn.*

One of the reasons that the relationships in both tables are highly significant (and the chance of a Type I error is so remote) is the large sample size. *The χ^2 statistic is directly proportional to sample size:* doubling the cell frequencies in a table will double the calculated value of χ^2, for example, although the *df* remain unchanged.

> The sensitivity of χ^2 to the sample size points to an important distinction between statistical significance and substantive importance in relationships between social variables. Two variables may covary only modestly, yet if the sample size is large enough, the null hypothesis can be rejected at almost any level of probability you choose.

4.5.1 Another Example Using Chi Square

The χ^2 distribution can also be used to investigate the interesting question of whether people change their political identifications after a presidential election year. Two kinds of changes might be expected. First, fewer individuals might declare themselves Independent and instead align themselves with one of the major political parties. And, second, those who changed their Independent identification might be more likely to align themselves with the party of the presidential victor.

The 1976 and 1977 GSS surveys can help shed light on these hunches. In 1976 Jimmy Carter—a Democrat—was elected president. What happened in 1977? Panel A of Table 4.8 suggests support for the first of these hypotheses; the data indicate 3.7 percent fewer Independents in 1977, following the 1976 election. There also are 2.5 percent more declared Democrats in the 1977 data, not very different from the 1.2 percent increase for the number of declared Republicans. The difference between these increases is in the predicted direction, however.

An important question is whether the difference between these two distributions is due to chance, that is, to mere sampling variability, or the distribution of political identification actually changed between 1976 and 1977. To examine this question we determine what the number of expected Democrats, Independents, and Republicans would have been under the assumption of statistical independence. If the 1976 and 1977 distributions are independent, then we should have expected the joint distribution shown in Panel B of Table 4.8. Are the observed frequencies significantly

TABLE 4.8

Distributions of Political Identification, GSS Respondents

A. OBSERVED DISTRIBUTIONS

	1976	*%*	*1977*	*%*
Democrat	629	42.1	675	44.6
Independent	553	37.1	505	33.4
Republican	310	20.8	333	22.0
N	1,492	100.0	1,513	100.0

B. EXPECTED DISTRIBUTIONS UNDER INDEPENDENCE

	1976	*1977*
Democrat	647.44	656.56
Independent	525.30	532.70
Republican	319.25	323.75
$\chi^2_2 = 4.476$		

Source: 1976 and 1977 General Social Surveys.

different from those expected on the assumption of independence? We can use χ^2 to find out.

First, convince yourself that in this case $df = 2$. Second, note that we subscript χ^2 with its df—a common practice. Then, $\chi^2_2 = (647.44 - 629)^2/647.44 + (525.30 - 553)^2/525.30 + (319.25 - 310)^2/319.25 + (656.56 - 675)^2/656.56 + (532.70 - 505)^2/532.70 + (323.75 - 333)^2/323.75 = 4.476$. A check of the chi-square table in Appendix B indicates that for $\alpha = .05$, an observed χ^2_2 value must be at least 5.99 to reject the null hypothesis. Hence we cannot reject the hypothesis that the distributions are the same. We conclude that more persons did not affiliate with the major parties in the year after the presidential election.

4.5.2 Chi Square as a Goodness-of-Fit Test

The chi-square statistic can also be used to test whether an observed frequency distribution could have been sampled from a *known* population distribution. For example, suppose you are shooting craps with a friend and suspect that your opponent is getting a lot more sevens than you would expect if the dice were fair. You expect that the dice are loaded in favor of ones and sixes; this would account for the large number of observed sevens when the two dice are thrown. If your hunch is true, you could throw one die, say, 600 times, record each outcome, and compute a chi-square test where the expected frequency for each of the outcomes is 100. That is, each outcome has a probability of 1/6 of occurring if the die is fair, so in 600 throws of a die, in theory, there should be 100 of each outcome (see Section 5.1.1 in Chapter 5). If the chi-square test is significant for a given level, you reject the hypothesis that the die is fair and confront your "friend."

equiprobable distribution—a probability distribution in which each outcome has an equal chance of occurrence

goodness-of-fit test—the chi-square statistic applied to a single discrete variable, with $K - 1$ degrees of freedom, where expected frequencies are generated on some theoretical basis

In this example the known distribution (which is the expected distribution) is called an **equiprobable distribution**, since each outcome has an equal probability of occurrence. In cases such as this, where χ^2 is used as a **goodness-of-fit test**, the degrees of freedom are $K - 1$, where K is the number of outcomes associated with the distribution being examined.

TABLE 4.9
Distribution of Births by Month, GSS Respondents

Month of Birth	*f*	*%*
January	116	7.7
February	120	7.9
March	113	7.5
April	124	8.2
May	127	8.4
June	126	8.3
July	126	8.3
August	117	7.7
September	149	9.9
October	141	9.3
November	122	8.1
December	131	8.7
$N =$	1,512	100.0
$\hat{f} =$	126	
$\chi^2_{11} =$	9.41	

Source: 1978 General Social Survey.

A second example using χ^2 as a goodness-of-fit statistic might concern a demographer interested in whether equal numbers of people are born each month or more births occur in some months than others. In this example the known theoretical distribution (the expected outcome) is again the equiprobable one. That is, the probability of being born in any given month is 1/12 if births occur randomly throughout the year (ignoring the slight differences among months with 31, 30, and 28 days).

We can use the data in the 1978 General Social Survey, when month of birth was coded, to test the hypothesis that the distribution of births could have been generated by an equiprobable population distribution. The distribution of births for the sample of $N = 1{,}512$ respondents is shown in Table 4.9. Since $N = 1{,}512$, the expected frequency for each outcome is $(1/12)(1{,}512) = 126$, under the assumption of an underlying equiprobable population distribution.

In this example there are $12 - 1 = 11$ *df*, and $\chi^2_{11} = (126 - 116)^2/126 + (126 - 120)^2/126 + \ldots + (126 - 131)^2/126 = 9.41$. If we set $\alpha = .05$, we can use Appendix B to determine that the critical value for 11 degrees of freedom is 19.7. Therefore we cannot reject the hypothesis that births in the population of adult Americans are distributed randomly across the months of the year, based on these sample results. Remember, however, that in reaching this conclusion we run some risk of Type II error, the false acceptance of the null hypothesis of equiprobable birth months.

4.6. Two Meanings of Statistical Inference

The examples presented in this chapter demonstrate nicely the different meanings that can be attached to the term *statistical inference* (see Chapter 5). Early in this chapter, when we dis-

BOX 4.2 Summary of Steps in Testing for Statistical Significance

Step 1. Choose an α level (probability of a Type I or false rejection error) for the null hypothesis.

Step 2. Examine the tabled values of the statistic to see how large the test statistic would have to be in order to reject the null hypothesis. This is called the critical value, or c.v.

Step 3. Calculate the test statistic.

Step 4. Compare the test statistic with the critical value. If it is as large or larger, reject the null hypothesis; if not, do not reject the null hypothesis.

cussed inference, we referred to generalizing from a given sample to a large population from which the sample was drawn. This is the most common use of the term. However, the examples in Sections 4.5.1 and 4.5.2 illustrate a second interpretation of inference—*a test of the hypothesis that the process which generated the data is a random rather than a systematic one.* If the process is random we could expect the two variables being studied to be statistically independent and a χ^2 test to be nonsignificant.

Keep these two notions of statistical inference in mind for the other tests of significance considered in the remaining chapters of this text. Both will be employed.

You may have noticed that all the tests for statistical significance we have computed follow a set pattern. This routine is summarized in Box 4.2. Since the procedure for testing statistical significance is the same for all of the statistics employed in this text, you would do well to commit the steps in this box to memory.

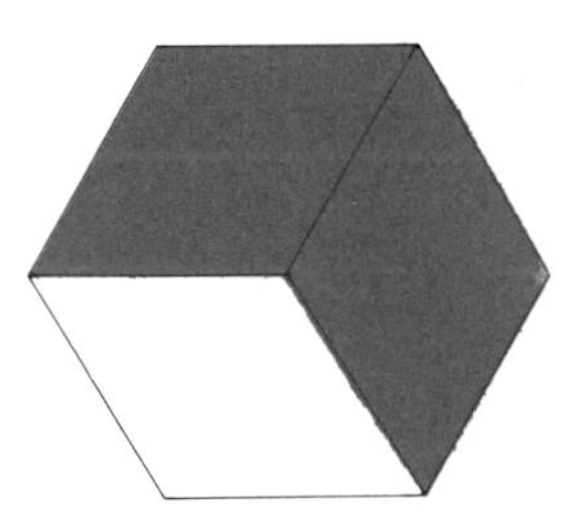

Review of Key Concepts

This list of the key concepts introduced in Chapter 4 is in the order of appearance in the text. Combined with the definitions in the margins, it will help you review the material and can serve as a self-test for mastery of the concepts.

crosstabulation (joint) contingency table)
cell
marginal distribution
row marginals
column marginals
statistical significance test
random sampling
null hypothesis
probability level
alpha (α)
Type I error (false rejection error)
Type II error (false acceptance error)
chi-square test (χ^2 test)
statistical independence
expected frequency
chi-square distribution
test statistic
sampling distribution
degrees of freedom *(df)*
critical value (c.v.)
equiprobable distribution
goodness-of-fit test

PROBLEMS

General Problems

1. In an introductory statistics course, the following students are enrolled: 25 senior sociology majors, 15 junior majors, and 10 sophomore majors; 5 senior nonmajors, 10 junior nonmajors, and 15 sophomore nonmajors. Set up both the raw-data and percentage crosstabulations to display the class-by-major relationship, treating major as the independent variable. Give a verbal interpretation.

2. A social scientist hypothesizes that religiosity was a significant factor in the 1980 election. She finds the following results in a sample: among "fundamentalist" respondents, 85 Reagan voters, 22 Carter voters, and 43 nonvoters; among "liberal" religious adherents, 55 Reagan voters, 35 Carter voters, and 10 nonvoters. Construct the appropriate percentage table and describe the relationship you observe.

3. If a simple random sample of 350 respondents is to be selected from the student body at Big State University, which has an enrollment of 45,600, what is Mary Jones's chance of being interviewed?

4. State the following research hypotheses as null hypotheses:
   a. Younger people are more sexually permissive than older people.
   b. Marital status is related to drinking behavior.
   c. Southern residents own more firearms than non-Southerners.
   d. The average price of suburban houses is greater than that of inner-city houses.

5. An experimental parole program is conducted at a state prison to test the research hypothesis that early-release prisoners will have fewer rearrests than those who serve out their full terms. State the null and the alternative hypotheses. If $\alpha = .01$ for a statistical test, what do Type I and Type II errors imply in this example?

6. For the data in Problem 1, what is the expected frequency of sophomore nonmajors under the hypothesis of statistical independence between class level and major?

7. Give the table of expected frequencies for the data in Problem 2 under the statistical independence hypothesis.

8. What are the chi-square test statistics for the data in Problems 1 and 2?

9. State the degrees of freedom in a table with *(a)* 3 rows and 4 columns, *(b)* 4 rows and 5 columns, and *(c)* 4 rows and 3 columns.

10. Find the critical values of chi square for the following:

	Rows	Columns	α
a.	3	5	.01
b.	2	2	.05
c.	4	8	.001
d.	3	5	.05
e.	7	4	.01

11. How large must χ^2 be in a 3×5 crosstabulation to reject the independence hypothesis *(a)* at $\alpha = .05$, *(b)* at $\alpha = .01$, and *(c)* at $\alpha = .001$?

12. In a juvenile court, 20 youths were sent to a youth home, and 40 were given probation for first offenses. Within a year, 60% of those sent to the home were arrested for another offense, while among those on probation, 40% were arrested for a second offense. Did probation significantly reduce the recidivism rate among this sample of juvenile delinquents? What would be your conclusion if you used a probability level of .01?

13. In a survey of 200 dormitory residents, a research team finds that among 150 undergraduates, 10% think the food is "excellent," 30% think it is "good," and the rest say it is "poor." Of the graduate students who comprise the remainder of the sample, 5 say the food is "excellent," 15 rate it "good," and the others answer "poor." Set up the appropriate raw-data and percentage crosstabulations. Can you reject the null hypothesis that undergraduate and graduate opinions about dormitory food are the same? (Use $\alpha = .05$.)

14. You obtain data that show the relationship between amount of studying (four categories) and grades on a final exam (five categories). The χ^2 for this table is 23.4. What is the chance

of Type I error if you decide to reject the hypothesis that studying has no impact on grades?

15. Public health scientists have test results showing that a new drug to treat morning sickness in pregnant women may also result in a slightly higher rate of birth defects, although the difference in birth defects between those using the drug and those not is significant only if $\alpha = .40$. What decision should be made about releasing the drug for public use? What criteria should enter into the scientists' decision?

16. Two kinds of classrooms in an elementary school—a traditional structure and an open structure—are compared on pupils' reading levels. Each class has 30 students, and there are 20 students in the high, 20 in the middle, and 20 in the low levels of reading for both classes combined. Construct two crosstabulations, one showing independence of structure and reading level, the second showing the highest possible association between traditional structure and high reading performance. Calculate χ^2 for both tables.

17. Here are the responses to the question in the 1978 and 1980 GSS whether the United States is spending enough on the military, armaments, and defense:

	1978	*1980*
Too little	413	825
About right	666	377
Too much	334	168

Decide whether a real change occurred after two years.

18. State the level of significance at which you can reject the hypothesis of no relationship between watching television and studying for 30 high school students who gave these responses:

		Do you watch too much TV?	
		Yes	*No*
Do you study hard enough?	*Yes*	5	5
	No	15	5

19. The catalog of Big State University states that its student body is recruited equally from the North, South, East, and West of the nation. Test this proposition with the following sample

data on 75 students' home states: North = 19, South = 14, East = 24, and West = 18.

20. Three coins are tossed 500 times, with the following results:

Heads	*Tails*	*f*
0	3	50
1	2	150
2	1	200
3	0	100

Do these results differ significantly from what you would expect if the coins were "fair" (i.e., the chance of a head or tail is one-half that for a given toss)?

Problems Requiring the 1980 General Social Survey Data

21. One question of interest to many Americans is the quality of life of people born under different Zodiac signs. Use $\alpha = .05$ and crosstabulate:
    a. HAPPY by ZODIAC, and calculate a chi-square test to determine whether persons born under different signs report themselves happier.
    b. HEALTH by ZODIAC, and calculate χ^2 to decide whether astrological sign is related to health.

22. Crosstabulate regularity of smoking (EVSMOKE) and alcoholic beverage drinking (DRINK). Are these two behaviors significantly related in the population? Set $\alpha = .01$.

23. Test the null hypothesis that a person's opinion about the amount of income tax paid (TAX) is unrelated to confidence in Congress (CONLEGIS). Set $\alpha = .001$.

24. Do people who favor the death penalty for murder (CAPPUN) also oppose laws requiring police permits before buying a gun (GUNLAW)? Set $\alpha = .05$.

25. Does belief in life after death (POSTLIFE) significantly increase with regularity of church attendance (ATTEND)? Use $\alpha = .001$.

Problems Requiring the 63-Cities Data

26. An urban sociologist hypothesizes that cities in different regions of the country experienced different amounts of fiscal stress in 1970 and 1974, measured as per capita long-term municipal debt. Dichotomize STRESS70 and STRESS74 at their median values, then crosstabulate each with REGION. Were significant differences in stress by region found in both years? Which region's cities had the largest percentages of high fiscal stress? Set $\alpha = .001$.

27. Decentralized decision-making structures are expected to covary with formal government structure, with decentralized power structures more prevalent in less reformed cities. Recode DECINDEX into three levels (under 6.33, 6.33 to 7.50, over 7.50) and crosstabulate it with REFORMGV. Is the hypothesis supported by the data for $\alpha = .05$?

28. Are reform government institutions more common in some regions of the country than in others? Crosstabulate REFORMGV and REGION, and determine whether the hypothesis of statistical independence can be rejected if $\alpha = .001$.

29. Divide urban crime rates (CRIMRATE) into three categories (33 and under, 34 to 51, 52 and over) and also trichotomize police expenditures (POLICEXP) (under 20, 20 to 26, and over 26). Do the data support the hypothesis that cities with higher crime rates spend more per capita for police protection? Set $\alpha = .01$.

30. Test whether the number of municipal functions (FUNCTION) performed by cities is an equiprobable distribution across its range, using the χ^2 goodness-of-fit test and $\alpha = .05$.

5

Statistical Inference and Hypothesis Testing

Some basic concepts necessary for understanding statistical inference, which makes it possible to generalize from a sample to a population in order to test a hypothesis, were introduced in Chapter 4. In this chapter we will address the problem of making inferences about a population from calculated sample data more precisely. First we will provide a general, elementary discussion of discrete and continuous probability distributions. Then we will consider the description of discrete probability distributions and present in detail two continuous probability distributions which are useful for drawing inferences about sample means: the normal distribution and the t distribution. A brief discussion of sample estimators and the properties that make sample statistics general estimators of population parameters is also included.

5.1. Probability Distributions

When a researcher chooses a random sample to study from some population, the goal is to be able to draw accurate inferences about characteristics of the population from which the sample has been drawn. The researcher hopes that the distribution of outcomes is a good approximation of the population distribution from which the sample was drawn. If it is a good approximation, the sample mean and standard deviation, for example, would also be expected to be good approximations of the population mean and standard deviation (these concepts were described in Chapter 3).

probability distribution —a set of outcomes, each of which has an associated probability of occurrence

Every population has a probability distribution associated with it. A **probability distribution** is defined by a set of K outcomes —$Y_1, Y_2, \ldots, Y_K$—each of which has an associated probability of occurrence—$p(Y_1), p(Y_2), \ldots, p(Y_K)$. There are two types of probability distributions, those where the outcomes of Y are discrete and those where they are continuous.

5.1.1. Discrete Probability Distributions

discrete probability distribution— a probability distribution for a discrete variable

As we defined the concept in Chapter 1, a *discrete variable* classifies persons, objects, or events according to the kind or quality of their attributes. In considering **discrete probability distributions**, however, we are concerned with another characteristic of discrete measures, and in this sense *a variable is discrete if its outcomes are countable.* Examples of discrete outcomes include the number of crimes per year in major U.S. cities, the number of blacks in high-level executive positions in *Fortune*'s 500 largest U.S. corporations, or the number of Democrats in Paducah, Kentucky, in 1981. The probability of observing an outcome in discrete probability distributions is simply the relative frequency of cases associated with that outcome. If the total population is of size T and the number of cases associated with outcome Y_i is n_i (where $(n_1 + n_2 + \ldots + n_K = T)$, then $p(Y_i) = n_i/T$. Furthermore, it follows from this definition that:

$$p(Y_1) + p(Y_2) + \ldots + p(Y_K) = \sum_{i=1}^{K} p(Y_i)$$
$$= 1.00$$

That is, the sum of the probabilities in the distribution is 1.00.

empirical probability distribution— a probability distribution for a set of empirical observations

Probability distributions can be either empirical or theoretical distributions. An example of an **empirical probability distribution** is the U.S. distribution of age in 1970, as shown in Table 5.1. Assuming there is no error in counting (which is an unrealistic assumption, of course), the data in Table 5.1 represent an empirical population, since the U.S. Census Bureau tries to count everyone in the nation at a given point in time. According to this table, for example, if we draw a name at random, the probability that the person with that name is under five years of age is .084.

TABLE 5.1
Empirical Probability Distribution for Age of U.S. Citizens in 1970 (in thousands)

Age	*N*	*p*
Under 5	17,154	.084
5–9	19,956	.098
10–14	20,790	.102
15–19	19,071	.094
20–24	16,371	.081
25–29	13,477	.066
30–34	11,431	.056
35–39	11,106	.055
40–44	11,981	.059
45–49	12,116	.060
50–54	11,104	.055
55–59	9,973	.049
60–64	8,617	.042
65–69	6,992	.034
70–74	5,444	.027
75–79	3,835	.019
80–84	2,285	.011
85 or over	1,510	.007
Total	203,213	1.000

Source: Adapted from U.S. Office of Budget and Management, *Social Indicators: 1973.*

An example of a **theoretical probability distribution** is the one associated with the outcomes of tossing a fair die, which was introduced in the χ^2 goodness-of-fit test in Chapter 4. By definition, if the die is fair the probability associated with all six possible outcomes is $1/6 = .167$; that is, $p(Y_i) = .167$ for $i = 1, 2, \ldots, 6$. This distribution is shown in Table 5.2. For this example as well

theoretical probability distribution—
a probability distribution for a set of theoretical observations

TABLE 5.2
Theoretical Probability Distribution for Numbers on Die Tosses

	Outcome					
Probability	*1*	*2*	*3*	*4*	*5*	*6*
$p(Y_i)$	.167	.167	.167	.167	.167	.167

as the previous one, $\sum_{i=1}^{K} p(Y_i) = 1.00$—the sum of the probabilities across all outcomes is unity. Other examples of discrete theoretical distributions are those associated with coin tossing and drawing cards of each suit from a deck.

FIGURE 5.1
A Continuous Probability Distribution

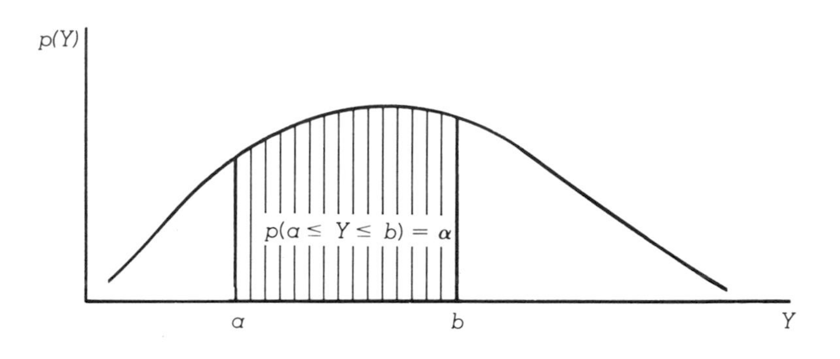

5.1.2. Continuous Probability Distributions

continuous probability distribution—
a probability distribution for a continuous variable, with no interruptions or spaces between the outcomes of the variable

Many of the variables in the social sciences can be thought of as continuous—grade point average, social status, and religiosity, for example. In Chapter 1 we noted that continuous variables are used to classify persons, objects, or events according to the magnitude or quantity of their attributes. When we are considering **continuous probability distributions**, however, *continuous* also means that *there are no interruptions or spaces between the outcomes of a variable.* In fact, of course, we must make some decision about how precisely to measure an underlying continuous variable. For example, grade point average is rarely computed beyond the nearest hundredth (e.g., 3.27), age is rarely measured beyond the nearest month, and so on.

Since the distribution of outcomes is continuous, the probabilities associated with the various outcomes can be connected by a single continuous line, as has been done in the hypothetical example in Figure 5.1. When a continuous variable is measured more and more precisely (such as a grade point average of 3.4276), the number of cases associated with an outcome and hence the probability of observing that outcome approach zero. For this reason, *when considering probabilities for continuous variables, we look at the area between two outcomes, a and b.* Notice that this probability of observing an outcome of Y that lies between points a and b is labeled alpha (α); that is, $p(a \leq Y \leq b) = \alpha$. (Read this as: "The probability of observing an outcome of Y that is greater than or equal to a and less than or equal to b is alpha.")

While many of the variables of interest in the social sciences can be thought of as continuous in theory, when they are actually measured they are, of course, discrete. Therefore, empirical probability distributions are always discrete rather than continuous.

There are several *theoretical continuous probability distributions* that are of great use to statisticians. Later in this chapter we will introduce two of them—the normal and t distributions. Another one is the χ^2 distribution discussed in Chapter 4. Still another—the F distribution—will be introduced in Chapter 6.

These four theoretical probability distributions are extremely important for the field of **inferential statistics**. This is the part of statistics that is concerned with generalizing from sample estimates to population parameters.

inferential statistics—numbers that represent generalizations, or inferences, drawn about some characteristic of a population, based on evidence from a sample of observations from the population

5.2. Describing Discrete Probability Distributions

In the same way researchers can describe and summarize a sample with statistics like the mean, as we noted in Chapter 3, they can also summarize and describe population distributions. The three major population descriptors are the mean, standard deviation, and variance. These descriptors are called **population parameters** because they are *constants* in the population. With a sample, in contrast, the mean and variance are *variables,* because they vary from sample to sample. For this reason, statistics are used as *estimators* of parameters.

population parameter—a descriptive characteristic of a population, such as a mean, variance, or correlation; usually designated by a Greek letter

5.2.1. The Expected Value and Mean of a Probability Distribution

The single outcome that best describes a probability distribution is its **expected value**, labeled $E(Y)$ and given by:

expected value—the number that best describes the typical observation in a population or sampling distribution; the expected value equals the mean of the distribution

$$E(Y) = \sum_{i=1}^{K} Y_i p(Y_i)$$

mean of a probability distribution—the expected value of a population of scores

The **mean of a probability distribution**, labeled μ_Y (Greek letter *mu*), is defined the same way as the expected value; that formula is:

$$\mu_Y = \sum_{i=1}^{K} Y_i p(Y_i)$$

The mean of the distribution of outcomes of a die toss shown by the data in Table 5.2 above, for example, is $\mu_Y = 1(1/6) + 2(1/6) + 3(1/6) + 4(1/6) + 5(1/6) + 6(1/6) = (1 + 2 + 3 + 4 + 5 + 6)1/6 = 21/6 = 3.5$. The average or typical outcome in this distribution, therefore, is 3.5.

In computing the mean of a probability distribution we simply weight each of the K outcomes by its probability of occurrence and add up the resulting terms. Note the similarity of this formula to that presented in Section 3.1.3 in Chapter 3 for computing the sample mean with grouped data. They differ only in that the sample outcome is weighted by (f_i/N), whereas the population outcome is weighted by $p(Y_i) = (n_i/T)$.

We can also compute the expected value of other functions of Y, which we will call $g(Y)$. In this case:

$$E[g(Y)] = \sum_{i=1}^{K} g(Y_i)p(Y_i)$$

One application of this general formulation for expected values is given in the following section.

5.2.2. The Variance of a Probability Distribution

variance of a probability distribution —the expected spread or dispersion of a population of scores

To define the **variance of a probability distribution**, which is labeled σ_Y^2 (Greek letter *sigma*), we first let $g(Y) = (Y - \mu_Y)^2$. Then it follows that:

$$\sigma_Y^2 = E(Y - \mu_Y)^2$$
$$= \sum_{i=1}^{K} (Y_i - \mu_Y)^2 p(Y_i)$$

As in a sample, *the variance of a probability distribution is a measure of spread or dispersion.*

The square root of the variance of a population is called the *standard deviation,* as it is with sample statistics (see Chapter 3). For a population, the standard deviation is symbolized σ_Y and is given by:

$$\sigma_Y = \sqrt{\sigma_Y^2}$$

Using the data on die tosses from Table 5.2, we can calculate that $\sigma_Y^2 = (1 - 3.5)^2 1/6 + (2 - 3.5)^2 1/6 + (3 - 3.5)^2 1/6 + (4 - 3.5)^2 1/6 + (5 - 3.5)^2 1/6 + (6 - 3.5)^2 1/6 = 17.50/6 = 2.92$, and $\sigma_Y = \sqrt{2.92} = 1.71$.

Since researchers do not ordinarily observe populations, the parameters μ_Y and σ_Y^2 are of largely theoretical interest. You need to understand the concept of expected value, however, in order to understand the discussion of inference in the sections and chapters that follow.[1] Box 5.1 provides a summary of symbols for the sample statistics introduced in Chapter 3 and the population parameters used thus far in this chapter.

Some theorems that can be used to determine how "deviant" or unlikely an outcome is in a distribution are described in the following sections. These preliminaries will be useful in thinking about how rare a given sample mean is in a population of sample means generated by taking all possible samples of size *N*. In drawing an inference we first hypothesize that a population has a mean equal to some value, μ_Y. If the discrepancy between the *observed* sample mean and the *hypothesized* population mean is too large—that is, if it is too "deviant" to have come from a population with a mean of μ_Y—we reject the hypothesis about the size of the population mean, μ_Y.

5.3. Chebycheff's Inequality

Observations which are distant from the mean of a distribution occur, on the average, with less frequency than those close to the mean. In all of the distributions we have examined, an outcome that is one-half standard deviation above the mean is more likely

1. We have not discussed the expected value, mean, and variance of continuous probability distributions because integral calculus is needed to do so. The meanings of the mean and variance for continuous probability distributions are identical to those for discrete probability distributions, however.

BOX 5.1 Population and Sample Symbols

Formulas can apply to data for an entire population and for a sample of observations drawn from a population. While the formulas are often the same or similar, the symbolic notation differs. Italic letters are used to stand for *statistics* calculated on sample data, while lower-case Greek letters stand for the population values, called *parameters.*

For some of the statistics and parameters we have introduced, the symbols are:

Name	*Sample Statistic*	*Population Parameter*
Mean	$\bar{Y}$	μ (mu)
Variance	s^2	σ^2 (sigma)
Standard deviation	s	σ (sigma)

than one that is, say, two standard deviations above the mean. In general, *the more distant an outcome is from its mean, the lower the probability of observing it.*

Chebycheff's inequality—a theorem which states that, regardless of the shape of a distribution, the probability of an observation being k standard deviations above (or below) the population mean is less than or equal to $1/k^2$

A Russian named Chebycheff first proved the relationship between the size of a deviation from the mean and the probability of observing it. **Chebycheff's inequality** theorem states that the probability of an observation being k standard deviations above (or below) the mean in a population is less than or equal to $1/k^2$. Stated more formally:

$$p(|Z| \geq k) \leq 1/k^2$$

In words, this equation says that the probability that the absolute value of an observation's standard or Z score (see Chapter 3) is equal to or greater than k standard deviations is equal to or less than 1 divided by the square of k standard deviations.

Another way to think of this idea is that at least $1 - (1/k)^2$ of

FIGURE 5.2
Probability Distribution Illustrating Chebycheff's Inequality

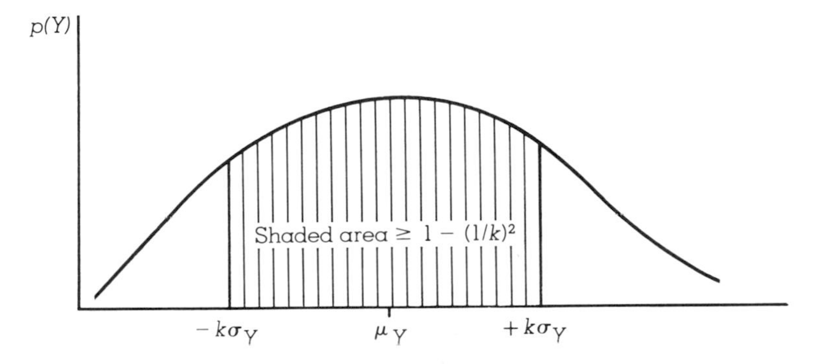

the observations in any population will lie in an interval bounded by $\mu_Y - k\sigma_Y$ and $\mu_Y + k\sigma_Y$, as illustrated in Figure 5.2. The theorem states, for example, that at least $1 - (1/2)^2 = 3/4$ of the observations will lie in the interval bounded by $\mu_Y - 2\sigma_Y$ and $\mu_Y + 2\sigma_Y$. Therefore, for observations drawn at random, an observation two or more standard deviations distant from the mean will occur less than 25% of the time. And an observation three or more standard deviations from the mean will occur *in any population* (regardless of its mean, standard deviation, and skew) less than 1/9, or 11%, of the time. *Chebycheff's inequality refers to outcomes in a population, not in a sample.* (We can always calculate the likelihood of observing an outcome in a sample once we have the data in hand.)

The power of Chebycheff's inequality is that it provides probabilities of observing outcomes in a population in the absence of any information about the shape of the population. If we can further assume that the population distribution is unimodal and symmetric about its mean, a derivative version of Chebycheff's inequality applies. This version is:

$$p(|Z| \geq k) \leq .4444(1/k)^2$$

For example, in a unimodal symmetric distribution an outcome two or more standard deviations from the mean will be observed $.4444(1/2)^2 = .4444(1/4) = 11\%$ of the time or less, compared to 25% of the time or less when no assumption about the shape of the distribution is made.

> The value of Chebycheff's inequality is to provide a way for deciding a priori how deviant a given observation in a distribution is. It also shows how altering an assumption about the shape of a distribution alters the probability of observing outcomes a given distance from the mean. This latter point will be critical in the sections that follow.

5.4. Normal Distributions

normal distribution— a smooth, bell-shaped theoretical probability distribution for continuous variables that can be generated from a formula

One family of unimodal, symmetric distributions which is especially important in statistics is comprised of **normal distributions**, which were introduced in Chapter 4. The term *normal* is really a misnomer, since normal distributions are rarely found in real data. Normal distributions are all described by a rather formidable equation:

$$p(Y) = \frac{1}{\sqrt{2\pi\sigma_Y^2}} e^{-(Y-\mu_Y)^2/2\sigma_Y^2}$$

The shape of any given normal curve is determined by two values, the population's mean, μ_Y, and variance, σ_Y^2. The theoretical curves are shown in Figure 5.3, one with $\sigma_Y^2 = 10$ and the other with $\sigma_Y^2 = 15$, and both with $\mu_Y = 0$. Notice that the smaller the variance, the closer the population scores are to the mean, and the "thinner" the tails of the normal distribution are. Although the tails of the normal curve appear in the figure to touch the horizontal axis, the theoretical distribution of values ranges from $-\infty$ to $+\infty$. Therefore the tails approach but never actually touch the horizontal axis.

We will refer to probabilities associated with outcomes of normal distributions very often in this text. Figuring out the probability of outcomes for different values of μ_Y and σ_Y^2 can be very tedious and time-consuming. But any distribution can easily be converted

FIGURE 5.3
Two Examples of Normal Distributions

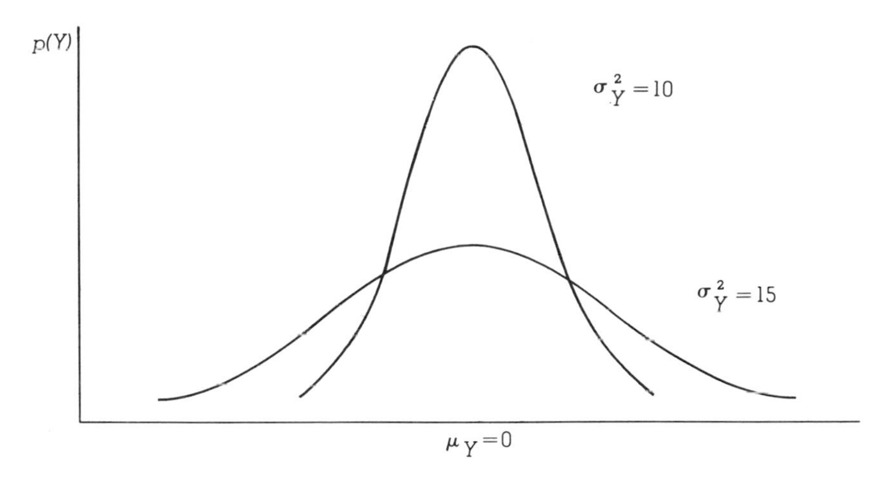

to Z scores; for population values, $Z = (Y - \mu_Y)/\sigma_Y$. Therefore, one table of probabilities associated with distributions will suffice —the one associated with Z scores. This table appears as Appendix C, "Area under the Normal Curve," and a schematic of it is shown in Figure 5.4.

The total area under the curve shown in Figure 5.4 is unity (1.0), since as we noted earlier in this chapter the sum of all probabilities in a distribution equals unity. Half the area lies left of its mean (which is zero, since the mean of Z scores is always zero), and the other half lies to the right of the mean. In Figure 5.4, Z refers to a given Z score for which we want to show the probability of occurrence. The shaded area refers to the probability of a value between zero and Z_α.

For example, suppose we want to know the probability of an outcome being at least 1.55 standard deviations above the mean, assuming a normal distribution. In this example, $Z_\alpha = 1.55\sigma = 1.55$, since the standard deviation of Z scores is always 1 (see Chapter 3 for a review of Z scores if you are confused). To determine the probability of this occurrence, turn to Appendix C. Look *down* the stub, or first column, of the table until you find 1.5. At that point look *across* the table, to the column labeled .05. The

FIGURE 5.4
Example of the Probability of Observing an Outcome in a Normal Distribution

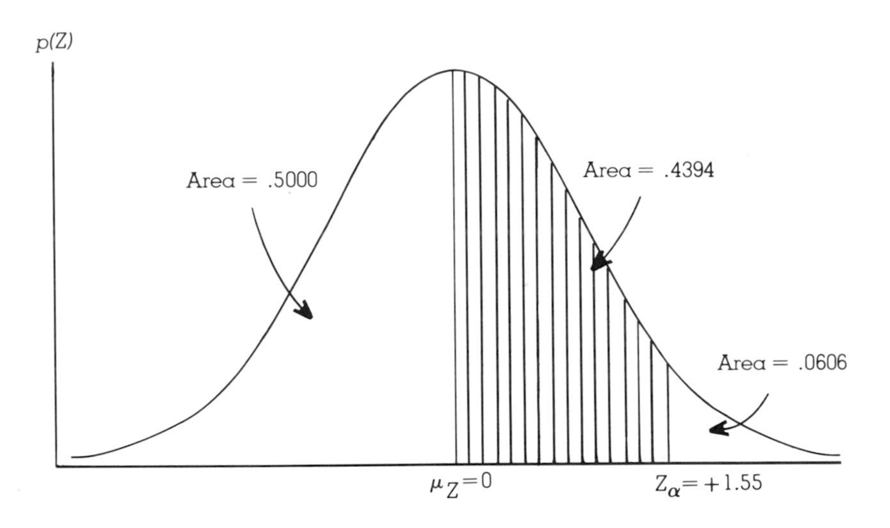

number you should see under this column is .4394. This is the probability of an outcome being between zero and Z_α (1.55 in this example). The probability of Z_α being 1.55 or greater is the *un*-shaded area in the right tail of the distribution. This probability is $.5000 - .4394 = .0606$, since the probability for the entire upper half of the distribution equals .5000. Since a normal distribution is symmetric, it should be clear that the probability of an observation being at least 1.55 standard deviations *below* the mean is also .0606. Thus, .8788 of the standard normal curve area lies between -1.55 and $+1.55$, while only .1212 of the area lies in both tails beyond Z scores of -1.55 and $+1.55$.

5.4.1. The Alpha Area

Although it will not be evident why until later in this chapter, the area from a given Z_α to the tail of a distribution is called alpha, or α. Stated mathematically:

$$p(|Z| \geq |Z_\alpha|) = \alpha$$

We have illustrated this in Figure 5.5, where the area between Z_α and ∞ is labeled α. This alpha is the same one introduced in

FIGURE 5.5
Probability Distribution for a Type I Error in the Right Tail

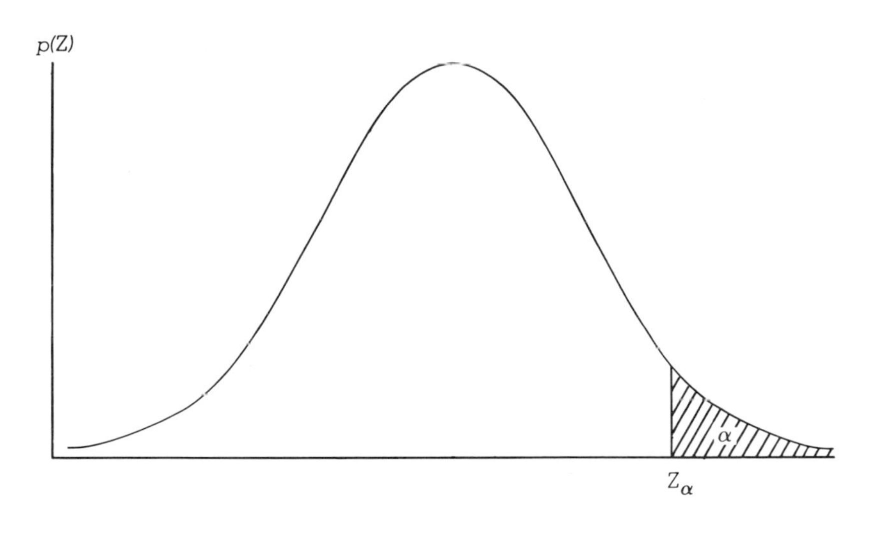

Chapter 4—the probability of a Type I or false rejection error (the probability of rejecting the null hypothesis when it is in fact true). We could also have chosen α to be in the left tail of the distribution. For this reason we use absolute values of Z and Z_α in the formula above.

Sometimes we want to split the probability of a Type I error between the two tails of a distribution. In this case the probability in the upper tail is $\alpha/2$, and the probability in the lower tail is also $\alpha/2$. The two Z scores that cut off these areas in the distribution are labeled $Z_{\alpha/2}$ and $-Z_{\alpha/2}$, respectively. Figure 5.6 should help clarify this discussion, as well as the two examples that follow.

Assume we have a problem where the use of the normal curve is appropriate, as will be the case later in this chapter. Suppose further that we choose $\alpha = .05$ and want it to be totally in the upper tail, as is the case in Figure 5.5. The strategy to follow is to ask what value of Z_α will cut off the upper 5% of the normal curve. We look up .4500 in Appendix C, since $.5000 - .4500 = .0500$. The two values closest to .4500 are .4495 and .4505. They are

FIGURE 5.6
Probability Distribution for a Type I Error, Split between Two Tails

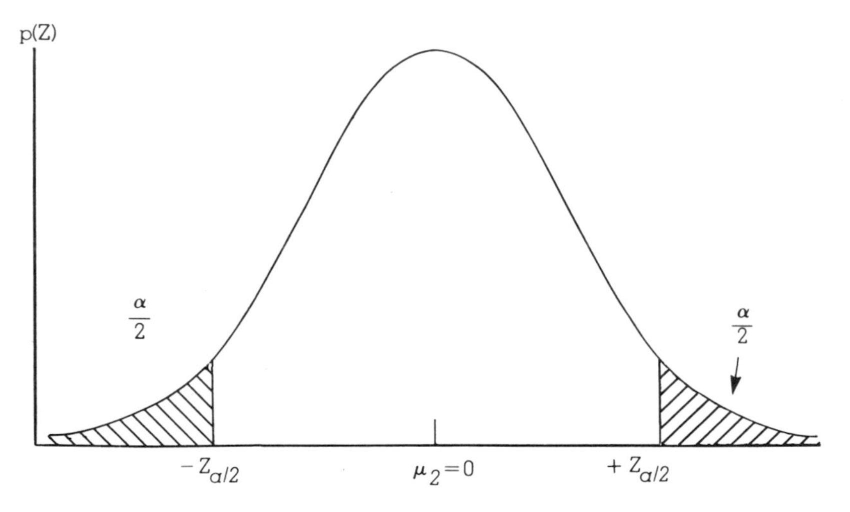

associated with the outcomes 1.64 and 1.65, respectively. We can just divide the difference and conclude that a Z_α of 1.645 standard deviations or more above the mean will occur only about 5% of the time in a normal distribution. Therefore $Z_\alpha = 1.645$ for this problem.

If we wish to split the probability of a Type I error between the two tails, as we have done in Figure 5.6, then for $\alpha = .05$, $\alpha/2 = .025$. To determine the Z_αs that place 2.5% of the area under the normal distribution in each tail, we calculate that $.5000 - .4750 = .0250$, and the Z_α associated with .4750 is 1.96. Since the normal distribution is symmetric, $-Z_\alpha = -1.96$ also.

Another way to think of this is as follows: 95% of the area under a normal curve lies between $-Z_\alpha = -1.96$ and $Z_\alpha = 1.96$. Now, convince yourself that about 68% of the outcomes in a normal distribution fall between standard deviations of -1 and $+1$, and about 99.7% of all outcomes will fall between standard deviations of -3 and $+3$. Assuming a normal distribution, an observation three standard deviations or more from the mean is a rare occurrence indeed.

Let's look at another example. The probability of an outcome two standard deviations or more above the mean of a normal distribution is .5000 − .4772 = .0228 or less. That is, only about 2% of the time will an outcome two standard deviations above the mean in a normal distribution be observed. As we noted in Section 5.3 on Chebycheff's inequality, when a distribution is symmetric the probability of an observation two or more standard deviations from the mean is .11 or less, and when no assumption about the shape of the distribution is made, the probability of such an observation is .25 or less. These differences make clear how assumptions about the shape of a distribution affect the probability of observing a given outcome. *The closer in shape the distribution is to normal, the smaller the probability of observing an outcome k standard deviations or more from the mean.*

The use of Z_α and α in using the normal distribution will become clearer when we discuss inference in Section 5.6.

5.5. The Central Limit Theorem

One very special and important application of the normal distribution follows from the **central limit theorem**, which states:

> If all possible random samples of N observations are drawn from a population with mean μ_Y and variance σ_Y^2, then as N grows large, the sample means will be approximately normally distributed, with mean μ_Y and variance σ_Y^2/N; that is:

$$\mu_{\overline{Y}} = \mu_Y$$
$$\sigma_{\overline{Y}}^2 = \sigma_Y^2/N$$

central limit theorem— a mathematical theorem that states that if repeated random samples of size N are selected from a normally distributed population with mean = μ and standard deviation = σ, then the means of the samples will be normally distributed with mean = μ and standard error = $\sigma/\sqrt{N}$ as N gets large.

According to the central limit theorem, *the mean of the distribution of all the sample means of a given sample size drawn at random will equal the mean of the population from which the samples were drawn. The theorem does not make any assumption about the shape of the population from which the samples are drawn.*

The population distribution of all possible means for samples

sampling distribution of sample means—the population distribution of all possible means for samples of size N selected from a population

of size N is the **sampling distribution of sample means**. Sampling distributions are hypothetical constructs. *A sampling distribution for means consists of all the sample means that would be obtained by forming all possible samples of the sample size, N, drawn at random from a population.* With billions and trillions of unique samples from any large population, no one is actually going to calculate a sampling distribution of means. Nevertheless, since the central limit theorem tells us how the population's two parameters (μ_Y and σ_Y^2) are related to the sampling distribution's mean and variance, we can readily determine the shape of the latter once we know the values of the former. The sampling distribution concept will be very useful when we discuss inferences about population means in the next section.

> The central limit theorem guarantees that a given sample mean can be made to come close to the population mean in value by simply choosing an N large enough, since the variance of the sampling distribution of means, σ_Y^2/N, becomes smaller as N gets larger.

standard error—the standard deviation of a sampling distribution

The standard deviation of a sampling distribution has a special name—the **standard error**. For means, this is:

$$\sigma_{\bar{Y}} = \sigma_Y/\sqrt{N}$$

Armed with the knowledge that the sampling distribution of sample means is normal, regardless of the shape of the population from which the samples are drawn, and assuming that N is large, we can make some powerful statements. Suppose a population has a mean of $\mu_Y = 100$ and a standard deviation of $\sigma_Y = 15$ and we take a random sample of 400 cases. We can immediately calculate the standard error of the sampling distribution of means for all samples of size 400 which can be drawn from the population. Using the formula above, the standard error of this specific sampling distribution is

$$\sigma_{\bar{Y}} = 15/\sqrt{400} = .750$$

Based on the information in the preceding section on normal distributions, we should expect that 95% of all sample means will fall within $\pm$ 1.96 *standard errors* of the population mean, or in this example, 95% of all sample means should fall in the interval between 98.53 and 101.47—that is, 100 $\pm$ (1.96)(0.75). Therefore, if we draw a random sample of size 400 its mean should be very close to the population mean in value; only 5% of samples drawn at random will lie outside the interval bounded by 98.53 and 101.47. If we increase the sample size from 400 to 1,000, the standard error of the sampling distribution becomes much smaller; specifically, $\sigma_{\bar{Y}} = 15/\sqrt{1,000} = .47$. For $N = 1,000$, we would expect 95% of all sample means to lie between 99.08 and 100.92; that is, 100 $\pm$ 1.96(.47). This result implies that we can have considerable confidence that *any* random sample of size 1,000 we choose will give an accurate estimate of the mean of the population from which it was drawn.

5.5.1. An Example: Occupational Prestige

When students are introduced to the central limit theorem they often have some trouble believing that the mean of the sampling distribution of all possible samples of a given size equals the mean of the population from which it was drawn. An example in which we treat a large sample as if it were the entire population and then compute sampling distributions for it should be convincing. The illustrations are drawn from data from the 1977 General Social Survey for a measure of occupational prestige which was developed at the National Opinion Research Center in 1963–65. Prestige scores were generated by asking a sample of respondents to estimate the social standing of occupations on a nine-step ladder. Occupational titles were printed on small cards, and the prestige ratings were collected by requesting respondents to sort the cards into boxes formed by the rungs of the ladder. The prestige scores eventually assigned to occupations were averages computed across all respondents in the three years.

These scores were applied to the occupations of the sample of 1977 GSS respondents. As Figure 5.7, a diagram of the distribution of NORC prestige scores for the occupations reported by this sample, shows, the mean of these scores is 38.4, and the standard deviation is 14.15. Clearly it is *not* a normal distribution, since it has a positive skew. Superimposed on the figure are two empirically derived "sampling distributions." Two sets of 10,000 different samples were drawn from the population of 1,416 respondents with prestige scores. In one case $N = 25$, and in the other, $N = 100$. The complete sampling distributions would require

FIGURE 5.7
Sampling Distributions for $N = 25$ and $N = 100$

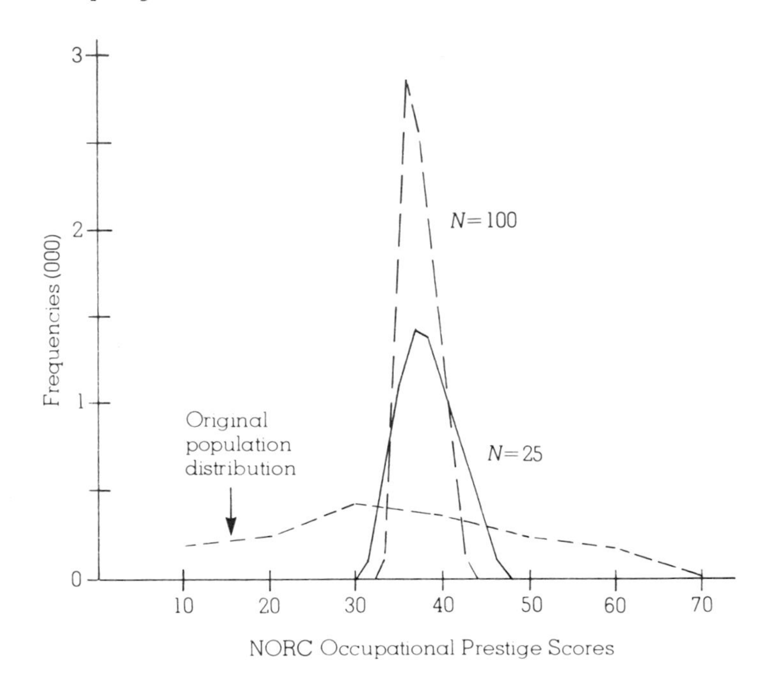

Source: 1977 General Social Survey.

many times more samples, so the two distributions are approximations. Nevertheless, they are very close approximations to the actual sampling distributions.

Four important facts can be noted about this example. First, while the mean of the GSS sample (which we are here treating as if it were a population of observation) is 38.4, the mean of the sampling distribution for a sample of size $N = 25$ is 38.4 and that for $N = 100$ is 38.3. These examples should make it clear that the mean of a sampling distribution does in fact equal the mean of the population from which the samples are drawn.

Second, even though the distribution of the original 1,416 observations from which the samples were drawn is somewhat skewed, both sampling distributions look somewhat like normal distributions. This is as they should be, according to the central limit theorem.

Third, since the central limit theorem states that the standard error of the sampling distribution equals $\sigma/\sqrt{N}$, using the statistics for the 1,416 observations we should expect the standard errors to equal $14.15/\sqrt{25} = 2.83$ and $14.15/\sqrt{100} = 1.42$ for the samples of 25 and 100, respectively. In fact they equal 2.79 and 1.37, which are very close to the theorem's guarantees.

Fourth, we would expect the standard error to be smaller for a larger N. That this is indeed the case is shown in Figure 5.7, where there is less variance or "spread" in the scores around the mean of the sampling distribution for $N = 100$ than for $N = 25$. If we choose cases at random from a population, the larger the sample size the more likely is the mean of any given sample mean to equal or nearly equal the mean of the population from which it was drawn. This example should make it clear to you that very accurate generalizations to even the largest of populations (such as that of the United States) can be made with samples as small as 1,416.

While we have said that N must be large for the central limit theorem to apply, we have not squarely addressed the question of how large it must be. Some textbooks say 30; others suggest 100. We do not use any hard-and-fast rule, but we suggest on the basis of experience that when a sample is the size of 100 we can safely assume that the sampling distribution of means closely approximates a normal distribution. And unless the underlying population is extremely odd in shape, we can be relatively confident that the central limit theorem applies for samples as small as 25 or 30.

In Section 5.9 of this chapter we will introduce the t distributions and some so-called *small-sample* estimation techniques. To make such decisions easier we will recommend that these techniques be used when N is less than 100. You should recognize this as only a loose rule of thumb, however, not as dogma. First we will be more explicit about how the central limit theorem can be used to make inferences with real data where we only have a single sample mean and variance (or standard deviation) and N is greater than or equal to 100.

5.6. Sample Point Estimates and Confidence Intervals

The fact that the mean of the sampling distribution of means equals the mean of the population from which the sample was drawn has an important corollary. This is that the sample mean for a randomly drawn sample is an unbiased estimator of the population mean from which the sample was drawn, that is,

$$E(\overline{Y}) = \mu_Y$$

When a random sample is drawn, therefore, the best single estimate of the population mean is the sample mean. Another term for the sample mean is the **point estimate** of the population mean.

point estimate—a sample statistic used to estimate a population parameter

confidence interval—a range of values constructed around a point estimate which makes it possible to state the probability that the interval contains the population parameter between its upper and lower confidence limits

lower confidence limit—the lowest value of a confidence interval

upper confidence limit—the highest value of a confidence interval

Once we choose a sample and compute a mean, we can construct a **confidence interval** around it. We will then be able to state the probability that the confidence interval contains the actual population mean. We would expect approximately 95% of the intervals constructed in repeated sampling of the same size to contain the population mean, μ_Y, where the interval is defined by boundaries approximately two standard errors below and two standard errors above the mean. The **lower confidence limit** (or LCL) is $\overline{Y} - 1.96\sigma_{\overline{Y}}$ and the **upper confidence limit** (or UCL) is $\overline{Y} + 1.96\sigma_{\overline{Y}}$ for a 95% confidence interval.

Based on the information on normal distributions in Section 5.4, we can determine that the LCL and UCL for a 68% confidence interval are $\overline{Y} - \sigma_{\overline{Y}}$ and $\overline{Y} + \sigma_{\overline{Y}}$, respectively. And for a 99% confidence interval, the LCL is $\overline{Y} - 2.58\sigma_{\overline{Y}}$ and the UCL is $\overline{Y} + 2.58\sigma_{\overline{Y}}$. In general, we can construct confidence intervals for any desired level of confidence, say $1 - \alpha$, by the formula:

$$\overline{Y} \pm Z_{\alpha/2}\sigma_{\overline{Y}}$$

We can now define the symbol $Z_{\alpha/2}$, which was introduced above, more precisely: *In Appendix C, $Z_{\alpha/2}$ is the value that has an area to the right which is equal to $\alpha/2$.* If we want to be 95% confident that a given interval contains μ_Y, $\alpha = .05$ by implication, and $\alpha/2 = .025$. As we saw in Section 5.4, $Z_{\alpha/2} = 1.96$ will give us the correct interval for a given $\sigma_{\overline{Y}}$.

Figure 5.8 illustrates the concept of a confidence interval. The solid vertical line represents the true population mean, which is, of course, a constant. The horizontal lines represent confidence

FIGURE 5.8
Example Illustrating the Concept of a Confidence Interval

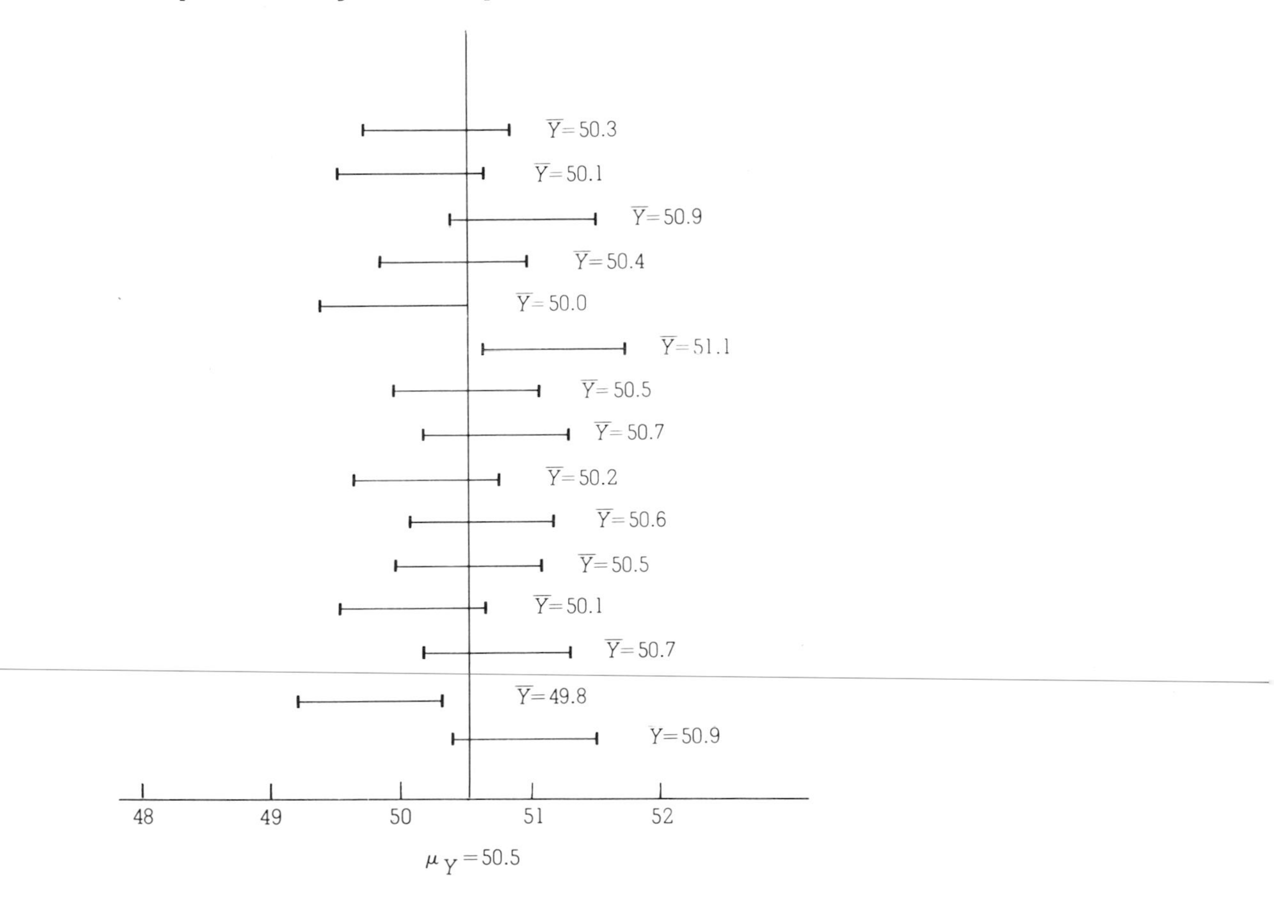

intervals constructed around 15 different sample means. Notice that all but two of them (the 6th and 14th from the top) contain the population mean, $\mu_Y = 50.5$. The point estimates (i.e., the sample means) are noted beside each interval.

In general, the larger the sample size, the smaller the interval around the sample mean for a given confidence interval. This can be easily shown. Suppose we know that $\sigma_Y = 15$. We observe $\overline{Y} = 100$ in a random sample of 100. If we want to be 95% confident that our constructed interval contains the population mean, our interval is bounded by $100 - 1.96\,(15/\sqrt{100})$, or 97.06, and $100 + 1.96\,(15/\sqrt{100})$, or 102.94. But now suppose we observe $\bar{Y} = 100$ for a random sample of 500. In this case, the interval is bounded by $100 - 1.96\,(15/\sqrt{500})$, or 98.69, and $100 + 1.96\,(15/\sqrt{500})$, or 101.31, for 95% confidence. *Thus, for a given α, increasing N decreases the size of the confidence interval.*

This is another way of saying that as N gets large, a given sample mean, $\bar{Y}$, is on average a better estimate of the population parameter μ_Y. This clearly fits our intuition, since we know that the sample mean equals the population mean as N approaches T. But, more important, you should convince yourself that for a sample of even $N = 1{,}500$, confidence intervals around a given sample mean are quite small relative to the size of the population standard deviation. This is why public opinion polling is such big business today!

A few summary statements about assumptions used in constructing confidence intervals may help you tie things together:

1. We assume that the sample for estimating μ_Y is drawn *randomly*.
2. We assume we have chosen $N \geq 100$.
3. We assume we know σ_Y.

To be able to compute the standard error, $\sigma_{\bar{Y}}$, requires knowledge of the population standard deviation, which we do not generally know. When N is "large" (i.e., 100 or more), however, we can be quite confident that the sample standard deviation, s_Y, is a good estimate of $\sigma_{\bar{Y}}$. A caret (^) is added to $\sigma_{\bar{Y}}$ to signify that it is estimated:

$$\hat{\sigma}_{\bar{Y}} = s_Y/\sqrt{N}$$

In Chapter 3, for example, we estimated average liking scores for Canada and Egypt, using data from Table 3.5. The point estimates on the 10-point scales (where 0 is extreme disliking and 9 is extreme liking) were 7.57 for Canada and 4.74 for Egypt, with standard deviations of 1.69 and 2.22, respectively. The sample sizes on which these statistics were based were 1,436 for Canada and 1,344 for Egypt. Using this information, we can construct confidence intervals. For example we can be 95% confident that the population mean for liking for Canada is contained in the interval defined by $7.57 \pm (1.96)(1.69/\sqrt{1{,}436})$. This computes out to a LCL of 7.48 and a UCL of 7.66. Now convince yourself that the LCL for liking for Egypt is 4.62 and the UCL is 4.86. Notice that with these quite-large samples, both confidence intervals are small relative to the size of the standard deviations. This shows the importance of N in making inferences.

Since these examples use real data from the 1977 General Social Survey, there is no way to check whether the confidence intervals do indeed contain the two population means. For a given sample size all we can do to increase our confidence is to increase the interval size by choosing $1 - \alpha$ larger. In these examples, the 99% confidence intervals would be defined by $7.57 \pm (2.58)(1.69/\sqrt{1{,}436})$ and $4.74 \pm (2.58)(2.22/\sqrt{1{,}344})$. That is, for Canada the LCL and UCL would be 7.45 and 7.69, respectively; for Egypt, 4.58 and 4.90, respectively.

Based on all these results, we are virtually certain that Americans liked Canada a lot more than they liked Egypt in 1977!

5.7. Desirable Properties of Estimators

To be useful, sample estimators should have certain desirable characteristics. The general notation system uses $\boldsymbol{\theta}$ (Greek *theta*) to designate any population parameter and $\hat{\boldsymbol{\theta}}$ for an estimator (sample statistic) of a population parameter. Since we use sample statistics to make inferences about the parameters, we want them to be "good" estimators. This section identifies the characteristics that make a good estimator.

5.7.1. Lack of Bias

An **unbiased estimator** is one that, on average, equals the population parameter. If $\hat{\boldsymbol{\theta}}$ is an unbiased estimate of $\boldsymbol{\theta}$, therefore,

unbiased estimator—an estimator of a population parameter whose expected value equals the parameter

$$E(\hat{\theta}) = \theta$$

That is, if the mean of the estimator of all possible samples of size

N from the same population equals the population parameter, the estimator (sample statistic) is said to be *unbiased.*

For example, as we indicated in Section 5.5 on the central limit theorem, $\mu_{\overline{Y}} = \mu_Y$, and so $E(\overline{Y}) = \mu_Y$. Hence the sample mean $\overline{Y}$ is an unbiased estimator of the mean of the population, μ_Y. In contrast, it can be proven the sample median is *not* an unbiased estimate of the population mean; that is, E (median) $\neq \mu_Y$. This should not be surprising, of course.

In Section 3.2.5 of Chapter 3, one way we defined the sample variance was:

$$s_Y^2 = \frac{\sum_{i=1}^{N} (Y_i - \overline{Y})^2}{N - 1}$$

This indicates that the summed and squared deviation is to be divided by $N - 1$ instead of by N, as is done when computing the mean. Suppose we use s_*^2 to symbolize the variance instead of s_Y^2, and define it:

$$s_*^2 = \frac{\sum_{i=1}^{N} (Y_i - \overline{Y})^2}{N}$$

It can be proven that

$$E(s_*^2) = \left(\frac{N - 1}{N} \right) \sigma_Y^2$$

That is, if we take all possible random samples of size N from some population and then we compute s_*^2 for each, the mean of the distribution will be smaller than the population variance by the factor $(N - 1)/N$. For s_*^2 to be unbiased, we need to multiply it by the reciprocal of the biasing factor, that is, by $N/(N - 1)$:

$$\left(\frac{N}{N - 1} \right) s_*^2 = \left(\frac{N}{N - 1} \right) \sum_{i=1}^{N} \left(\frac{Y_i - \overline{Y}}{N} \right)^2$$

$$= \sum_{i=1}^{N} \frac{(Y_i - \overline{Y})^2}{N - 1}$$

$$= s_Y^2$$

Therefore the statistic for the variance of a sample, s_Y^2, is an unbiased estimator of the variance of the population, σ_Y^2.

5.7.2. Efficiency

In addition to an unbiased estimator, the distribution of all possible sample estimates from some population should have *minimum variance.* An **efficient estimator** is one where $E(\hat{\theta} - \theta)^2$ is as small as possible.

efficient estimator—the estimator of a population parameter among all possible estimators that has the smallest sampling variance

We can illustrate this by again considering the median and sample means as estimators of μ_Y, the population mean. It can be shown that the variance of the sampling distribution for the median is $(\pi/2)(\sigma_Y^2/N)$ for a large N, and we have shown that the variance of the sampling distribution for the mean is σ_Y^2/N. To compare their relative efficiencies we can compute the *ratio* of the sampling variance of the median to that of the mean, and we find that $\pi/2 = 1.57$. That is, the mean is roughly 1.5 times more efficient than the median as an estimator of μ_Y.

5.7.3. Consistency

A **consistent estimator** is one that approximates the population parameter more closely as N gets large. More formally, $\hat{\theta}$ is a consistent estimator if as $N \rightarrow \infty$, $E(\hat{\theta} - \theta)^2 \rightarrow 0$. That is, as N gets larger, the variance of the distribution of estimates gets smaller, thereby ensuring that the sample estimate is closer and closer to the population parameter. In this case, *both* the sample mean and the median are consistent estimates of the population mean, μ_Y. As N gets large, the variance of the sampling distribution for each of them gets smaller. Since the mean is more efficient than the median, however, it is a better estimator of the population mean.

consistent estimator—an estimator of a population parameter that approximates the parameter more closely as N gets large

Because the sample mean, $\bar{Y}$, and the variance, s_Y^2, are unbiased, efficient, and consistent, they are desirable estimates of the population mean, μ_Y, and variance, σ_Y^2, respectively.

5.8. Testing Hypotheses about Single Means

One type of hypothesis testing involves a research hypothesis about the specific value of a population mean. Considerable research has shown, for example, that when people are asked to describe their own physical attractiveness, most of them say their looks are above average.[2] Because physical attractiveness helps

2. E. Berscheid and E. Walster, "Physical Attractiveness," in L. Berkowitz (Ed.), *Advances in Experimental Social Psychology,* Vol. 7 (New York: Academic Press, 1974).

determine how others treat and judge them, individuals tend to bias estimates of their own appearance upward. In research done by one of the authors, a sample of 2,013 *Psychology Today* readers were asked to indicate how physically attractive they considered themselves to be on a seven-point scale of 7 for "much more attractive than others" through 1 for "much less attractive than others." The middle score of the scale, 4, was "about the same as others."[3]

null hypothesis about a single mean—a null hypothesis that the population mean is equal to or unequal to a specific value

We can test our hypothesis that people tend to bias their ratings of physical attractiveness in the positive direction by positing a **null hypothesis about a single mean;** that, is, in the population, the mean score is 4 or lower (i.e., one's physical attractiveness is about the same or less than others).

alternative hypothesis—a secondary hypothesis about the value of a population parameter that often mirrors the research or operational hypothesis

Choosing an alternative hypothesis is more difficult. We can choose either an exact alternative (e.g., $\mu_Y = 5.0$) or an inexact alternative (e.g., $\mu_Y > 4.0$). (A second inexact alternative is $\mu_Y \neq 4.0$.) Suppose, for the sake of example, past research suggests that $\mu_Y = 4.85$. Then the null hypothesis (H_0) and the **alternative hypothesis** (H_1) can be stated as follows:

$$H_0: \mu_Y = 4.00$$
$$H_1: \mu_Y = 4.85$$

Since our sample is large, we can safely assume that the sampling distributions under H_0 and H_1 are roughly normal. Therefore we can calculate the probability of observing a given mean, given $\mu_Y = 4.0$ versus $\mu_Y = 4.85$. We also know that s_Y can be taken as a reasonable estimate of σ_Y because we have a large sample size. To test which of the two hypotheses is the more viable (*both* may be wrong, of course) we follow these steps:

1. Choose an α level—the probability of making a Type I error and falsely rejecting a true null hypothesis.
2. Compute the test statistic under H_0 and determine the probability that the observed mean could have come from the sampling distribution implied by H_0. Call this test statistic Z_0.
3. Compute the test statistic under H_1 and determine

3: E. Berscheid, E. Walster, and G. Bohrnstedt, "The Happy American Body: A Survey Report," *Psychology Today,* December 1973, pp. 119–31.

the probability that the observed mean could have come from the sampling distribution implied by H_1. Call this test statistic Z_1.

4. Calculate the probabilities associated with Z_0 and Z_1 in steps 2 and 3 above.
5. Reject H_0 or not, or withhold final judgment if the results do not clearly support either hypothesis.

To compute the test statistic for steps 2 and 3, we compute Z scores (in Appendix C it is assumed that the variables are Z standardized). Since the standard deviation of the sampling distribution is $\sigma_{\overline{Y}}$, according to the central limit theorem, the two Z scores for accomplishing steps 2 and 3 are:

$$Z_0 = \frac{\overline{Y} - \mu_{Y_0}}{\sigma_{\overline{Y}}}$$

$$Z_1 = \frac{\overline{Y} - \mu_{Y_1}}{\sigma_{\overline{Y}}}$$

Where:

μ_{Y_0} = The population means hypothesized under H_0.
μ_{Y_1} = The population means hypothesized under H_1.

For the physical attractiveness example, suppose we choose $\alpha = .05$. Since the alternative states that μ_Y is larger than 4.00, the probability of a Type I error is in the right tail (as in Figure 5.5). That is, we will reject H_0 if $Z_0 \geq 1.65$, the Z score which cuts off the upper 5% of the distribution in Appendix C. In the sample of $N = 2{,}013$ readers, the observed $\overline{Y} = 4.90$ and $s_Y = 1.153$.[4] Therefore the Z scores under the null and alternative hypotheses are:

$$Z_0 = \frac{4.90 - 4.00}{1.153/\sqrt{2{,}013}} = 35.02$$

$$Z_1 = \frac{4.90 - 4.85}{1.153/\sqrt{2{,}013}} = 1.95$$

4. Unpublished data from Berscheid, Walster, and Bohrnstedt, "Happy American Body."

Since Z_0 is far larger than 1.65, the probability that $\mu_Y = 4.00$ is virtually zero. Hence we *reject* the null hypothesis. However, the alternative hypothesis that $\mu_Y = 4.85$ is *also* unlikely, since (as Appendix C shows) the probability of observing a Z score of 1.95 or greater is about .03. Thus, while the likelihood of H_1 being true is greater than the likelihood of H_0 being true, neither hypothesis seems very likely. Therefore instead of accepting either hypothesis, **suspending judgment** pending further research is a better procedure.

suspending judgment—a position taken by a researcher when the results of a statistical test permit neither clear rejection nor clear acceptance of the null or alternative hypotheses

A rational next step would be to state that *the best estimate of the population mean is the sample mean,* $\overline{Y} = 4.90$. We might follow this up by saying we can be 95% confident that the interval bounded by $4.90 - 1.65(1.153/\sqrt{2{,}013}) = 4.86$ and $4.90 + 1.65(1.153/\sqrt{2{,}013}) = 4.94$ contains the true population mean. Thus, we have strong evidence for our general research hypothesis that on average, people say they are above average in looks.

Theory and past research in the social sciences rarely provide enough information to state an exact alternative hypothesis, as was done above. A more typical approach to test the hypothesis that people rate themselves as above average on looks would use these null and alternate hypotheses:

$$H_0\colon \mu_Y = 4.00$$

$$H_1\colon \mu_Y > 4.00$$

When the alternative is inexact, the steps taken to test the hypotheses for statistical significance are identical to those given in Box 4.2 (Chapter 4):

1. Choose an α level.
2. Examine Appendix C to see how large the critical value has to be to reject the null hypothesis.
3. Compute the test statistic under H_0.
4. Compare the test statistic with the critical value. If it is as large or larger, reject the null hypothesis; if not, do not reject the null hypothesis.

As noted above, if we choose $\alpha = .05$ for the physical attractiveness example, the test statistic is $Z_0 = 35.02$. Since the c.v. is 1.65 for $\alpha = .05$, the null hypothesis that $\mu_Y = 4.00$ must be rejected. In this case, however, when accepting the alternative hypothesis all we are saying is that, on average, people tend to rate their physical attractiveness above average, without saying how much above average.

As the above examples make clear, if we choose an N large enough, almost no exact hypothesis will ever be accepted, since tests of statistical significance are very sensitive to sample size. For this reason it makes good sense to report the point estimate of a parameter and confidence intervals, in addition to or instead of tests of statistical hypotheses.

5.9 Hypothesis Testing When the Standard Error Is Unknown: The *t* Distribution

In the preceding examples we have assumed we know the standard error of the sampling distribution of the mean, $\sigma_{\overline{Y}}$, or that N is large enough so that we can use the sample variance, s_Y^2, as a reasonable estimate of σ^2. There is another family of distributions which are very similar to normal distributions, but for which we do not need to know $\sigma_{\overline{Y}}$. These are referred to as ***t* distributions.**

***t* distribution**—one of a family of test statistics used with small samples selected from a normally distributed population or, for large samples, drawn from a population with any shape

A ***t* variable**, or *t score,* is given as:

$$t = \frac{\overline{Y} - \mu_Y}{s_Y/\sqrt{N}}$$

***t* variable (*t* score)**—a transformation of the scores of a continuous frequency distribution derived by subtracting the mean and dividing by the estimated standard error

The similarity to Z used for hypothesis testing is evident. The only difference is that t involves s_Y, whereas Z assumes knowledge of σ_Y. The t statistic was first introduced by W. S. Gossett, who signed his research article "Student." For this reason these sampling distributions are often called Student's t distributions.

There are many t distributions, and their shape varies with the sample size and the sample standard deviation. All t distributions, like Z-transformed normal distributions, are bell-shaped and have a mean of zero. But there are two important differences between a t distribution and a normal distribution. These are:

1. The use of a t distribution to test hypotheses assumes that the sample is drawn from a normally distributed population.

2. A t distribution for a given sample size has a larger variance than a normal Z distribution. Therefore, the standard error of a t distribution is larger than that of a normal Z distribution.

The second statement needs to be qualified, however. As N gets large (i.e., in the range of 100), a t distribution becomes increasingly similar to a *normal* Z distribution in shape. Therefore, as N gets large, the standard error of the t distribution approaches that of the normal Z distribution. This means that for $N = 100$, the probabilities associated with outcomes in the two distributions are virtually identical. This can be verified by comparing probabilities of Z values in Appendix C with t values in Appendix D for $N = 100$.

The assumption that a sample is drawn from a normally distributed population may seem to be restrictive. Research has shown, however, that violations of this assumption have only minor effects on the computation of a test statistic. Therefore, unless we are certain that the underlying population from which the sample is drawn is grossly nonnormal, we can use a ***t* test** to test a hypothesis even when N is small.

t test—a test of significance for continuous variables where the population variance is unknown and the sample is assumed to have been drawn from a normally distributed population

FIGURE 5.9

Examples of t Distributions for Different Degrees of Freedom

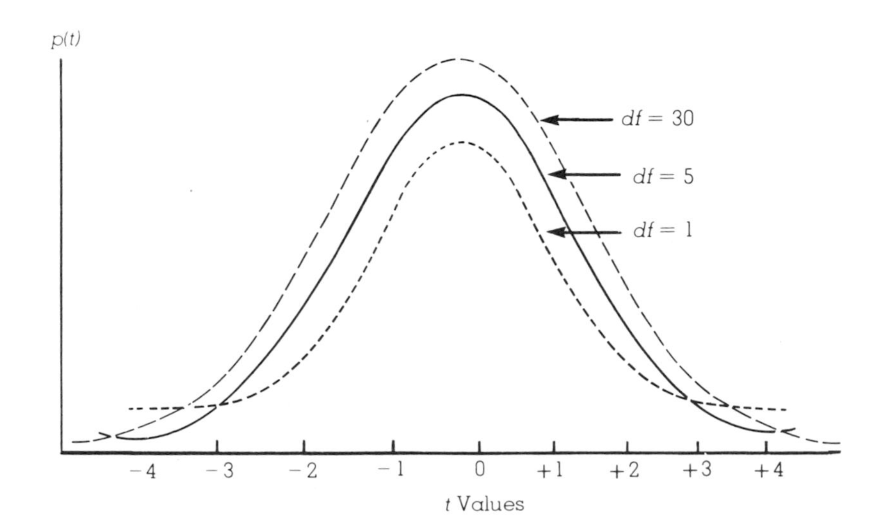

FIGURE 5.10
Tabled Values of Student's *t* Distribution

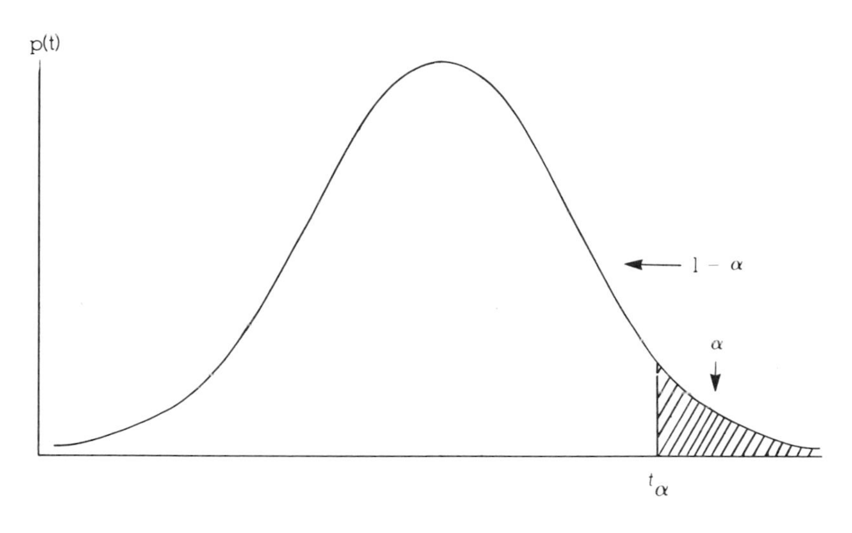

As noted in Section 5.7.1, the numerator of the formula for the sample variance, s_Y^2, is divided by $N - 1$. With a *t* test, the value of $N - 1$ is called the *degrees of freedom* (see Chapter 4) associated with the *t* test. *In a set of N observations with a given mean, N − 1 scores can assume any value, but the Nth score is not free to vary for a given mean.* For example, if we know the mean of a distribution is 3, $N = 4$, and three observations equal 1, 2, and 4, the fourth value is constrained to be 5. Once we know the mean and $N - 1$ values, the *N*th value is constrained. For this reason, we say there are $N - 1$ degrees of freedom *(df).*

Sampling distributions, as we noted at the beginning of the chapter, are bell-shaped, symmetric curves with means of 0 and differing standard deviations, depending on the population standard deviation and the degrees of freedom. The smaller the *df,* the flatter the curve, and the more slowly the tails approach the base line, as can be seen in Figure 5.9.

Since the values of a *t* distribution are the same as the values of a *Z*-normal distribution when *N* is large, we can *always* use a *t* test to test hypotheses about means. Therefore, *when results involving tests of means are presented in published research, they are almost always evaluated using t statistics.* If N is small, however, you need to remember that the t distribution requires random sampling from a normal population, whereas Z does not assume a normal parent population. As suggested above, unless the parent population is grossly nonnormal, violations of this assumption do not cause serious problems in drawing inferences.

Figure 5.10 shows how to use Appendix D. The values tabled in Appendix D cut off the upper α percent of a t distribution for each given degree of freedom. Thus for $\alpha = .05$, the critical value, t_α, is found by looking down the column labeled .05. For example, for $\alpha = .05$ and $df = 30$, $t_\alpha = 1.70$. As we found in Section 5.6, for $\alpha = .05$, $Z = 1.645$. This illustrates that *when N is small, a larger critical value of t is needed to reject the null hypothesis.*

5.9.1. An Example Using the t Test

Demographic data, which some social science researchers use in their work, indicate a huge growth in U.S. cities since World War II. More recently there appears to have been a reversal of this trend, however, as more people seem to be moving to less densely populated areas. We can use the 63-cities data (see Box 2.4 in Chapter 2) to test whether there has been a decline in the size of U.S. cities between 1960 and 1970. Our two hypotheses are:

$$H_0: \mu_Y = 0$$

$$H_1: \mu_Y < 0$$

where Y is percent change in population between 1960 and 1970. If we choose $\alpha = .05$, the relevant sample statistics to compute the t test are $\bar{Y} = -1.26\%$ and $s_Y = 6.32$. The notation t_{62} refers to the fact that the t test has $63 - 1 = 62$ degrees of freedom associated with it. Therefore:

$$t_{62} = \frac{-1.26 - 0}{6.32 / \sqrt{63}} = -1.58$$

We examine Appendix D and see that the critical value for a t test with 62 degrees of freedom for $\alpha = .05$ is 1.67. Therefore, we *cannot reject* the null hypothesis in this case.

All the hypothesis tests we have presented in this chapter involve single means. In practice, however, we are rarely interested in hypotheses about a single mean. Instead we wish to compare two or more means. For example, given equal education and experience for professional men and women, are professional men's salaries the same as or higher than professional women's salaries in the same occupations? A similar question could be asked to compare the salaries of blacks and whites in the same profession, or we might want to ask whether premarital sexual activity is greater for teenage boys than for teenage girls. All of these questions involve comparing two means and asking whether they are significantly different from each other. The evaluation of such types of questions is undertaken in Chapter 6.

Review of Key Concepts

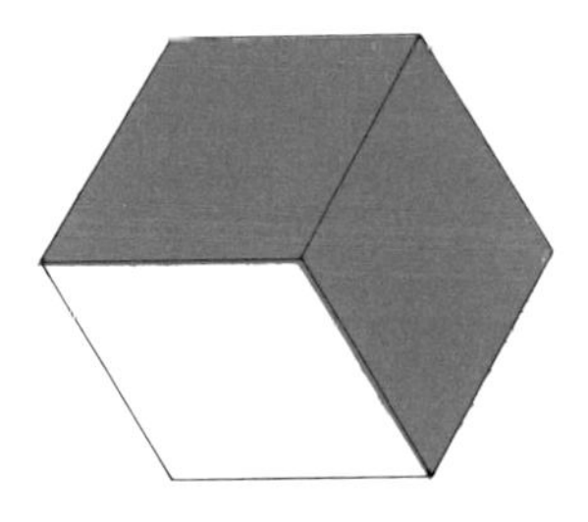

This list of the key concepts introduced in Chapter 5 is in the order of appearance in the text. Combined with the definitions in the margins, it will help you review the material and can serve as a self-test for mastery of the concepts.

probability distribution
discrete probability distribution
empirical probability distribution
theoretical probability distribution
continuous probability distribution
inferential statistics
population parameters
expected value
mean of a probability distribution
variance of a probability distribution
Chebycheff's inequality
normal distribution
central limit theorem
sampling distribution of sample means
standard error
point estimate
confidence interval
lower confidence limit (LCL)
upper confidence limit (UCL)
unbiased estimator
efficient estimator
consistent estimator
null hypothesis about a single mean
alternative hypothesis
suspending judgment
t distribution
t variable (*t* score)
t test

Problems

General Problems

1. Assuming that birthdays are equally distributed throughout the year, what is the theoretical probability of a sample respondent chosen at random who is *(a)* born on March 4, *(b)* born in the month of May, or *(c)* born on a Tuesday?

2. Using the empirical probability distribution for age in 1970 shown in Table 5.1, state the probability that a person chosen at random that year would *(a)* be in his or her 30s; *(b)* be less than 20 or more than 64; *(c)* not be in his or her 50s.

3. Find the expected value of the following probability distributions:

(a) Y_i	$p(Y_i)$	*(b)* Y_i	$p(Y_i)$
10	.05	10	.05
20	.20	20	.05
30	.50	30	.15
40	.20	40	.60
50	.05	50	.15

4. Find the variances and standard deviations of both probability distributions in Problem 3.

5. Use Chebycheff's inequality to find the probabilities that an observation is *(a)* 4 or more standard deviations away from the population mean, *(b)* within 2.5 standard deviations of the mean, *(c)* 1.75 standard deviations or further from the mean, and *(d)* 1.33 or fewer standard deviations from the mean. Make no assumptions about the shape of the population distribution.

6. The mean income of residents of Bigtown is known to be $15,000, with a standard deviation of $4,000. If we know nothing about the shape of the income distribution in Bigtown, how often, at the very most, could we expect to find people with incomes of $23,000 or more? With incomes of $5,000 or less?

7. Find the areas under the normal distribution between the mean and the following *Z*-scores: *(a)* -1.57, *(b)* $+3.12$, *(c)* $+0.99$, and *(d)* -0.23.

8. Find the *Z*- scores for the normal distribution that correspond to the following alphas:
   a. α = .03, one-tailed.
   b. α = .15, one-tailed.
   c. α = .02, one-tailed.
   d. α = .02, two-tailed.
   e. α = .20, two-tailed.
   f. α = .001, two-tailed.

9. Find the areas of the normal distribution between the following *Z*-scores:
   a. $-0.50 \leq Z \leq +0.50$.
   b. $-1.25 \leq Z \leq +2.00$.
   c. $-2.33 \leq Z \leq +2.33$.
   d. $-.25 \leq Z \leq +1.50$.

10. According to the central limit theorem, find the means and standard errors of sampling distributions with the following characteristics:

	μ_Y	σ_Y^2	N
a.	12.5	40	50
b.	40	100	100
c.	0	100	500
d.	14	160	80
e.	200	200	200

11. The mean number of hours students at Big State University devote to study each day is known to be 4.5, with a variance of 4.0. If the sampling distributions of sample means for all samples of size N = 25, N = 50, and N = 100 were constructed, what would be the values for the means and the standard errors?

12. In a population of high school girls, the average number of dates per month is 5.5, with a standard deviation of 2.5.
    a. Find the standard error for the sampling distribution of these sample sizes: (1) N = 20, (2) N = 60, and (3) N = 200.
    b. What is the probability, in a sample of N = 100, that the observed sample mean would be six or more dates per month?

13. Find the upper and lower confidence limits for the following:

	$\bar{Y}$	$\sigma_{\bar{Y}}$	*Confidence Interval*
a.	75	15	95
b.	75	15	99
c.	3.6	1.4	99
d.	14.5	3.5	90

14. From a population of automobile assembly workers, a sample of 100 are randomly selected for a study of the effects of a new training program on productivity. At the end of the program, the mean productivity score of the sample is 6.66. If you know that the population standard deviation is 35, what is the 95% confidence interval for the sample? Can you reject the null hypothesis that the mean productivity in the population is zero?

15. A hypothesis asserts that "The average college graduate has a higher intelligence quotient (IQ) than that of the general population." Rewrite this statement in null and alternative hypothesis form, using symbolic notations. (IQ scores are standardized at a mean of 100.)

16. Rewrite the statement, "The average college graduate has an IQ of 110," in formal notation for null and alternative hypotheses, with the population mean score of 100 as the null hypothesis.

17. In a sample of 500 college graduates, the variance of IQ scores is 225 and the mean is 107. Can you reject either null hypothesis in favor of the alternative in Problem 15 or 16, if $\alpha = .05$?

18. Find the areas under the t distribution between the mean and the following values:

	df	t
a.	14	2.62
b.	20	−2.85
c.	8	3.36
d.	30	2.04
e.	17	−2.90
f.	500	2.58

19. Find the critical values of t that correspond to the following:

	N	α
a.	12	.05, one-tailed
b.	12	.01, one-tailed
c.	19	.05, one-tailed
d.	19	.05, two-tailed
e.	28	.01, one-tailed
f.	40	.05, two-tailed
g.	8	.001, two-tailed
h.	500	.001, two-tailed

20. Find the areas in the tails of the *t* distributions for the following:
    a. $-1.80 \leq t_{11} \leq +3.11$. *c.* $-2.07 \leq t_{23} \leq +2.50$.
    b. $-2.10 \leq t_{18} \leq +2.10$. *d.* $-1.70 \leq t_{30} \leq +2.75$.

21. Find the *t* scores for a test of the null hypothesis that $\mu_Y = 0$, given the following information:

	$\overline{Y}$	s_Y	N
a.	−3.47	2.40	15
b.	+15	9	8
c.	−500	1600	26

22. In a sample of 125 experimental subjects, the mean score on a postexperimental measure of aggression was 55, with $s = 5$. If subjects scoring 47 points or lower are judged to be completely nonaggressive, what number of subjects fall into this category, assuming that the distribution of aggression scores is approximately normal?

23. In a study of perception accuracy under stress, one group of 36 persons in a stressful condition had a mean score of 90 on a perception scale, with a standard deviation of 24. Construct the *(a)* 95% and *(b)* 99% confidence intervals around this point estimate.

24. After a kindergarten class spends eight months with a new curriculum, the 15 pupils are tested for reading readiness. The average score is 15, with a variance of 9. If the minimum score for entry into the first grade is 12, can we conclude that the new curriculum has boosted reading readiness significantly above the minimum requirement? (Use $\alpha = .01$.)

25. The standard deviation of political tolerance in a sample of 27 adults is 3.60 on a 10-point scale. The sample mean is 7.40. Test the null hypothesis that this sample is drawn from a population with a mean of 8.20 on the political tolerance measure.

Problems Requiring the 1980 General Social Survey

26. Find the empirical probability distribution for the number of children a respondent has ever had (CHILDS) and determine the probability that he or she has had four or more.

27. As the American population changes, the proportion having only a grade school education is declining. Using the EDUC variable to obtain the mean and standard deviation for years of formal schooling, find the Z- score for 8 years or less. What is the corresponding area of the tail for this Z score?

28. Using the SPSS selection procedure to select only the black respondents, test the null hypothesis that they feel the nation is spending about the right amount on improving the conditions of blacks (NATRACE). Use $\alpha = .05$.

29. Test the hypothesis that the GSS respondents were drawn from a population in which satisfaction with the city or place they live in (SATCITY) is 2.5 on the 7-point scale. Use $\alpha = .001$.

30. Selecting only women (SEX) who work part time during the week (WRKSTAT), test the hypothesis that the mean number of hours they work (HRS1) is 20 or less, setting $\alpha = .01$.

Problems Requiring the 63-Cities Data

31. In the post-World War II baby boom, many suburbs grew while the central cities aged. Test the hypothesis that the mean percentage of the present housing stock (HOUSEPCT) built before 1950 is greater than 55%. Use $\alpha = .01$.

32. Looking at retail sales (RETAIL) in the 63 cities, how likely is it that this sample of cities came from a population of cities in which the mean retail sales were exactly $500 million? Use $\alpha = .01$.

33. Find the standard errors of the sampling distributions for samples of size 63 for: *(a)* the percentage in nonmanual occupation (WCPCT), *(b)* per capita income (INCOMEPC), *(c)* percent population change (POPCHANG), and *(d)* percent female labor force (FEMLABOR).

34. Test the hypothesis that the 63-cities sample was selected from a population of cities whose mean size (POPULAT) was smaller than 300,000 residents. Use $\alpha = .05$.

35. Construct the *(a)* 95% and *(b)* 99% confidence intervals around the sample mean for the age of cities (CITYAGE). Does a population mean of 75 years fall within either interval?

6

Testing for the Difference between Two Means

The information on statistical inference and hypothesis testing presented in Chapter 5 was limited to inferences about a single mean. But because social science researchers are interested in relationships, they almost always want to compare two or more means. The hypothesis might be, for example, that the average income of high school graduates is higher than that of those who do not reach this level of education, or that people in the South are more religious than those in other parts of the United States, or that college students who attend church or temple regularly engage in premarital sex less than those who do not. All these hypotheses involve comparing two means and deciding whether one is larger than the other. The statistical procedures presented in this chapter provide ways to test whether two hypothesized population means differ from one another. This procedure is called a **mean difference test.**

mean difference test—a statistical test of whether two means differ in the population

6.1 Testing for the Difference between Two Means When the Standard Error Is Known

The research efforts of sociologists have often been directed at improving our understanding of the stratification of society, or social status. Traditionally, studies of social stratification have centered on the prestige conveyed by jobs in the occupational structure, which for years was almost exclusively a male concern. The relation between fathers' and sons' educational and occupational achievements was a popular research topic. Blau and Duncan, for example, suggested two variables that contribute to a man's social status: his own educational achievement and his parents' educations.[1] They argued that the more education the parents had, the more education they were likely to provide for their male offspring. Furthermore, education was seen as an important factor in learning the general and specific skills necessary for occupational achievement.

Because the data to test the hypotheses presented in this section are derived from male respondents in the 1977 General Social Survey, we will use men in our examples. As the boundaries between conventional sex roles for men and women are weakening, however, the conclusions could also apply to people in general.

We have greatly oversimplified the Blau-Duncan model of occupational achievement, in terms of both the variables in occupational achievement and the level of explanation. The specification in the example is sufficient, however, to allow us to state two general propositions:

P1: The higher the educational achievement of a man's father, the higher his educational achievement.

P2: The higher a man's educational achievement, the higher his occupational status.

A derivative proposition follows from P1 and P2, namely that men with highly educated fathers should have higher occupational status than those who have less-well-educated fathers.

6.1.1. Stating the Operational Hypotheses

We can use the 1977 GSS data set to test these two propositions, since respondent's education, father's education, and the prestige of respondent's occupation were all measured. Our analysis dichotomizes continuous measures, that is, divides them into two groups. To test P1 we recode father's education into two categories—a high school degree or less, versus at least some college

1. Peter M. Blau and Otis Dudley Duncan, *The American Occupational Structure* (New York: John Wiley & Sons, 1967).

completed—and the respondent's own education is measured by asking how many years of formal schooling he had completed. The hypothesis to be used to test P1 is:

H1: Men whose fathers completed at least some college have more years of schooling themselves, on average, than those whose fathers have a high school education or less.

The second proposition (P2) can be tested by recoding the respondent's education into the same two categories used for father's education in setting up H1—a high school degree versus at least some college completed. The respondent's occupational prestige is measured by the NORC prestige score described in Section 5.5.1. of Chapter 5. Using these variables, P2 can be tested with:

H2: Men who have completed at least some college have occupations with higher average prestige scores than those who have a high school education or less.

To evaluate the first research hypothesis, we will use the following strategy. As we noted in Chapter 1, hypotheses cannot be proven, only rejected by the data in favor of their opposites. Since we believe the evidence will favor the alternative hypothesis, we need to state a null hypothesis, H_0, that can be tested and, presumably, rejected by the data. Thus, the nature of the alternative or research hypothesis, H_1, gives us the form of the null hypothesis, H_0, that we will test. In this example, H_1 states that the mean years of schooling for the population of men whose fathers had some college education is greater than the mean of the population of those whose fathers did not attend college. In symbolic form:

$$H_1: \mu_2 > \mu_1$$

(read "mu two is greater than mu one"), where μ_2 is the mean of the population of men whose fathers had some college, and μ_1 is the mean of the population of men whose fathers did not attend college. The contrasting null hypothesis, also in symbolic terms, is:

$$H_0: \mu_2 \leq \mu_1$$

(read "mu two is less than or equal to mu one").

If, based on a statistical test of H_0 with data from two **independent random samples** of men, we conclude that H_0 can be rejected with a low probability of Type I error, we automatically establish evidence in favor of H_1, the alternative hypothesis. That

independent random samples—samples drawn according to random selection procedures so that the choice of one observation for a sample does not affect the probability of another observation being chosen for a different sample

is, we do not directly test the hypothesis that $\mu_2 > \mu_1$. Rather, we expect to reject the null hypothesis whose form is given by the way the research hypothesis is formed. Box 6.1 describes other forms of null and alternative hypotheses.

6.1.2. Test Procedures

The central limit theorem (see Section 5.4 in Chapter 5) guarantees that if the sample size is large enough, the distribution of sample means (i.e., the sampling distribution) will be normal, with a mean equal to the mean of the population from which the samples were drawn. A corollary of the central limit theorem which is very useful in testing for the difference between two population means states:

> The distribution of differences between two sample means generated by taking random samples of N_1 and N_2 from populations with mean μ_1 and μ_2 and variances σ_1^2 and σ_2^2 follows a normal distribution, with mean $\mu_2 - \mu_1$ and standard deviation (standard error) $\sqrt{\sigma_1^2/N_1 + \sigma_2^2/N_2}$; that is:

$$\mu_{(\bar{Y}_2 - \bar{Y}_1)} = \mu_2 - \mu_1$$
$$\sigma_{(\bar{Y}_2 - \bar{Y}_1)} = \sqrt{\sigma_1^2/N_1 + \sigma_2^2/N_2}$$

Thus, the mean of the sampling distribution of differences between the sample means equals the difference between the two population means. Note that *no assumption is made about the shape of the original population distributions.*

mean difference hypothesis test—a statistical test of a hypothesis about the difference between two population means

With this corollary, we can use the table of probabilities associated with the normal distribution to perform a **mean difference hypothesis test.** That is, we can test hypotheses about two population means, assuming the N_1 and N_2 are large (as defined in Chapter 5) and we can estimate the variances of both populations, σ_1^2 and σ_2^2.

■ *Estimating the Standard Error.* In social research based on sampled data, we will never know the *true* values of the population variances. Indeed, if we had population data available to us, there

BOX 6.1 Forms of Null and Alternative Hypotheses

Theory and past research clearly indicate that one population mean can be expected to be greater than another, although the amount of the difference is unclear. The alternative hypothesis, therefore, can be stated as a general *range* of values, while any difference in the opposite direction is consistent with the null hypothesis. The example in the text uses the most common form of the null and alternative hypotheses about two population means:

$$H_0: \mu_2 \leq \mu_1, \text{ or } \mu_2 - \mu_1 \leq 0$$

$$H_1: \mu_2 > \mu_1, \text{ or } \mu_2 - \mu_1 > 0$$

At times, however, prior knowledge or belief is an insufficient basis for deciding which population mean should be larger, although the researcher expects they will not be equal. Since the substantive hypothesis is that the two means are unequal, the null hypothesis must be that the means are equal. Hence:

$$H_0: \mu_1 = \mu_2 \text{ or } \mu_1 - \mu_2 = 0$$

$$H_1: \mu_1 \neq \mu_2 \text{ or } \mu_1 - \mu_2 \neq 0$$

Thus, evidence from two samples that the population means are not equal—regardless of whether the first mean or the second mean is the larger of the two—permits rejection of the null in favor of the alternative hypotheses.

A third form of the test hypotheses is used whenever precise values of the two population means can be stated. For example, if the IQ mean of one group is hypothesized to be 100 and that of a second group is hypothesized to be 115, the researcher can construct null and alternative hypotheses about this precise 15-point IQ difference. Hence:

$$H_0: \mu_2 - \mu_1 = k$$

$$H_1: \mu_2 - \mu_1 \neq k$$

where k is the hypothesized amount of difference in the two means; that is, $k = 15$ in this example. The null hypothesis

(Continued next page)

Box 6.1 (continued)

is rejected if the sample means do not differ significantly from each other by the specified amount *k*, because the difference observed in the samples is either greater than *k* or less than *k*.

You should recognize, of course, that rejecting the null hypothesis does not imply that the alternative hypothesis is true. The results may be consistent with the alternative hypothesis, but as we have noted several times, hypothesis testing never leads to absolute truth.

Most social theory is so imprecise that explicit values for population means can seldom be stated in advance. Hence, in their most common form, the null and alternative hypotheses in social research state either a general range of differences or merely that the two groups differ, without indicating which has the larger mean value.

would be no reason to perform significance tests. Instead, we must use information from the two samples to *estimate the population standard error* for subsequent use in significance tests (see Section 5.4 in Chapter 5). When N_1 and N_2 are large (i.e., $N_1 + N_2 \geq 100$), the appropriate test is the one discussed in this section. When N_1 and N_2 are small, the appropriate significance test is a *t* test. This concept was introduced in Section 5.8 of Chapter 5, which made a more precise distinction between large and small sample sizes. Use of the *t* test for the difference between two means is discussed in Section 6.4.

We can estimate the standard error of the sampling distribution, $\sigma_{(\bar{Y}_2 - \bar{Y}_1)}$, by substituting for the unknown population variances, σ_1^2 and σ_2^2, with the known sample variances, s_1^2 and s_2^2, *if N is large.* Our estimated standard error for the sampling distribution of the difference between two means is given by:

$$\hat{\sigma}_{(\bar{Y}_2 - \bar{Y}_1)} = \sqrt{\frac{s_1^2}{N_1} + \frac{s_2^2}{N_2}}$$

■ *Testing the Null Hypothesis.* To test the null hypothesis that $\mu_2 \leq \mu_1$, which implies that $\mu_2 - \mu_1 \leq 0$, we refer back to the summary of steps for testing hypotheses given in Box 4.2 (Chapter 4). First we need to choose an α level (the probability of making a Type I error). Second, we calculate the test statistic. Third, we calculate the critical value (c.v.). Fourth, we compare the test statistic with the critical value.

In this example we choose $\alpha = .05$. Computing the test statistic involves ascertaining how large the observed difference between the two sample means is against the null hypothesis. Since Appendix C is in standard-score or *Z*-score form, we compare the observed hypothesized difference as a *Z* score, which is, in general, $Z = (Y - \mu_Y)/\sigma_Y$. In generating the difference between two means, the numerator of a *Z* score is $(\overline{Y}_2 - \overline{Y}_1) - (\mu_2 - \mu_1)$ and the denominator is $\sigma_{(\overline{Y}_2 - \overline{Y}_1)}$.

Under the null hypothesis, $\mu_2 - \mu_1 \leq 0$, that is, the mean from the second population is less than or equal to the mean from the first population. Therefore, when testing the null hypotheses, we need only to determine whether the observed differences in sample means $(\overline{Y}_2 - \overline{Y}_1)$ could be generated from two populations in which the true difference in means is exactly zero. The outcome which would be least favorable yet still consistent with the null hypothesis occurs when $\mu_2 - \mu_1 = 0$. The *Z* score actually used to test the null hypothesis will not contain a term involving the population means because the value of their difference under the null hypothesis is zero. Therefore the numerator of *Z* is $(\overline{Y}_2 - \overline{Y}_1) - 0$, and it follows that:

$$Z_{(\overline{Y}_2 - \overline{Y}_1)} = \frac{\overline{Y}_2 - \overline{Y}_1}{\sigma_{(\overline{Y}_2 - \overline{Y}_1)}} = \frac{\overline{Y}_2 - \overline{Y}_1}{\sqrt{\dfrac{\sigma_1^2}{N_1} + \dfrac{\sigma_2^2}{N_2}}}$$

Or, using our large sample estimate for $\sigma_{(\overline{Y}_2 - \overline{Y}_1)}$, which we label $\hat{\sigma}_{(\overline{Y}_2 - \overline{Y}_1)}$:

FIGURE 6.1
Two Examples of Outcomes When the Null Hypothesis about Mean Differences Is True

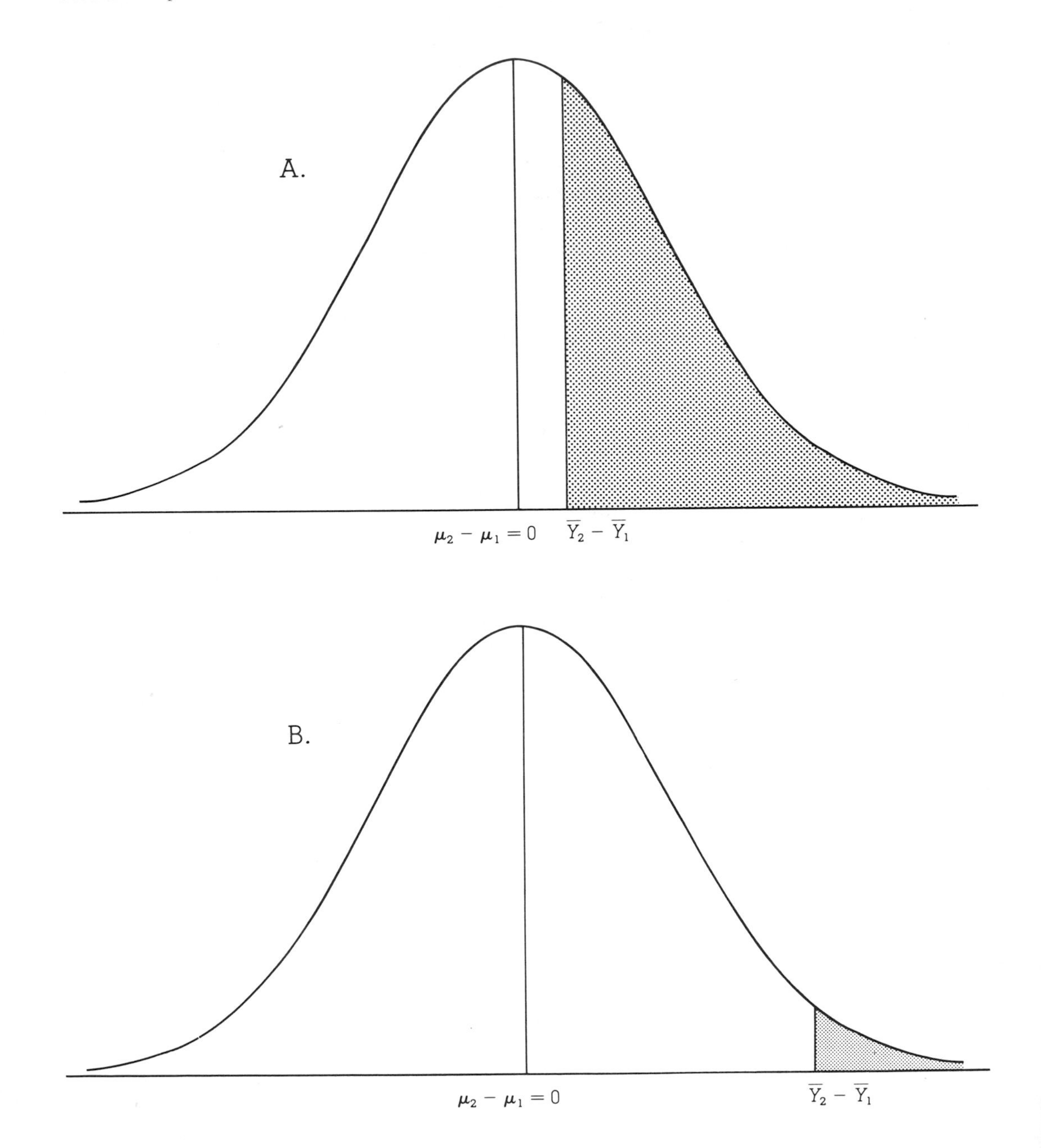

$$Z_{(\overline{Y}_2 - \overline{Y}_1)} = \frac{\overline{Y}_2 - \overline{Y}_1}{\hat{\sigma}_{(\overline{Y}_2 - \overline{Y}_1)}}$$
$$= \frac{\overline{Y}_2 - \overline{Y}_1}{\sqrt{\dfrac{s_1^2}{N_1} + \dfrac{s_2^2}{N_2}}}$$

■ *Diagraming the Distribution.* To understand what happens when the difference in two sample means is tested, it is useful to diagram the sampling distribution. In Figure 6.1, two situations are depicted for a sampling distribution where the null hypothesis that $\mu_2 - \mu_1 = 0$ is in fact true. In Panel A, the difference in observed sample means, $\overline{Y}_2 - \overline{Y}_1$, is a small positive value which lies close to the hypothesized population difference of zero. Since this sample difference has a high probability of occurrence in a population where $\mu_2 - \mu_1 = 0$, we would not reject the null hypothesis at conventional levels of α.

In Panel B of Figure 6.1, however, the observed sample difference is substantial. It would be found only in a small proportion of all the sample mean differences if the true difference in population means is zero. If the observed means are in fact very different, as they are in Panel B of Figure 6.1, we would probably reject the null hypothesis that the two samples came from populations where $\mu_1 = \mu_2$, since the outcome $\overline{Y}_2 - \overline{Y}_1$ is a highly unlikely one in the sampling distribution (note it occurs in the far right tail). Instead we would conclude that the $\overline{Y}_1$ and $\overline{Y}_2$ came from populations where $\mu_2 > \mu_1$.

■ *Conclusions.* We rarely specify exactly what μ_2 and μ_1 are in stating the alternative hypothesis. Suppose, however, that we assume our alternative hypothesis is H_1: $\mu_2 - \mu_1 = k$, where k is some number derived from previous research. In Figure 6.2, the sampling distribution to the left is the same as it is in Figure 6.1. The sampling distribution to the right, however, is the one associated with H_1: $\mu_2 - \mu_1 = k$. Note that the mean difference $(\overline{Y}_2 - \overline{Y}_1)$ observed in this sample is a highly unlikely outcome if in fact the null hypothesis is true, that is, if $\mu_2 - \mu_1 = 0$. However, the observed $\overline{Y}_2 - \overline{Y}_1$ is a very likely outcome if the alternative hypothesis, H_1: $\mu_2 - \mu_1 = k$, is true.

FIGURE 6.2
Likelihood of the Same Sample Outcome under Two Different Hypotheses about Mean Differences

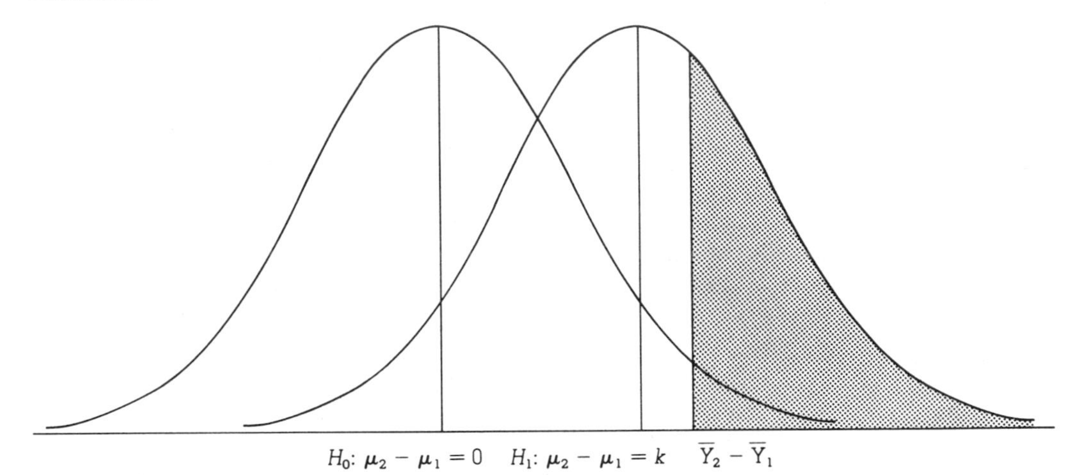

For this reason we reject the null hypothesis in favor of the alternative. *This decision in no way implies that the true difference in population means is k, only that this hypothesis is more compatible with the observed sample outcome than the hypothesis that the true difference is zero.* In other words, we are quite confident that the true difference is *not* zero, and instead it is some positive number.

6.1.3. Testing the Hypotheses

We can demonstrate these principles by testing the hypotheses about educational achievement and occupational status. Using the GSS data for 1977, we estimate that the mean number of years of education for male respondents is 11.92 for those whose fathers completed at least some college, compared to 11.89 for those whose fathers had a high school education or less. The variances for the two groups are 15.02 and 11.54, respectively, based on samples of 267 and 423. The test statistic is computed to be:

$$Z = \frac{11.92 - 11.89}{\sqrt{\frac{15.02}{267} + \frac{11.54}{423}}}$$

$$= \frac{.03}{.289} = .104$$

Now for $\alpha = .05$, the critical value is that Z_α which cuts off the upper 5% of the normal distribution. As we found in Section 5.4, $Z_{.05} = 1.645$. Finally, we compare the test statistic with the critical value. Since the test statistic is .104 and the critical value is 1.645, the null hypothesis cannot be rejected. That is, the 1977 GSS data do *not* provide evidence for the proposition (P1) that fathers' education is positively related to men's own education.

The second hypothesis states that men who have completed at least some college will have occupations with more prestige than those who have had a high school education or less. Using the NORC prestige score as the dependent variable, we find that the average prestige score of men who have had at least some college education is 49.66 (with a variance of 15.57), compared to an average score of 33.95 (with a variance of 10.89) for those who have a high school education or less. The sample sizes on which these estimates are based are 444 and 236, respectively. We compute the test statistic:

$$Z = \frac{49.65 - 33.95}{\sqrt{\dfrac{15.57}{444} + \dfrac{10.89}{236}}}$$

$$= \frac{15.70}{.285} = 55.09$$

Since we know from the preceding example that c.v. = 1.645 for $\alpha = .05$, the null hypothesis is rejected. That is, we find evidence for the proposition (P2) that a man's occupational status is positively related to his educational achievement.

Notice from these examples that the alternative hypothesis reflects the substantive hypothesis of interest. Remember that *we can never prove a hypothesis; we can only offer indirect evidence for it—by rejecting the null hypothesis of no relationship.* Hypothesis testing in statistics refers to the practice of deciding whether the null hypothesis can be rejected as a means of garnering support for an alternative or research hypothesis.

> By itself, hypothesis testing is obviously not a powerful or necessarily a convincing way to test a hypothesis. For this reason, it is important to replicate the findings with other independently drawn samples. It is also important to estimate the strength of relationships, a concept considered in Chapters 8–11.

The role that sample size plays in hypothesis testing should be emphasized: *If we choose N_1 and N_2 large enough, virtually any difference between $\overline{Y}_2$ and $\overline{Y}_1$ will be statistically significant.* This can be seen clearly by examining the formula for the estimated standard error for the sampling distribution of the difference between two means, $\hat{\sigma}_{(\overline{Y}_2 - \overline{Y}_1)}$, given above. As N_1 and N_2 get large, the standard error goes to zero. While this is a virtue in making a single point estimate of $\mu_2 - \mu_1$ or in constructing confidence intervals around $\overline{Y}_2 - \overline{Y}_1$ (as we will do in Section 6.6), in hypothesis testing the results can be extremely misleading. For this reason more conservative strategies should be employed.

We might argue, for example, that unless a mean difference of at least one-quarter standard deviation is found, the difference is not *substantively* important, regardless of whether or not the result is statistically significant. An even better approach, in our opinion, is to use exact methods for estimating the strength of a relationship (as discussed in Chapters 8–11). The magnitude of a relationship that is considered to be important will vary from research problem to research problem, so general rules are really quite useless. The important point is that you should *not* be misled by simple statistical significance when hypothesis testing.

6.2 Hypothesis Testing with Proportions

As we defined the term in Chapter 2, a *proportion* is simply the frequency of cases of a given type, divided by the total number of cases. And as we noted in Chapter 3, the sample mean is given by: $\overline{Y} = \sum_{i=1}^{N} Y_i/N$. But if Y takes on only the values 1 and 0, the numerator term, $\sum_{i=1}^{N} Y_i$, equals f_1. *Hence, the mean of a dichotomous variable is the proportion of cases with the value of 1.*

A simple example should make this clear. Assume there are 10 children in a class, 4 of whom are boys. If the boys are coded 1 and the girls 0, the mean is given by:

$$\overline{Y} = (1 + 1 + 1 + 1 + 0 + 0 + 0 + 0 + 0 + 0)/10$$
$$= 4/10 = 0.40$$

Notice that $\sum_{i=1}^{10} Y_i = 4$, which is the number of boys in the class, or f_b. But f_b/N equals 4/10, which is exactly equal to the mean of the variable. To use an example from the 1976 General Social

Survey, 286 of 1,499 respondents said the United States should withdraw from the United Nations. The mean response in favor of withdrawal is thus $p = 286/1{,}499 = 0.191$.

Since the proportion of a variable with only two outcomes is the mean of that variable, we can do **significance testing with proportions**, applying the same formulas used in Section 6.1 to test the hypotheses. *The variance of a variable with only two outcomes is simply pq, where $q = 1 - p$.* Thus in the example sampling attitudes toward withdrawing from the UN, the sample variance is simply $s_Y^2 = (0.191)(1 - 0.191) = (0.191)(0.809) = 0.155$. And the standard error of a sampling distribution of proportions, s_p, is:

significance testing with proportions—using statistical tests to determine whether the observed difference between sample proportions could occur by chance in the populations from which the samples were selected

$$s_p = \sqrt{pq/N}$$

In this example $s_p = \sqrt{.155/1{,}499} = 0.010$.

Suppose we hypothesize that Republicans, who are conventionally considered more chauvinistic and conservative, are more likely to favor withdrawal from the UN than other U.S. citizens are. Then our two hypotheses are:

$$H_0: p_R \leq p_O$$

$$H_1: p_R > p_O$$

where p_R is the proportion of Republicans favoring withdrawal and p_O is the proportion of other citizens favoring withdrawal in the population. To test whether we can reject the null hypothesis, we test whether p_R equals p_O. We again draw on the 1976 GSS data, setting $\alpha = .05$, and find that $p_R = 0.216$ and $p_O = 0.184$. Now to test whether this difference is statistically significant, we calculate the test statistic, using the formula for the difference between two means presented in the preceding section. That is:

$$Z_{(p_R - p_O)} = \frac{p_R - p_O}{\sqrt{\dfrac{p_R q_R}{N_R} + \dfrac{p_O q_O}{N_O}}}$$

In the UN attitude example, this would be:

$$Z_{(p_R - p_O)} = \frac{0.216 - 0.184}{\sqrt{\frac{(0.216)(0.784)}{310} + \frac{(0.184)(0.816)}{1,189}}}$$

$$= \frac{0.032}{\sqrt{0.007}} = 1.23$$

Since the critical value is 1.645, we cannot reject the null hypothesis. We conclude that there is no significant difference between Republicans and other citizens with respect to support for U.S. membership in the UN.

6.3 Two-Tailed Hypothesis Tests

one-tailed hypothesis test—a hypothesis test in which the alternative is stated in such a way that the probability of making a Type I error is entirely in one tail of a probability distribution

All the hypotheses we have tested in this chapter to this point have been **one-tailed hypothesis tests.** By *one-tailed* we mean that *the alternative is stated in such a way that the probability of making a Type I or false rejection error is entirely in one tail of a probability distribution.* (See Box 4.1 if you do not remember what a Type I error is.) Or, stated slightly differently, the alternative hypotheses have all followed the pattern of $H_1: \mu_1 > \mu_2$ or $H_1: \mu_2 > \mu_1$.

Sometimes, however, the researcher does not have a very-good guess about the alternative hypothesis. For example, there may be no reason to believe that women are any likelier than men to favor withdrawing from the UN. In this case the null and alternative hypothesis can be stated as:

$$H_0: p_m = p_f$$
$$H_1: p_m \neq p_f$$

where: p_m is the proportion of men favoring withdrawal and p_f is the women's proportion. Note that the alternative hypothesis is simply that the two means (proportions in this case) are not equal. Therefore there will be no alternative hypothesis about which mean is greater if the results direct us to reject the null hypothesis that the two means are equal.

two-tailed hypothesis test—a hypothesis test in which the region of rejection falls equally within both tails of the sampling distribution

Where the alternative hypothesis states simply inequality of means, the region of rejection falls within both tails of the sampling distribution, rather than in just one tail, as was true for the preceding examples. For this reason, such a test is called a **two-tailed hypothesis test.**

FIGURE 6.3
Area of Rejection for a Null Hypothesis, Using a Two-Tailed Hypothesis Test

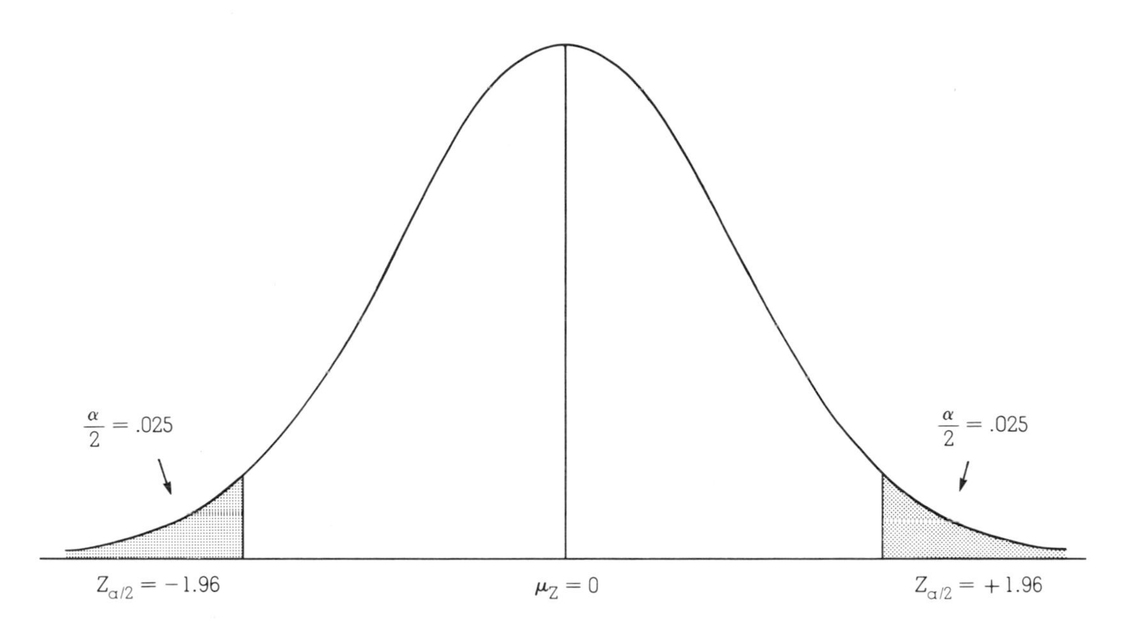

Suppose we set $\alpha = .05$. Now we want half the area associated with a Type I error to be above the mean and half below. In Figure 6.3 we show a sampling distribution for a two-tailed test with $\alpha = .05$. Note that half the area of rejection, or $\alpha/2 = .05/2 = .025$, is in the upper tail, and half is in the lower tail. We look up $Z_{\alpha/2}$ associated with $.5000 - .0250 = .4750$ in Appendix C and find it is 1.96. Since the sampling distribution is symmetric, it follows that $-Z_{\alpha/2} = -1.96$. *If our test statistic is greater than 1.96 or less than −1.96, we reject the null hypothesis. Otherwise we do not.*

We can use the 1976 GSS data to test the null hypothesis stated above, that the proportion of males and females favoring withdrawal from the UN is equal. The alternative hypothesis, also stated above, is that the two proportions are unequal. We choose $\alpha = .01$, compute the two proportions, and find that $p_m = 0.209$ and $p_f = 0.173$, with $N_m = 669$ and $N_f = 830$. This suggests that males may favor withdrawal more than females do. Now we compute the test statistic:

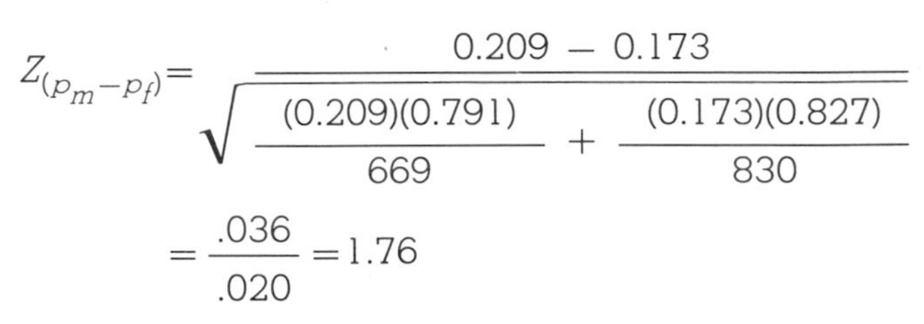

$$Z_{(p_m - p_f)} = \frac{0.209 - 0.173}{\sqrt{\frac{(0.209)(0.791)}{669} + \frac{(0.173)(0.827)}{830}}}$$

$$= \frac{.036}{.020} = 1.76$$

For $\alpha = .01$ we find that $Z_{\alpha/2} = 2.58$ and $-Z_{\alpha/2} = -2.58$. Since the test statistic is only 1.76, we cannot reject the null hypothesis. We therefore conclude that there is no difference between the sexes in the population in their support (or lack of support) for U.S. membership in the United Nations, in spite of the sex difference found in the sample.

In most cases we have a good idea, based on theory or past research, whether one mean should be larger or smaller than a second. Therefore, in almost all instances, one-tailed hypothesis tests are used. But in those cases where theory is absent or past research is unclear about the equality or inequality of two means, a two-tailed hypothesis is appropriate.

6.4. Testing for the Difference between Two Means When the Standard Error is Unknown: The *t* Test

When we do not know the variance of a variable in each of two populations, $\sigma^2_{Y_1}$ and $\sigma^2_{Y_2}$, as a result we also do not know the standard error of the sampling distribution of mean differences, $\sigma_{(\bar{Y}_2 - \bar{Y}_1)}$. In such a case, when N_1 and N_2 are small (i.e., $N_1 + N_2 \leq 100$), it is necessary to test hypotheses about two means by using a *t* test. As we showed in Chapter 5, the only statistic necessary to compute a *t* test is the sample variance. The use of the *t* distribution to test the difference between two means is illustrated here with an example involving hypotheses about urban change.

Since the end of World War II, the urban population of the United States has redistributed itself within extended metropolitan areas, as many residents of the central cities have moved to suburbs and smaller satellite towns ringing the industrial core areas. At the same time, population shifts across geographic regions have brought the decline of the older Northeastern states and the growth of the Sunbelt states, below a line stretching from North Carolina to the West Coast. Urbanologists hypothesize that these types of population change have numerous causes: technological developments such as superhighways, cheap motor fuels, and industrial automation; economic factors like inexpensive, low-tax suburban industrial parks and nonunion Southern plants; the amenities of suburban living—low-density housing, quality schools, and a means of escape from blight and crime in the inner city; and Southern climate and lifestyle.[2]

We will use the 63-cities data (see Box 2.4 in Chapter 2) to examine population change in response to two factors. First, the pull of the Sunbelt region presumably draws population away from the cities of the North and Midwest (the Frostbelt) as industries and commerce move their operations to the South and Southwest. Second, cities located in the Sunbelt states (roughly those below the 37th parallel) should gain population or at least not lose population as fast as the Frostbelt cities during a single period. These considerations lead to the following proposition about urban population change:

P1: Frostbelt cities are losing population at a higher rate than cities located in the Sunbelt.

Changing this proposition into a hypothesis testable with the 63-cities data is relatively easy. Sunbelt cities are located in California, Arizona, New Mexico, Texas, Oklahoma, Arkansas, Louisiana, Mississippi, Alabama, Tennessee, North and South Carolina, Georgia, and Florida. All other cities are in the Frostbelt region. For population change in these cities we have two figures, for populations in 1960 and in 1974. The analysis of proposition P1, therefore, will be based on the percentage change in population from 1960 to 1974. The operational or testable hypothesis is:

H1: Frostbelt cities lost population from 1960 to 1974 at a higher rate than did cities located in the Sunbelt states.

We use the t distribution to test the hypothesis that Sunbelt cities did not lose population as fast as Frostbelt cities did from 1960 to 1974. The raw data to test this hypothesis appear in Table 6.1, which shows the percentage (to nearest tenth) by which a community's 1960 population grew or declined over the next 15 years. The average for all 63 cities was a -1.26% loss (s_Y = 6.32). But the means are markedly different for the two regions.

2. Thomas M. Guterbock, "The Push Hypothesis: Minority Presence, Crime, and Urban Deconcentration," in Barry Schwartz (Ed.), *The Changing Face of the Suburbs* (Chicago: University of Chicago Press, 1976).

The 26 Sunbelt cities grew an average of +2.8%, while the 37 Frostbelt cities declined by an average of −4.1% of their 1960 populations. Only 2 of the total of 17 cities that grew at all were located outside the Sunbelt states. The standard deviations show a substantially greater dispersion in the Sunbelt sample, reflecting in part the extremely high growth of three Sunbelt cities (San Jose, Phoenix, and Charlotte). The smaller value of $s_2 = 3.16$ for the Frostbelt cities compared to $s_1 = 7.59$ for the Sunbelt cities implies that the sample of Frostbelt cities is clustered more closely about the mean.

TABLE 6.1
Percent Population Change, 1960–1974, in Sunbelt and Frostbelt Cities

Sunbelt Cities		*Frostbelt Cities*		
18.1	2.2	2.5	−2.7	−5.1
17.8	0.7	1.4	−2.9	−5.2
15.6	−0.8	−1.0	−3.0	−5.8
9.6	−1.7	−1.2	−3.2	−5.9
8.7	−2.5	−1.3	−3.6	−7.3
8.4	−2.9	−1.8	−3.7	−7.6
7.2	−3.3	−1.9	−3.8	−7.9
7.1	−3.4	−2.0	−3.8	−8.1
6.9	−3.6	−2.0	−3.9	−8.3
5.7	−3.9	−2.1	−3.9	−9.6
3.6	−4.4	−2.5	−4.2	−10.3
3.4	−8.6	−2.5	−4.5	−12.0
3.2	−9.2	−2.6		
$N_1 = 26$		$N_2 = 37$		
$\bar{Y}_1 = 2.84$		$\bar{Y}_2 = -4.14$		
$s_1 = 7.44$		$s_2 = 3.16$		

Source: NORC Permanent Community Sample (63-cities data).

Although the means of the two sets of cities differ in the direction predicted by the hypothesis, the data are from a sample of cities. To reject the null hypothesis of no difference in change in population size between regions for all U.S. cities, we must perform a significance test to decide whether the observed sample difference in means has a low probability of occurring in a population of cities where no difference exists. Since the samples are small (26 and 37), the appropriate test of significance is the *t* test. To use the *t* test we assume that independent random samples N_1 and N_2 are drawn from the two normally distributed populations. The null hypothesis states that the Sunbelt cities' population loss between 1960 and 1974 is equal to or greater than the Frostbelt cities' population loss. The alternative hypothesis is that the mean population loss from 1960 to 1974 is greater for Frostbelt than for Sunbelt cities. That is,

$$H_0: \mu_2 - \mu_1 \geq 0$$
$$H_1: \mu_2 - \mu_1 < 0$$

To use the *t* distribution to test hypotheses about means, we need to make the following assumptions:

1. Random samples are drawn from two independent, normally distributed populations.
2. The two population variances are homoscedastic, or equal; that is: $\sigma_1^2 = \sigma_2^2 = \sigma^2$.

With two sample variances, s_1^2 and s_2^2, either could be used to estimate σ^2. Since we want the best possible estimate of σ^2, we use a *weighted average* of s_1^2 and s_2^2, rather than arbitrarily choosing one of them as the estimate. That is, we *pool* the information on variation from both samples into a single estimate. The formula for doing this is:

$$s^2 = \frac{(N_1 - 1)s_1^2 + (N_2 - 1)s_2^2}{N_1 + N_2 - 2}$$

where $N_1 + N_2 - 2$ are the degrees of freedom that are associated with s^2.

Note that the degrees of freedom associated with s_1^2 and s_2^2 sum to equal the degrees of freedom associated with s^2. Specifically, $(N_1 - 1) + (N_2 - 1) = N_1 + N_2 - 2$.

The test statistic for the difference between two means under the null hypothesis using small samples is:

$$t_{(N_1 + N_2 - 2)} = \frac{(\overline{Y}_2 - \overline{Y}_1) - (\mu_2 - \mu_1)}{s_{(\overline{Y}_2 - \overline{Y}_1)}}$$

$$= \frac{\overline{Y}_2 - \overline{Y}_1}{\sqrt{\dfrac{s^2}{N_1} + \dfrac{s^2}{N_2}}}$$

and it follows that:

$$= \frac{\bar{Y}_2 - \bar{Y}_1}{s\sqrt{\frac{1}{N_1} + \frac{1}{N_2}}}$$

To test H_1 in the urban population-change example, we first need to calculate s^2 and then take its square root to obtain s:

$$s^2 = \frac{(26 - 1)7.44^2 + (37 - 1)3.16^2}{26 + 37 - 2}$$

$$= 28.58$$

and then

$$s = 5.35$$

For the t test of the null hypothesis we choose $\alpha = .05$, and $df = N_1 + N_2 - 2 = 26 + 37 - 2 = 61$. Then t is calculated as:

$$t_{61} = \frac{(-4.14) - (2.84)}{5.35\sqrt{\frac{1}{26} + \frac{1}{37}}}$$

$$= \frac{-6.98}{1.37} = -5.10$$

When $\alpha = .05$ with 61 degrees of freedom, the critical value is -1.67 for a one-tailed test. Since the test statistic is -5.10, we reject the null hypothesis. Therefore, the alternative hypothesis that Frostbelt cities are losing population faster than Sunbelt cities is supported.

Technically, you should use a t rather than a Z distribution in hypothesis testing (or constructing confidence intervals) whenever the standard error of the sampling distribution is unknown. For all intents and purposes, we never know the standard error, so we should always employ a t test. However, as we saw in Chapter 5, as N gets large (or $N_1 + N_2$ when considering two samples, as we are in this chapter), the t distribution approaches a normal distribution.

You can check this assertion by examining and comparing

Appendices C and D. In Appendix D (*t* distribution), look at *df* equals infinity for a given α and compare the *t* value to the *Z* value in Appendix C for the same α. This convergence has another practical implication: As *N* gets large, the requirement that the sample be drawn from a population that is normal becomes less important. When $N_1 + N_2$ is 100 or greater, we can invoke the central limit theorem. Then we need not worry whether the population is normal, since we can be confident that the sample variances will be good estimators of the population variances.

6.5. New Methods for Diagraming Continuous Data

The diagrams used to display data in preceding chapters have been designed as charts, histograms, or polygons, which we introduced in Chapter 2. This section provides a brief introduction to some recent innovations developed by John Tukey to display continuous data, which are especially useful for showing differences between two or more groups.[3] We will illustrate these techniques with the population-loss hypothesis for Sunbelt and Frostbelt cities.

One method for displaying frequency distributions of grouped continuous data is called a **stem-and-leaf diagram.** Figure 6.4 shows two stem-and-leaf diagrams of percent change in popula-

stem-and-leaf diagram —a type of graph that displays the observed values and frequency counts of a frequency distribution

FIGURE 6.4
Stem-and-leaf Diagrams of Percent Change, 1960–1974, in Sunbelt and Frostbelt Cities

SUNBELT CITIES			***FROSTBELT CITIES***		
1	0688	(4)	1		(0)
0	12334677789	(11)	0	13	(2)
−0	12333344499	(11)	−0	11122222333333344444445556678888	(32)
−1		(0)	−1	002	(3)

Note: Data rounded to nearest whole percent.
Source: Data derived from Table 6.1.

3. John Tukey, *Exploratory Data Analysis* (Reading, Mass.: Addison-Wesley Publishing Co., 1977).

tion for Sunbelt and Frostbelt cities, based on the data in Table 6.1 above. A stem-and-leaf diagram somewhat resembles a histogram turned on its side. The stem of the diagram on the left is the leading (first) digit of the population-change data. To the right of the stem are the "leaves" consisting of the second digits of the observed scores. In this example we rounded the rates to the nearest whole number, although with more observations we might have preserved the decimal as well. Each successive entry on the leaf stands for an observation, so the longer the leaf, the more cases there are in the given interval. The leaf totals can be written in parentheses at the extreme right of the diagram, as we have done.

As an example, look at the bottom leaf of the Frostbelt stem. At a glance, you can tell the actual scores of the three cities in this interval (in nearest whole percent) were −10, −10, and −12 percent population loss. If you compare the stem-and-leaf diagram for Sunbelt cities with that for Frostbelt cities, it should be clear that Frostbelt cities experienced greater population loss between 1960 and 1974 than Sunbelt cities did.

A second innovation of Tukey's displays the distribution by

FIGURE 6.5

Box-and-Whisker Diagram for Sunbelt-Frostbelt Population Change

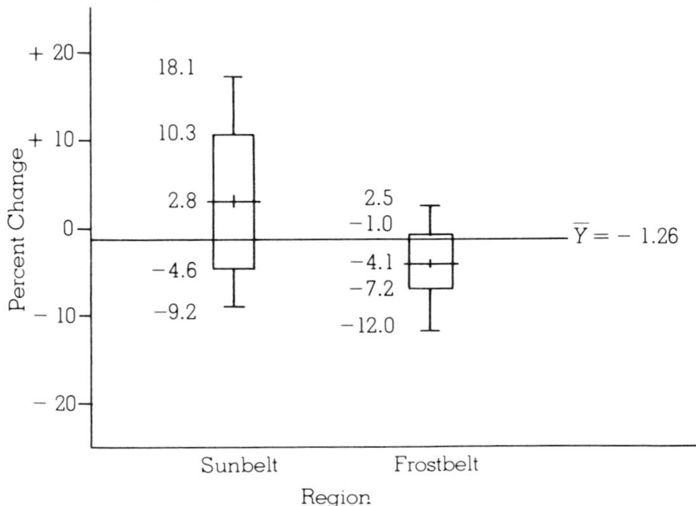

Source: Data drawn from Table 6.1.

using the mean, the standard deviation, and the range. This diagram, which is called a **box-and-whisker diagram**, is shown for the same population-change data in Figure 6.5. The scores are marked on the vertical axis. The mean of the sample is indicated by a short horizontal line at the appropriate level, and a rectangular box is drawn about the mean to a height of exactly 1 standard deviation in both directions. Then, the upper and lower scores which mark the range are designated by short horizontal lines and connected to the box by thin "whiskers."

box-and-whisker diagram—a type of graph for discrete and continuous variables in which boxes and lines represent central tendency and variability

The virtues of such box-and-whisker diagrams are clear when we consider the distribution of a continuous variable within categories of an independent variable, as in Figure 6.5. Even a casual inspection of the diagram should make it clear that the Frostbelt cities have, on the average, experienced more population loss than the Sunbelt cities. You should also be able to see that there is considerably more variability in population changes for the Sunbelt cities.

6.6. Confidence Intervals and Point Estimates for Mean Differences and Proportions

Confidence intervals can be constructed around the difference between two means in much the same way as for a single mean, as we did in Section 5.6 in Chapter 5. The interpretation also is similar. We can state the probability that the **confidence interval for mean differences** contains the difference between the population means, that is, $\mu_2 - \mu_1$. More formally, we can construct confidence intervals for any desired level of confidence, such as $1 - \alpha$, with the formula:

confidence intervals for mean differences—a confidence interval constructed around the point estimate of the difference between two population means

$$(\overline{Y}_2 - \overline{Y}_1) \pm t_{\alpha/2}\, s_{(\overline{Y}_2 - \overline{Y}_1)}$$

Suppose we want to compute a 95% confidence interval around our estimate of difference in percent population change between Sunbelt and Frostbelt cities. In Section 6.4 we found that $\overline{Y}_2 - \overline{Y}_1 = -6.98$, and we calculated $s_{(\overline{Y}_2 - \overline{Y}_1)} = 1.37$. For the 95% confidence level, $t_{\alpha/2}$ is 2.00 with $N_1 + N_2 - 2 = 61$ *df*, as can be verified from Appendix D. Therefore, the lower confidence limit (LCL) is $-6.98 - (2.00)(1.37) = -9.72$, and the upper confidence limit (UCL) is $-6.98 + (2.00)(1.37) = -4.24$. In other words, we can be 95% confident that the interval bounded by -9.72 and -4.24 contains the true difference in percent population change between Sunbelt and Frostbelt cities during the 1960–74 period. And our **point estimate for mean differences**, or the change in population, is simply $\overline{Y}_2 - \overline{Y}_1 = -6.98$.

point estimate for mean differences—the difference between the sample means used to estimate the difference between two population means

We can also construct confidence intervals around differences between *proportions*. In this case the intervals are given by:

$$(p_2 - p_1) \pm t_{\alpha/2} s_{(p_2 - p_1)}$$

In the example where we examined support for withdrawing from the UN, according to political preference, $p_R - p_O = 0.032$ and $s_{(p_R - p_O)} = \sqrt{.0007} = 0.026$. Now suppose we want to construct a 99% confidence interval around the point estimate, .032. Since in this sample the number of Republicans was $N_R = 310$ and the number of others was $N_O = 1{,}189$, there are $310 + 1{,}189 - 2 = 1{,}497$ *df*. And we can see from Appendix D that $t_{\alpha/2} = 2.58$.

We can use all of this information to determine that the LCL is $.032 - (2.58)(.026) = -.035$ and the UCL is $.032 + (2.58)(.026) = .099$. That is, we can be 99 percent confident that the true difference between the proportions of Republicans versus others who favor withdrawal from the UN lies between $-.035$ and .099. Furthermore, our best estimate of the true difference is the point estimate, $p_R - p_O = .032$. In other words, according to the GSS data, the difference between Republicans and others on the issue of withdrawing from the UN was very small indeed in 1976.

6.7 Reporting *p* Values

Our strategy for testing hypotheses, as summarized in Box 4.2 in Chapter 4, has been as follows: We choose an α level, compute a test statistic based on the null hypothesis, determine a critical value, and, finally, compare the test statistic to the critical value, rejecting the null hypothesis if the test statistic is larger than the critical value. There is another approach that goes a step further and has considerable support among researchers. After following all four steps, this research reports what is called the *p value.*

A ***p* value** is the probability of observing the test statistic under the assumption that the null hypothesis is true. Suppose we set $\alpha = .05$ and then observe a t statistic of 2.60 for $N = 500$. Assuming a one-tailed test, the probability of observing a t of 2.60 or more under the null hypothesis is less than .005. Instead of merely reporting that the null hypothesis can be rejected, we report that it is rejected with $p < .005$.

***p* value**—the probability of observing a test statistic under the assumption that the null hypothesis is true

As another example, again suppose $\alpha = .05$, but we observe $t = 2.54$ with 20 *df* while doing a one-tailed test. An examination of Appendix D indicates that for 20 *df* a t of 2.54 would occur less than 1% of the time. Therefore we would report that the null hypothesis is rejected with $p < .01$.

When a p value is reported, the researcher is stating not only that the null hypothesis can be rejected, but that it can be rejected at an even more stringent α level. That is, the probability of a Type I error is even smaller than the chosen α level. And, furthermore, there is an implication that a hypothesis which can be rejected at the .001 level represents a *stronger* relationship than one that is rejected at the .05 level.

While it is true that for a given sample size the relationship between two variables is even stronger for a p value of .001 than for one equal to .05, we believe there are much better and more accurate ways to assess strength of relationship. These form much of the substance of Chapters 8–12.

> In our view, the practice of reporting p values focuses on demonstrating statistical significance rather than substantive significance. And almost any difference can be made statistically significant by choosing samples large enough. Therefore, when p values are reported, the strength of the relationship should be assessed using other tools as well.

6.8 Comparing Means from the Same Sample Across Time

Often the researcher is interested in whether persons have *changed* from time 1 to time 2 on some variable, Y. In such a case, $\mu_2 - \mu_1$ is seen as a measure of change. Since the two samples being compared are the *same* across time, the methods introduced in this chapter cannot be used without violating the assumption that the two random samples are *independently* drawn.

To be able to measure change by comparing the same observations across time requires knowledge of correlations—a topic to be introduced in Chapter 8. For this reason, we will return to the topic of testing for difference in means in nonindependent populations again in that chapter.

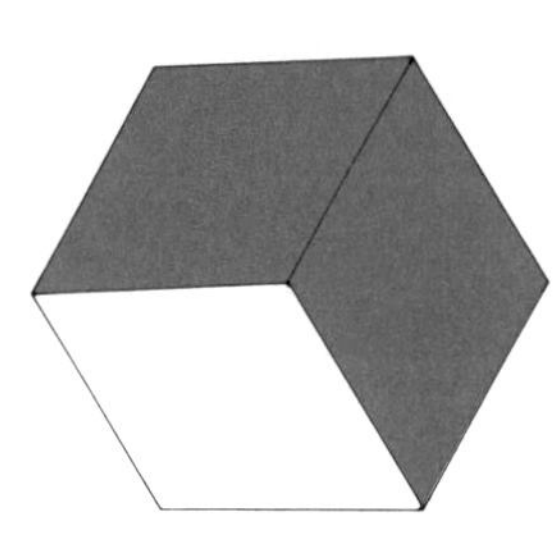

Review of Key Concepts

This list of the key concepts introduced in Chapter 6 is in the order of appearance in the text. Combined with the definitions in the margins, it will help you review the material and can serve as a self-test for mastery of the concepts.

mean difference test
independent random samples
mean difference hypothesis test
significance testing with proportions
one-tailed hypothesis test
two-tailed hypothesis test.
stem-and-leaf diagram
box-and-whisker diagram
confidence interval for mean differences
point estimate for mean differences
p value

Problems

General Problems

1. A theoretical proposition in the sociology of religion states that "Societies in which decision-making groups extend beyond the local community are more likely to be monotheistic than societies that lack nonlocal decision-making groups." Restate this proposition as a pair of null and alternative hypotheses, using symbolic notation. Does the hypothesis require a one- or two-tailed test of significance?

2. A proposition in stratification holds that "The average annual incomes of white workers is greater than that of black workers." Rewrite this proposition as a pair of null and alternative hypotheses, using symbolic notation. Does the hypothesis call for a one- or two-tailed test of significance?

3. What are the standard errors for testing the difference between two means, given the following:

	σ_1^2	σ_2^2	N_1	N_2
a.	14	12	25	75
b.	14	12	200	300
c.	42	37	60	50
d.	240	360	400	600

4. Find the Z scores for the following and state whether the null hypotheses should be rejected:

	$\overline{Y}_1$	$\overline{Y}_2$	s_1^2	s_2^2	N_1	N_2	α
a.	24	20	225	360	60		.05, one-tailed
b.	24	20	225	360	600		.05, one-tailed
c.	3.50	6.40	20	30	40		.01, two-tailed
d.	73	68	15	150	150		.001, two-tailed

5. In 1972 the proportion of Americans agreeing that "an admitted communist" should be allowed to make a speech in the community was 0.54. By 1980, the proportion was 0.57. If these two independently drawn samples' sizes are 1,556 and 1,430, respectively, test the null hypothesis of no change over the period, using $\alpha = .05$ and a two-tailed significance test.

6. In 1972, 73.6% of 1,533 respondents interviewed by GSS said they would vote for a woman for president if she were qualified for the job. By 1980, this figure had risen to 81.6%, based on 1,492 respondents. Test the null hypothesis of no difference in population percentages, using $\alpha = .01$ and a one-tailed test.

7. Given that H_0: $\mu_A - \mu_B = 0$ and $\alpha = .02$, what Z score(s) are the critical values defining the region of rejection?

8. Given that H_0: $\mu_A - \mu_B \geq 0$ and $\alpha = .01$, what Z score(s) are the minimum values required to reject the null hypothesis?

9. One sample has 16 respondents, a second has 13. In a test of the mean difference, how many degrees of freedom has the sampling distribution of mean differences?

10. Find the critical values of the *t* score or *Z* score for the following:
    a. $\alpha = .05$, one-tailed, $N_1 = 8$, $N_2 = 6$.
    b. $\alpha = .05$, two-tailed, $N_1 = 80$, $N_2 = 60$.
    c. $\alpha = .01$, two-tailed, $N_1 = 16$, $N_2 = 16$.
    d. $\alpha = .01$, one-tailed, $N_1 = 16$, $N_2 = 8$.
    e. $\alpha = .001$, two-tailed, $N_1 = 11$, $N_2 = 7$.

11. An anthropologist measures the level of monotheism in two types of societies, and finds the following scores:

Societies with Only Local Decision-Making Groups	Societies with Nonlocal Decision-Making Groups
6	6
5	12
8	15
12	13
6	9
4	12
7	

Is the difference in monotheism by societal political structure hypothesized in Problem 1 supported by this evidence? Set $\alpha = .05$.

12. Construct the 95% and 99% confidence intervals around the point estimate of the mean population difference estimated in Problem 11.

13. Give the 95% and 99% confidence intervals for *(a)* the difference in proportions in Problem 5 and *(b)* the difference in proportions in Problem 6.

14. An experimental study of the hypothesis that small monetary rewards will increase group participation yields the following sample statistics:

Control group: $\bar{Y}_1 = 14$, $s_1^2 = 6$, $N_1 = 10$
Experimental group: $\bar{Y}_2 = 19$, $s_2^2 = 15$, $N_2 = 12$

If α is set at .01, should you reject the null hypothesis?

15. A stratification researcher hypothesizes that businessmen have higher social origins than academics. From data on 12 businessmen and 8 academics, the researcher finds that the mean occupational status of the businessmen's fathers was 58.2, and that of the academics' fathers was 51.7. The standard error of the mean difference is 3.2. Carry out the test of

the hypothesis, using $\alpha = .05$, and state your decision about the null hypothesis.

16. One class of 10 pupils scored an average 84 on the statistics midterm exam, while a second class of 10 averaged 77. The estimated standard error is 4. Calculate the confidence interval limits for the 95% confidence interval. Can you be 95% confident that the first class scored significantly higher than the second class?

17. A poll by the student paper at Big State University on the abolition of foreign language requirements found that 43% of the 150 sampled underclassmen favored the abolition, while 68% of the upperclassmen ($N = 150$ also) favored it. The sample point estimate of the difference is thus 25%. Construct the 95% and 99% confidence intervals around this point estimate. Can you be 99% confident that upperclassmen are more in favor of abolition of the requirements than underclassmen are?

18. A researcher is uncertain whether hungry or not-hungry experimental subjects can perform a hand-eye coordination task with fewer mistakes. Data from the experiment show that 75 hungry subjects had a mean mistake rate of 6.0, while the 80 not-hungry subjects had a mean mistake rate of 3.0. The estimated standard error was 2.0. If the researcher sets $\alpha = .05$, should the null hypothesis of no difference in rates of mistakes be rejected?

19. Construct a box-and-whisker diagram for the data shown in Problem 11.

20. Construct a stem-and-leaf diagram for the following scores of 30 subjects on a measure of test anxiety:

12	14	43	11	32
14	22	22	28	19
26	27	16	8	25
37	33	25	23	26
18	19	31	24	10
32	20	28	29	25

Problems Requiring the 1980 General Social Survey Data

21. A basic proposition in political sociology holds that voters for different political parties have different political ideologies.

Test the hypothesis that in 1976 Ford voters described themselves as more conservative (POLVIEWS) than did Carter voters. (Use IF76WHO only for respondents voting for either major party candidate.) Set $\alpha = .01$.

22. Test the hypothesis that persons living in rural areas are less afraid to walk alone at night near their homes (FEAR) than are residents living in large central cities. Note that you must select only respondents in the two extremes categories of the XNORCSIZ variable. Set $\alpha = .01$.

23. Do people who have ever received some form of aid from a government agency (GOVAID) have more favorable attitudes toward increasing welfare spending (NATFARE)? Set $\alpha = .01$.

24. Because television did not become widely available until after World War II, the TV-watching habit may be more strongly engrained in those born during that era. Test this hypothesis by comparing the amount of TV viewing (TVHOURS) of persons born after 1945 and those born in 1945 or earlier. (Note: you must recode AGE into two categories.) Set $\alpha = .01$.

25. Determine whether whites or blacks (RACE) object more strenuously to interracial marriage (RACMAR). Exclude respondents of other racial or ethnic groups from your analysis. Set $\alpha = .01$.

Problems Requiring the 63-Cities Data

26. The municipal reform movement was supposed to make city government more responsive to citizens' demands. Compare those cities having all three reform characteristics to all other cities (REFORMGV) to determine whether the former have higher per capita spending (EXPENDIT). Set $\alpha = .01$.

27. Perform the same comparison as in Problem 26, this time using number of municipal workers per 1,000 residents (GOVTEMPL) as the dependent variable. Set $\alpha = .05$.

28. Do cities located in the South (REGION) have more centralized decision making (DECINDEX) than cities elsewhere? Set $\alpha = .05$.

29. Dichotomize the city data into a set with fewer women in the labor force (FEMLABOR) and a set with a larger female labor force than the median (40%). Test the hypothesis that in cities where more women work, the birth rates are significantly lower (BIRTHS). Set $\alpha = .05$.

30. Do middle-class cities (WCPCT of 55% or higher) have significantly larger per capita incomes (INCOMEPC) than cities with smaller middle classes (WCPCT of less than 55%)? Set $\alpha = .05$.

Statistics for Social Data Analysis

7

Testing for the Difference between Several Means

How to test the hypothesis that two sample means come from two different populations, rather than being drawn from the same population, was examined in Chapter 6. This chapter shows how to test the hypothesis that the sample means of two or more, or *J*, groups come from the same rather than different populations. The technique used is called the **analysis of variance**, or ANOVA for short.

analysis of variance— a statistical test of the difference of means for two or more groups

7.1 The Logic of Analysis of Variance: An Example

We will use fictitious data to demonstrate the logic behind analysis of variance. Suppose we are interested in evaluating whether a person's performance on a task suffers when it is observed or monitored by another person. We hypothesize that individuals who are being observed will perform less well on a task than individuals who perform the same task alone, on the assumption that having one's performance monitored by another person creates anxiety which interferes with efficient behavior. The null hypothesis is that monitored and unmonitored groups do not differ on performance.

To test this hypothesis, we randomly assign 20 persons to each experimental condition. These subjects are asked to try to complete a set of 20 puzzles, with two minutes allowed for each puzzle before a bell rings to signify that they should move on to the next one. The order in which the puzzles are presented is randomized across subjects to rule out order of presentation as a possible explanation of results. In the *monitored* condition a confederate is instructed to watch intently the subjects' attempts to complete the puzzles, and the subjects and the confederate are instructed not to talk with one another during the experiment. There are two unmonitored conditions. In the *not-monitored-together* condition two subjects are put together but are instructed to work independently on the puzzles. In the *not-monitored-alone* condition the subjects work on the task alone.

The dependent variable is the number of puzzles solved correctly. The research hypothesis of interest is that the subjects in the not-monitored-alone condition *(na)* will complete more puzzles correctly than those in the not-monitored together condition *(nt)*, who in turn will complete more puzzles correctly than those in the monitored condition *(m)*. If there is *no* effect of monitoring on performance we would expect the three population means, labeled μ_{na}, μ_{nt} and μ_m, to be equal to one another and hence equal to the overall mean, or **grand mean**, of the population, μ. We would also expect the sample means, $\overline{Y}_{na}$, $\overline{Y}_{nt}$ and $\overline{Y}_m$, to be roughly equal. However, if the alternative or research hypothesis is true, we would expect to observe that $\overline{Y}_{na} > \overline{Y}_{nt} > \overline{Y}_m$. We will look at some fictitious data to test these hypotheses, but first we will describe analysis of variance more fully.

grand mean—in analysis of variance, the mean of all observations

An ANOVA model provides a way to test the *null hypothesis* that *all J sample means are drawn from the same population and therefore are all equal.* Formally, the null hypothesis is:

$$H_0: \mu_1 = \mu_2 = \ldots = \mu_J.$$

The *alternative hypothesis* is that *one or more of the sample means are drawn from populations with different means.* The possibility that the null hypothesis can be rejected implies one of several alternative possibilities. These are:

1. All the population means are different from each other, i.e., $\mu_1 \neq \mu_2 \neq \cdots \neq \mu_J$.
2. Some subsets of the population means differ from

one another (e.g., μ_1 differs from μ_2 but not from μ_3 and μ_4).

3. Some combination of the means is different from some single mean or from some other combination of means (e.g., μ_2 differs from the average of the means of three other populations).

In other words, if we reject the null hypothesis, we still have to explain *how* the means differ from one another.

7.2 Effects of Variables

To examine the effects of the variables in an analysis of variance we will consider a population with mean μ. The null hypothesis, as noted above, states that the J group means are all equal, that is, $\mu_1 = \mu_2 = \ldots = \mu_J$. But if they all are equal to each other, clearly they will all equal the overall or grand mean, μ. If, for example, the mean number of hours of television watched is $\mu = 2.80$ per day, and the rate of watching is the same for men and women, then clearly $\mu = \mu_m = \mu_f = 2.80$. To measure the **effect** of an independent variable on a dependent variable, we can take advantage of this fact.

effect—the impact on a dependent variable of being in a certain treatment group

An effect of being in a subgroup or one of a number of groups, labeled j, is defined as:

$$\alpha_j = (\mu_j - \mu)$$

Note that if being in group j has no effect on the dependent variable, Y, then $\alpha_j = 0$.[1] But if being in a group does have an effect, α_j will be positive or negative, depending on whether a group's mean is above or below the grand mean, μ.

Suppose young people watch television an average of 5.0 hours per day (μ_Y), middle-age people watch it 1.5 hours per day (μ_M), and older people watch it 1.9 hours daily (μ_O), and the overall mean equals 2.80. Using the notation for age-group ef-

1. The use of alpha with a subscript (α_j) is not to be confused with α without a subscript, which refers to the probability of a Type I error (see Chapter 4).

fects, $\alpha_Y = (5.00 - 2.80) = 2.2$, $\alpha_M = (1.50 - 2.80) = -1.3$, and $\alpha_O = (1.90 - 2.80) = -0.9$ hours per day.

7.3 The ANOVA Model

The term *analysis of variance* is not used accidentally. When we do an ANOVA we are asking how much of the total variation in Y can be explained by the independent or "treatment" variable(s) and how much is left unexplained. The difference between an observed score and a score predicted by the model is called the **error term**, or *residual term*. The general model for ANOVA with one independent variable is:

error term—the difference between an observed score and a score predicted by the model

$$Y_{ij} = \mu + \alpha_j + e_{ij}$$

Where:

e_{ij} = The error, or residual, term.

This formula indicates that *the score of observation i, which is also a member of group j (hence Y_{ij}), is a function of a group effect, α_j, plus the population mean, μ, and random error, e_{ij}.* We need the error term to take into account that not every observation in subgroup j has the same Y_{ij}. For example, every young person does not watch exactly five hours of TV per day; some watch more and some watch less. The error term, e_{ij}, reflects this fact.

7.4 Sums of Squares

To be able to estimate the proportion of variance in Y_{ij} due to group effects (the α_j) and due to error, *we partition the numerator of the sample variance into two independent additive components.* We begin with:

$$\sum_{i=1}^{N} (Y_i - \bar{Y})^2$$

and divide it into two components. Before doing so, however, we need to reexpress this term to take into account that each of the N observations belongs to one of the J groups.

If the number of cases in the jth group is n_j, then it follows that $n_1 + n_2 + \ldots + n_J = N$. *That is, the sum of observations across the J subgroups or treatments equals the total sample size, N.* Furthermore:

$$\sum_{i=1}^{N} (Y_i - \overline{Y})^2 = \sum_{j=1}^{J} \sum_{i=1}^{n_j} (Y_{ij} - \overline{Y})^2$$

As Appendix A notes, when there is a double summation operator, the inside one is indexed faster.

Suppose we have $N = 5$ observations. We assign each observation either to group 1 or group 2 (i.e., $J = 2$), and $n_1 - 3$ and $n_2 = 2$. Then if we expand the terms on the right above, we see that:

$$\sum_{j=1}^{2} \sum_{i=1}^{n_j} (Y_{ij} - Y)^2 = [(Y_{11} - Y)^2 + (Y_{21} - \overline{Y})^2 + (Y_{31} - \overline{Y})^2] + [(Y_{12} - \overline{Y})^2 + (Y_{22} - \overline{Y})^2]$$

The sum in the first line is for $j = 1$, and in the second it is for $j = 2$. And the result is the same as if we had used $\sum_{i=1}^{5} (Y_i - \overline{Y})^2$, except that *we have tagged the group to which each observation belongs as well.* The term $\sum_{j=1}^{J} \sum_{i=1}^{n_j} (Y_{ij} - \overline{Y})^2$ is called the **total sum of squares**, or SS_{TOTAL}.

total sum of squares—a number obtained by subtracting the scores of a distribution from their mean, squaring, and summing these values

> The object of one-way analysis of variance is to divide or partition the total sum of squares into two components: the sum of squares lying between the means of the categories, called the **between sum of squares**, or SS_{BETWEEN}, and the sum of squared deviations about the category means, called the **within sum of squares**, or SS_{WITHIN}.

between sum of squares—a value obtained by subtracting the grand mean from each group mean, squaring these differences for all individuals, and summing them

within sum of squares—a value obtained by subtracting each subgroup mean from each observed score, squaring, and summing

To make this partition, we can add and subtract the same value to any deviation without changing that deviation. Suppose, in a sample, we subtract and add the mean score of the independent variable category j into which observation i falls (denoted $\overline{Y}_j$) to the deviation of that observation from the grand mean:

$$Y_{ij} - \overline{Y} = Y_{ij} + (\overline{Y}_j - \overline{Y}_j) - \overline{Y}$$

and, regrouping terms:

$$Y_{ij} - \overline{Y} = (Y_{ij} - \overline{Y}_j) + (\overline{Y}_j - \overline{Y})$$

Notice that the second term in the second equation $(\overline{Y}_j - \overline{Y})$, is an *estimate* of group effect $\alpha_j = (\mu_j - \mu)$, the effect of being in category j. And the first term, $(Y_{ij} - \overline{Y}_j)$ is an estimate of the error term, e_{ij}. If both sides of this equation are squared and summed over all scores, after some algebra is applied we obtain the following:

$$\sum_{j=1}^{J} \sum_{i=1}^{n_j} (Y_{ij} - \overline{Y})^2 = \sum_{j=1}^{J} \sum_{i=1}^{n_j} (Y_{ij} - \overline{Y}_j)^2 + \sum_{j=1}^{J} n_j(\overline{Y}_j - \overline{Y})^2$$

This equality shows that the total sum of squares can always be partitioned into:

$$SS_{\text{BETWEEN}} = \sum_{j=1}^{J} n_j(\overline{Y}_j - \overline{Y})^2$$

and

$$SS_{\text{WITHIN}} = \sum_{j=1}^{J} \sum_{i=1}^{n_j} (Y_{ij} - \overline{Y}_j)^2$$

It then follows that:

$$SS_{\text{TOTAL}} = SS_{\text{BETWEEN}} + SS_{\text{WITHIN}}$$

An interpretation of this partitioning in ordinary English is that whenever observations differ from one another, the variance is greater than zero. Further, some part of this variance is due to the effects of the groups to which the observations belong. In other words, the sum of squares *between* groups captures the effects of the independent variable under study. However, individuals *within* the same group can still differ from one another, due to the operation of chance factors such as sampling variability or omitted causal variables. The sum of squares within groups reflects the operation of these other, unmeasured factors. Thus the within-group sum of squares (or *error sum of squares,* as the term is used in Chapter 8) implies that by assuming each group member has the same score, we make an *error* in estimating the individual score.

7.4.1. Sums of Squares in the Problem-solving Example

To return to the example of the monitored and unmonitored puzzle-solving experiment, suppose the null hypothesis is true, and we observe that $\overline{Y}_{na} = \overline{Y}_{nt} = \overline{Y}_{m}$. *If the null hypothesis is true, the term $SS_{BETWEEN}$ will equal zero, and hence $SS_{TOTAL} = SS_{WITHIN}$.* That is, all of the variation is random error variance. In this case the general ANOVA model reduces to:

$$Y_{ij} = \mu + e_{ij}$$

Suppose, however, that being monitored *does* have an effect on performance, and in particular that $\overline{Y}_{na} > \overline{Y}_{m}$, and $\overline{Y}_{nt} > \overline{Y}_{m}$; that is, subjects who are not monitored complete more puzzles than those who are monitored. Suppose further that those in the monitored group all score at the *same* performance level, and so do those in the not-monitored groups. Then the scores for the members of the three treatment groups are:

$$Y_{i,m} = \overline{Y} + \alpha_{m}$$

$$Y_{i,nt} = \overline{Y} + \alpha_{nt}$$

$$Y_{i,na} = \overline{Y} + \alpha_{na}$$

In this case $SS_{TOTAL} = SS_{BETWEEN}$, meaning that *all of the variation in performance is due to the monitoring vs. not-monitoring manipulations.* In reality, this never happens, for several reasons. First, *performance is likely to vary systematically as a function of other variables* such as intelligence, previous experience with solving puzzles, and so on, which were not included as part of the experiment. Second, *there will be random influences* such as luck, fatigue, and the order in which the puzzles are presented, and these also will create variation in performance scores. Since we know that variation in performance is due to both systematic factors not included in the research design and random uncontrollable factors, the scores we observe will be a function of both an effect, α_{j}, and error, e_{ij}.

According to the hypothetical set of data for this experiment given in Table 7.1, the average number of puzzles completed for

TABLE 7.1
Hypothetical Data Showing Performance on a Task as a Function of Being Monitored or Not Monitored

Not Monitored Alone		*Not Monitored Together*		*Monitored*	
Observation	$Y_{i,na}$	*Observation*	$Y_{i,nt}$	*Observation*	$Y_{i,m}$
1	13	1	9	1	8
2	14	2	11	2	6
3	10	3	10	3	9
4	11	4	8	4	7
5	12	5	10	5	8
6	10	6	12	6	10
7	12	7	11	7	8
8	12	8	10	8	9
9	13	9	9	9	6
10	11	10	10	10	11
$\sum_{i=1}^{10} Y_{i,na} = 118$		$\sum_{i=1}^{10} Y_{i,nt} = 100$		$\sum_{i=1}^{10} Y_{i,m} = 82$	
$\overline{Y}_{na} = 11.8$		$\overline{Y}_{nt} = 10.0$		$\overline{Y}_{m} = 8.2$	
$s^2_{na} = 1.733$		$s^2_{nt} = 1.333$		$s^2_{m} = 2.622$	
		$\overline{Y} = 10.0$			
		$s^2_Y = 4.0$			

the monitored group is 8.2, compared to 10.0 for the not-monitored-together group and 11.8 for the not-monitored-alone group. By adding the sums of the observations for the three groups together (118 + 100 + 82) and dividing by the total $N = 30$, we can determine that the grand mean is $\overline{Y} = 10.0$.

The data from Table 7.1 are plotted in Figure 7.1 for two groups: not-monitored alone and monitored. By drawing in the grand mean as well as the two subgroup means, we can demonstrate how observations can be thought of as due to the grand mean, plus an effect, plus error.

To show this we will focus on two observations—one from the monitored group and one from the not-monitored-alone group. Observation 1 in the latter group is 13. We will use the fact that in general

$$Y_{ij} = \mu + \alpha_j + e_{ij}$$

and, in general

$$\hat{\alpha}_j = (\overline{Y}_j - \overline{Y})$$

(Note that the caret (^) is used on the α_j to indicate that it is a sample estimate and not a population parameter.) We can see that for these data,

$$13 = 10.0 + (11.8 - 10.0) + e_{1,na} = 10.0 + 1.8 + e_{1,na}$$

Now by solving for $e_{1,na}$, we can see that:

$$e_{1,na} = 13 - (10 + 1.8) = 1.2$$

The first observation in the not-monitored-alone group, therefore, can be thought of as due to the grand mean, 10.0, plus an effect due to working without being observed (1.8), plus an error component (1.2). The fact that observation 1 can be seen as the sum of these three components is shown in Figure 7.1. We have done the same kind of decomposition of observation 2 in the monitored group. Since $\hat{\alpha}_m = (8.2 - 10.0) = -1.8$, it is easy to show that $6 = 10.0 - 1.8 + e_{2,m}$, and $e_{2,m} = -2.2$.

FIGURE 7.1
Distributions of Outcomes for Puzzles Completed, Not Monitored Alone or Monitored

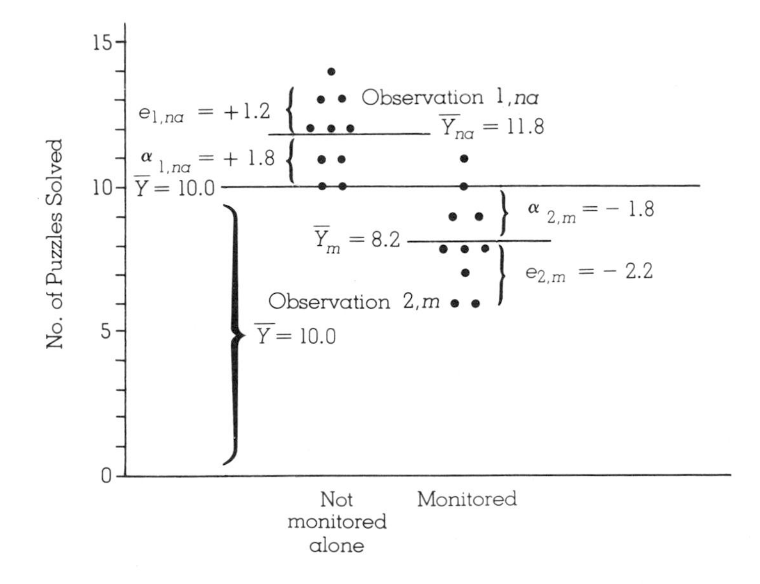

Notice in Table 7.1 that there is within-group variation (error) in all three groups. And the fact that all three subgroup means differ from one another suggests that there is between-group variation due to treatment effects. That is, the variation in performance scores (SS_{TOTAL}) can be seen in this simple example to be due to both SS_{BETWEEN} and SS_{WITHIN}.

To this point we have not actually computed the three sums-of-squares components in this example. To compute SS_{TOTAL}, we square the observations and add them up: $SS_{\text{TOTAL}} = (13 - 10)^2 + (14 - 10)^2 + \ldots + (11 - 10)^2 = 116$. And SS_{BETWEEN} is directly calculated by the formula:

$$SS_{\text{BETWEEN}} = \sum_{j=1}^{J} n_j(\overline{Y}_j - \overline{Y})^2$$

which in this case is simply $10(11.8 - 10.0)^2 + 10(10.0 - 10.0)^2 + 10(8.2 - 10.0)^2 = 64.8$.

Now SS_{WITHIN} can be computed in two ways. The direct calculation formula is:

$$SS_{\text{WITHIN}} = \sum_{j=1}^{J} \sum_{i=1}^{n_j} (Y_{ij} - \overline{Y}_j)^2$$

which directs us to subtract observations from their *subgroup* mean, square them, and sum them. If we do this we find that $SS_{\text{WITHIN}} = (13 - 11.8)^2 + (14 - 11.8)^2 + \ldots + (11 - 8.2)^2 = 51.2$. The other formula says that since $SS_{\text{TOTAL}} = SS_{\text{BETWEEN}} + SS_{\text{WITHIN}}$, it follows that:

$$SS_{\text{WITHIN}} = SS_{\text{TOTAL}} - SS_{\text{BETWEEN}}$$

In other words, once we know SS_{TOTAL} and SS_{BETWEEN}, we can simply subtract the latter from the former to obtain SS_{WITHIN}. This checks with the longer method of computation shown above. We suggest that you calculate all three components in the example directly, as a check on your understanding and the accuracy of your arithmetic.

7.5 Mean Squares

The next step in an analysis of variance is to compute the mean squares for SS_{BETWEEN} and SS_{WITHIN}. When we compute **mean squares**, we are computing two variances—one due to treatment effects and one due to error. If in fact treatment effects exist, we expect that the between-group variance, which we call **mean square between,** or MS_{BETWEEN}, will be significantly larger than the within-group variance, which we call **mean square within**, or MS_{WITHIN}.

mean square—estimate of variance use in the analysis of variance

mean square between—a value in ANOVA obtained by dividing the between sum of squares by its degrees of freedom

mean square within—a value in ANOVA obtained by dividing the within sum of squares by its degrees of freedom

As we noted in Chapter 3, a variance is an average or mean in the sense that it is derived by dividing a sum of squared deviations about the mean by their degrees of freedom. Thus we showed that, in general, the variance of a set of sample scores is:

$$s_Y^2 = \frac{\sum_{i=1}^{N} (Y_i - \overline{Y})^2}{N - 1}$$

where $N - 1$ is the degrees of freedom. *The degrees of freedom associated with the between-group variance are simply $J - 1$,* since once we know the grand mean and $J - 1$ group means, the jth group mean can be determined automatically. This is explained in Box 7.1.

Since there are in general $J - 1$ degrees of freedom in computing the variance due to treatment, we know that:

$$MS_{\text{BETWEEN}} = \frac{\sum_{j=1}^{J} n_j (\overline{Y}_j - \overline{Y})^2}{J - 1} = \frac{SS_{\text{BETWEEN}}}{J - 1}$$

Therefore in the puzzle-solving example:

$$MS_{\text{BETWEEN}} = 64.8/(3 - 1) = 32.40$$

The degrees of freedom associated with the within squares are $N - J$. This is also easy to comprehend. Each group has $n_j - 1$ degrees of freedom, so we can just add them up over all J groups.

Box 7.1 Determining Degrees of Freedom for J Groups: $J-1 = df$

If we know the overall or grand mean and any of the $J-1$ subgroup means, then the jth subgroup mean can be determined automatically. This relationship is true, since:

$$\bar{Y} = \sum_{j=1}^{J} n_j \bar{Y}_j / N$$

That is, the grand mean is a weighted sum of the subgroup means, where the weight for the jth subgroup mean is n_j/N.

It therefore follows directly that, where J is the last group:

$$N\bar{Y} = \sum_{j=1}^{J-1} n_j \bar{Y}_j + n_J \bar{Y}_J$$

or

$$n_J \bar{Y}_J = N\bar{Y} - \sum_{j=1}^{J-1} n_j \bar{Y}_j$$

and

$$\bar{Y}_J = (N\bar{Y} - \sum_{j=1}^{J-1} n_j \bar{Y}_j)/n_J$$

This derivation looks more complex than it is. In the puzzle-solving experiment described in the text, there are $J = 3$ groups and $\bar{Y} = 10.0$. Furthermore, $n_m = n_{nt} = n_{na} = 10$ and $N = (n_m + n_{nt} + n_{na}) = (10 + 10 + 10) = 30$. Suppose we know that $\bar{Y}_{na} = 11.8$ and $\bar{Y}_{nt} = 10.0$. With this information, we can figure out what $\bar{Y}_m$ must equal. In particular:

$$\bar{Y}_m = \frac{30(10.0) - [10(11.8) + 10(10.0)]}{10}$$

$$= \frac{300 - (118 + 100)}{10}$$

$$= \frac{(300 - 218)}{10}$$

$$= 8.2$$

This shows that there are $J-1$ degrees of freedom associated with SS_{BETWEEN}.

That is:

$$
\begin{aligned}
&(n_1 - 1) + (n_2 - 1) + \ldots + (n_J - 1) \\
&= (n_1 + n_2 + \ldots + n_J) - (1 \underbrace{+ 1 + \ldots + 1}_{J \text{ of these}}) \\
&= N - J
\end{aligned}
$$

Therefore, to compute MS_{WITHIN}, we divide SS_{WITHIN} by $N-J$. That is,

$$MS_{\text{WITHIN}} = \frac{SS_{\text{WITHIN}}}{N - J}$$

Using the data from the example, $MS_{\text{WITHIN}} = 51.2/(30 - 3) = 1.90$.

The variance due to treatment appears to be considerably larger than that due to error (32.40 vs. 1.90), which should be the case if treatment effects exist. We need to examine how hypotheses are tested in ANOVA, however, before drawing any firm conclusions about treatment effects in this example.

It can be shown that MS_{WITHIN} is an unbiased estimate of the population error variance:

$$E(MS_{\text{WITHIN}}) = \sigma_e^2$$

It can also be shown that the MS_{BETWEEN} is an unbiased estimate of the population variance plus another term reflecting treatment effects. That is,

$$E(MS_{\text{BETWEEN}}) = \sigma_e^2 + \frac{\sum_{j=1}^{J} n_j \alpha_j^2}{J - 1}$$

Notice that when there are *no* treatment effects, the numerator of the second component in this formula will be zero, since α_j^2 will be zero for all j. And in this case MS_{BETWEEN} will also be an unbiased estimate of the population error variance. That is, when all $\alpha_j = 0$:

$$E(MS_{\text{BETWEEN}}) = \sigma_e^2$$

If the null hypothesis is true and there are no treatment effects, we should expect MS_{BETWEEN} and MS_{WITHIN} to be roughly equal to each other, and their ratio will be roughly 1.0. (We say *roughly* because it is unlikely, due to sampling variability, that MS_{WITHIN} and MS_{BETWEEN} will exactly equal σ_e^2 in any given sample, even when the null hypothesis is true.) If the ratio is far greater than 1.0, this suggests that MS_{BETWEEN} is estimating more than σ_e^2, and the null hypothesis should be rejected. In other words, when MS_{BETWEEN} is "substantially" larger than MS_{WITHIN}, this suggests that treatment effects exist.

The next problem to be considered is how much larger MS_{BETWEEN} must be relative to MS_{WITHIN} before we can reject the hypothesis of no treatment effects.

7.6 The *F* Distribution

***F* ratio**—a test statistic formed by the ratio of two mean-square estimates of the population error variance

A variable called the ***F* ratio** is simply the ratio of MS_{BETWEEN} to MS_{WITHIN}. That is:

$$F = \frac{MS_{\text{BETWEEN}}}{MS_{\text{WITHIN}}}$$

Any *observed* F ratio can be tested against the assumption that it came from a population where the null hypothesis is true—that is, against the hypothesis that none of the observed sum of squares is due to treatment effects. The F ratio has a known sampling distribution under the null hypothesis if certain assumptions can be met. We must assume that *the J samples are independently drawn from a normally distributed population.* We must also assume that *the variance in the population is the same for all J treatment categories.* The second assumption is sometimes called **homoscedasticity.** (If the J population variances differ, they are said to be *heteroscedastic.*)

homoscedasticity—
a condition in which the variances of two or more population distributions are equal

If these assumptions are met, an F ratio is distributed according to the ***F* distribution,** *with* $J - 1$ *degrees of freedom associated with the numerator and* $N - J$ *with the denominator.* Since the alternative hypothesis in ANOVA implies that the between-group variance is larger than the within-group variance *in the population,* we usually do a one-tailed test of significance. We choose an α level, and if the observed F ratio is larger than the critical value associated with α, we reject the null hypothesis and instead conclude that there indeed are treatment effects.

***F* distribution**—
a theoretical probability distribution for one of a family of F ratios having $J - 1$ and $N - J$ *df* in the numerator and denominator, respectively

We will not show what F distributions look like, since they vary somewhat as a function of both degrees of freedom. However, we can show the use of the F table with the results of the hypothetical puzzle-solving experiment. Suppose we set $\alpha = .05$. Since $J = 3$ and $N = 30$ in this example, there are 2 degrees of freedom associated with the numerator and 27 with the denominator. Appendix E provides tables of the F distribution for $\alpha = .05$, .01, and .001. The degrees of freedom for the numerator run across the top rows of the tables and are labeled ν_1 (Greek letter *nu*), and the degrees of freedom associated with the denominator run down the first columns and are labeled ν_2.

To use the first table, for $\alpha = .05$, find the cell where $\nu_1 = 2$ and $\nu_2 = 27$. The value in that cell is the critical value (c.v.) which the observed F ratio must equal or exceed to reject the null hypothesis of no treatment effects. In this case c.v. = 3.35. Since $MS_{\text{BETWEEN}} = 32.40$ and $MS_{\text{WITHIN}} = 1.90$,

$$F_{2,27} = \frac{32.40}{1.90} = 17.05$$

This means that we reject the null hypothesis. We conclude that performance on tasks is reduced as a result of being observed by another person.

7.7 Reporting an Analysis of Variance

Procedures have been established for summarizing analysis of variance in research reports. A common practice is to summarize

TABLE 7.2

Summary Table for One-Way Analysis of Variance of Hypothetical Data on Task Performance as a Function of Monitoring

Source	*SS*	*df*	*Mean Square*	*F*
Between groups	64.80	2	32.40	17.05*
Within groups (error)	51.20	27	1.90	
Total	116.00	29		

*$p < .05$.

ANOVA summary table—a tabular display summarizing the results of an analysis of variance

the results of analysis of variance in an **ANOVA summary table** of the sort shown in Table 7.2. This table is very useful for the reader of the report, since it provides easy access to all the relevant information needed to interpret the research—sums of squares, degrees of freedom, and the F ratio.

7.8 The Relationship of *t* to *F*

You might have wondered what the relationship is between the t test discussed in Chapter 6 and the F test described in this chapter when comparing the means for only two *groups,* that is, when $J = 2$. Will the t test and an ANOVA give the same results, and how do we decide which test to use?

> The t test and ANOVA do yield identical results. Indeed, the square root of an F test with 1 and ν_2 degrees of freedom equals a t test with ν_2 *degrees of freedom for the same set of data:*
>
> $$t_{\nu_2} = \sqrt{F_{1,\nu_2}}$$

Usually when $J = 2$ the researcher reports a t test, and when $J > 2$, the results of an ANOVA are always reported.

7.9 Determining the Strength of a Relationship: Eta Squared

After the researcher has rejected the null hypothesis in an ANOVA, the question of how strong the relationship is remains to be answered. The problem with hypothesis testing is that *almost any difference between and among means will be statistically significant if we choose N to be large enough.* For that reason, when the null hypothesis has been rejected we should also assess the strength of the relationship found. This can be done by computing **eta squared,** or η^2 (Greek letter *eta*).

eta squared—a measure of nonlinear covariation between a discrete and a continuous variable, the ratio of SS_{BETWEEN} to SS_{TOTAL}

In Section 7.4 we showed that:

$$SS_{\text{TOTAL}} = SS_{\text{BETWEEN}} + SS_{\text{WITHIN}}$$

If we divide both sides of the equation by SS_{TOTAL}, we get:

$$1.0 = \frac{SS_{\text{BETWEEN}}}{SS_{\text{TOTAL}}} + \frac{SS_{\text{WITHIN}}}{SS_{\text{TOTAL}}}$$

$$= \text{"Explained" } SS + \text{"Unexplained" } SS$$

We can think of the ratio of SS_{BETWEEN} to SS_{TOTAL} as the sum of squares explained by the treatment or the independent variable(s) under study. Similarly, the ratio of SS_{WITHIN} to SS_{TOTAL} can be thought of as the unexplained sum of squares.

Notice that the two components add to 1.0. Therefore, to determine the proportion of variation in a dependent variable accounted for by a treatment variable, we define:

$$\eta^2 = \frac{SS_{\text{BETWEEN}}}{SS_{\text{TOTAL}}}$$

Eta squared (η^2) *is always a positive number ranging between zero and 1.00.* It measures the proportion of the variance in the dependent variable which is "explained" (in a statistical rather than in a causal sense) by the independent variable. Since η^2 is technically a population parameter, when it is estimated by sample data we place a caret over it (i.e., $\hat{\eta}^2$). The more the sample means differ from each other and the more the sample variances exhibit small dispersions, the higher the SS_{BETWEEN}, and hence the larger the value of $\hat{\eta}^2$.

Using the results from the hypothetical set of data for the problem-solving experiment,

$$\hat{\eta}^2 = \frac{64.80}{116.00} = .559$$

That is, 55.9% of the variation in task performance can be accounted for by the monitored/unmonitored manipulations in the experiment.

As social science experiments go, this η^2 is a large proportion of variance explained, and therefore it should not be taken as a guideline against which to gauge the strength of findings in general. Typically, a single independent variable in the social sciences will account for no more than 25% to 30% the variance in a dependent variable, and often as little as 5% or 10%.

7.10 Testing for Differences between Individual Treatment Means

The alternative hypothesis for an ANOVA (and therefore the test for a significant F) can take several forms, as we noted in Section 7.1. The three possibilities are: (1) All the population means differ from one another, (2) some subsets of the population means differ from one another, or (3) some combination of means is different from some single mean or from some other combination of means. By itself, the F value is mute on which of these alternatives is true in any given situation. You might think that the thing to do is to test for differences among all possible pairs of means—three in the above example. It can be shown, however, that *for J means, only $J - 1$ of the comparisons among means (or combinations of means) will be independent of each other.* For this reason, *individual t tests are not to be used after obtaining a significant F.*

planned comparison—hypothesis test of differences between and among population means carried out before doing an analysis of variance

post hoc comparison—hypothesis test of the differences among population means carried out following an analysis of variance

There are two approaches to the problem of how to compare means. The first is called *a priori,* or **planned comparison**, and the second is called *a posteriori,* or **post hoc comparison**. Both approaches are guided by the researcher's hypotheses about expected differences between and among the treatment groups. While planned comparisons have more statistical power and hence are to be preferred over post hoc comparisons, the treatment of this important topic requires more mathematics than we can assume you will have as a student in this course. Therefore (with some regret) we will not consider this topic. Post hoc comparisons are less statistically powerful, but they are useful and are often used in social science research. A brief introduction to one of them—the Scheffé test—follows.

7.10.1. Multiple Means Comparison Using Contrasts

To make multiple comparisons among J means, we can utilize the notion of a contrast. A **contrast**, labeled ψ (Greek letter *psi*), among J population means is defined as:

contrast—a set of weighted population means that sum to zero, used in making post hoc comparisons of treatment groups

---

$$\psi = c_1\mu_1 + c_2\mu_2 + \ldots + c_J\mu_J$$

where the c_j are weights under the constraint that $c_1 + c_2 + \ldots + c_J = 0$; *that is, the c_j when summed add to zero.*

An example should make this approach clearer. In the problem-solving experiment we had two hypotheses. First we hypothesized that the number of puzzles completed when subjects are monitored will be fewer than when they are not monitored. We suggested that the *monitored condition* can be contrasted with the *unmonitored condition,* whether the subjects were working alone or together. Then, we hypothesized that those who were not monitored but who worked in the presence of another person would complete fewer puzzles than those who were not monitored and worked at the task alone. This second hypothesis contrasts the *not-monitored-together* and *not-monitored-alone* groups. The contrast weights, or c_j, for the first hypothesis might be chosen as follows: $c_m = 1$ and $c_{nt} = c_{na} = -½$. Note that the c_j add to zero: $c_m + c_{nt} + c_{na} = 1 - ½ - ½ = 0$.

The contrast for this first hypothesis, call it ψ_1, is given by:

$$\begin{aligned}\psi_1 &= (1)\mu_m + (-½)\mu_{nt} + (-½)\mu_{na} \\ &= \mu_m - \frac{(\mu_{nt} + \mu_{na})}{2}\end{aligned}$$

That is, we are contrasting the monitored condition with the *average* of the two unmonitored conditions. (We could have let $c_m = 2$ and $c_{nt} = c_{na} = -1$; this contrast would yield the same result when tested for statistical significance using the procedure detailed below.)[2]

---

2. Indeed, when using SPSS (see Preface) on analysis of variance problems, you must use either integer contrasts or single-place decimal contrasts. So 0.5 and −0.5 are permissible, but contrasts like 0.25 would be rounded incorrectly to 0.3.

The second hypothesis can be tested by letting $c_{nt} = 1$ and $c_{na} = -1$ (and, by implication, $c_m = 0$). The contrast for the second hypothesis, labelled ψ_2, is given by:

$$\begin{aligned}\psi_2 &= (1)\mu_{nt} + (-1)\mu_{na} + (0)\mu_m \\ &= \mu_{nt} - \mu_{na}\end{aligned}$$

This contrast compares the two unmonitored conditions, as required by the hypothesis. It also satisfies the condition that the c_j sum to zero.

> Once we form a contrast using sample means we need to compare it in size to its standard error, much as we do when computing a t test. If the ratio is large enough, we conclude that the comparison between the means is significant. If not, we cannot reject the null hypothesis that the true difference is in fact zero.

An *unbiased estimate* of a contrast using sample data is given by:

$$\hat{\psi} = c_1\overline{Y}_1 + c_2\overline{Y}_2 + \cdots + c_J\overline{Y}_J$$

And the *estimated variance* of a contrast is given by:

$$\hat{\sigma}^2_{\hat{\psi}} = MS_{\text{WITHIN}}\left(\frac{c_1^2}{n_1} + \frac{c_2^2}{n_2} + \ldots + \frac{c_J^2}{n_J}\right)$$

Where:

MS_{WITHIN} = The mean square within obtained from the ANOVA.

The *test statistic* is given by the absolute value of $\hat{\psi}$ divided by its standard error:

$$t = \frac{|\hat{\psi}|}{\hat{\sigma}_{\hat{\psi}}}$$

The *critical value* against which the test statistic is compared is given by:

$$\text{c.v.} = \sqrt{(J - 1)(F_{J-1,\,N-J})}$$

Where:

$F_{J-1,\,N-J}$ = The critical value for the α level chosen to test the null hypothesis in the ANOVA.

Therefore, if:

$$|\hat{\psi}|/\hat{\sigma}_{\hat{\psi}} \geq \sqrt{(J - 1)(F_{J-1,\,N-J})}$$

we reject the null hypothesis. *There are N − J degrees of freedom associated with this test.*

7.10.2. Mean Comparison in the Problem-Solving Example

An example from the problem-solving experiment should help clarify how to do a multiple comparison. The first contrast (based on the first hypothesis) is:

$$\hat{\psi}_1 = 8.2 - \frac{(10.0 + 11.8)}{2} = -2.70$$

and

$$\hat{\sigma}^2_{\hat{\psi}1} = 1.90\left[\frac{1^2}{10} + \frac{(-1/2)^2}{10} + \frac{(-1/2)^2}{10}\right]$$
$$= 1.90(.15)$$
$$= 0.29$$

Taking the square root:

$$\hat{\sigma}_{\hat{\psi}_1} = \sqrt{.29} = .53$$

Therefore, the test statistic is $|-2.7|/.53 = 5.06$. The critical value for $\alpha = .05$ is $\sqrt{(3-1)(3.35)} = 2.59$. Since 5.0 is larger than 2.59, we conclude that being monitored does affect task performance negatively, compared to performing the task without monitoring.

The second contrast and its statistical test of significance are given by:

$$\hat{\psi}_2 = (1)(10.0) + (-1)(11.8) + (0)(8.2) = -1.8$$

and

$$\hat{\sigma}^2_{\hat{\psi}_2} = 1.90\,[(\frac{1^2}{10}) + (\frac{-1^2}{10}) + (\frac{0^2}{10})] = 1.90\,(\frac{2}{10})$$
$$= .38$$

Taking the square root:

$$\hat{\sigma}_{\hat{\psi}_2} = \sqrt{.38} = .62$$

The test statistic for the second contrast is $|-1.8| / .62 = 2.90$, and for $\alpha = .05$, the critical value is 2.59. Therefore, we also can reject the null hypothesis associated with the second contrast. We conclude that working in the presence of a second person decreases performance compared to working alone, even if the other person is not monitoring performance directly.

You should now know how to do an ANOVA and how to make post hoc comparisons among means. This means you should be able to undertake meaningful tests of hypotheses involving several means.

7.11 The Use of ANOVA in Nonexperimental Research

The term *treatment* has a long history in discussions of the analysis of variance. In fact, ANOVA was designed originally for the analysis of data generated by experiments and the treatments used in them. But this technique is very useful for the analysis of nonexperimental data as well.

> The term *independent variable* can be freely substituted for treatment when the analysis involves nonexperimental data. This will often be the case in the nonexperimental social sciences such as sociology and political science. The term *effect* can be substituted for *treatment effect* in the same way. However, it is much more difficult to draw *causal* inferences from nonexperimental designs.

An example in which some actual nonexperimental data are examined is given in the following section.

7.11.1. An Example: Crime Rate and Population Loss

When, in Chapter 6, we tested the proposition that Frostbelt cities lose population at a higher rate than cities located in the Sunbelt, we found support for the hypothesis. Now we will consider a second proposition related to population loss. The proposition is that the push of crime should cause cities with higher crime rates to lose population faster than cities with lower crime rates:

P2: The higher a city's crime rate, the faster its population loss.

We will test this proposition with the 63-cities data (see Box 2.4 in

TABLE 7.3
Percentage Distribution for Population Change, 1970–1974, in Cities Classified by 1970 Serious Crime Rate per 1,000 Residents

Population Change			
Cities with 0–30 Rate	***Cities with 31–45 Rate***	***Cities with 46–60 Rate***	***Cities with 61–90 Rate***
−1.0	17.8	18.1	3.2
−1.8	15.6	9.6	1.4
−1.9	8.7	7.1	−2.5
−2.0	8.3	0.7	−3.0
−2.1	7.2	−0.8	−3.2
−2.5	6.9	−3.4	−3.4
−2.6	5.7	−3.6	−3.6
−3.7	3.5	−3.8	−3.9
−3.8	3.4	−4.4	−3.9
−3.9	2.5	−5.0	−4.2
−4.5	2.2	−5.2	−8.2
−5.9	−1.2	−7.3	−10.3
−7.6	−1.3	−7.9	
	−1.7	−9.2	
	−2.0	−9.6	
	−2.5	−12.0	
	−2.9		
	−2.9		
	−3.3		
	−5.8		
	−8.1		
	−8.6		
$\bar{Y}_1 = -3.33$	$\bar{Y}_2 = 1.89$	$\bar{Y}_3 = -2.29$	$\bar{Y}_4 = -3.47$
$s_1 = 1.85$	$s_2 = 6.94$	$s_3 = 7.89$	$s_4 = 3.56$
$n_1 = 13$	$n_2 = 22$	$n_3 = 16$	$n_4 = 12$

Source: NORC Permanent Community Sample (63-cities data).

FIGURE 7.2
Population Change in Cities Grouped by Crime Rate

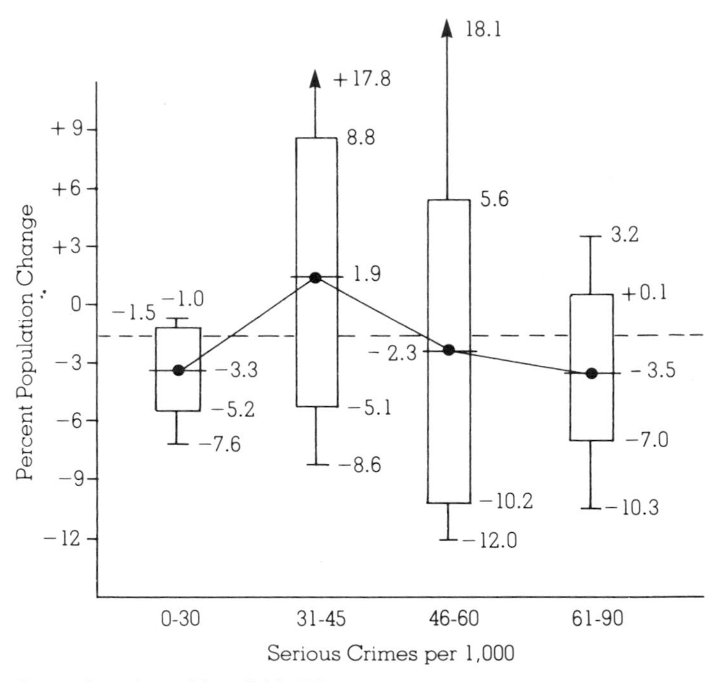

Source: Data derived from Table 7.3

Chapter 2). *Crime rate* is defined as the number of serious crimes per 1,000 persons, as reported by the Federal Bureau of Investigation in 1970. Serious crimes include murder, robbery, burglary, assault, auto theft, rape, and grand larceny. Population change is based on changes that occurred in the 63 cities between 1970 and 1974. The hypothesis to be tested is:

H2: The higher the serious crime rate in 1970, the greater the population loss between 1970 and 1974.

Since both variables are measured as continuous variables, we could use statistical tests appropriate to bivariate relationships between continuous variables (see Chapter 8). However, we want to illustrate how to analyze the relationship between a discrete independent and a continuous dependent variable, and so we have recoded the 1970 serious crime rate into four categories. These four new classifications are for cities with 0 to 30, 31 to 45, 46 to 60, and 61 to 90 reported serious crimes per 1,000 residents. As shown in Table 7.3 and Figure 7.2, this gives us four groups of cities with unequal sample sizes.

Contrary to the hypothesis, the relationship between crime rate and subsequent population loss appears not to be linear but rather appears to form a *curvilinear* pattern for the four group means. As the box-and-whisker diagram, Figure 7.2, shows, population loss is greatest in the two groups of cities experiencing the highest and the lowest levels of crime. The two groups of cities with intermediate levels of 1970 crime later appear to have experienced somewhat lower mean population declines. Indeed, the three higher crime-rate categories conform to the hypothesized pattern of decreasing population with increasing crime; only the first category mean appears to be substantially out of line. The reason for this curvilinear pattern is not obvious. Further research with other independent variables would be required to determine whether the low- and high-crime-rate cities are similar in other respects that would cause them to lose population during this period more rapidly than middle-crime-rate communities did.

We will continue with this example to illustrate how to perform a statistical test on the observed relationship. We need to determine the sums of squares, degrees of freedom, and mean squares in order to compute an F ratio.

Using the formula in Box 7.1, we can compute the grand mean from the subgroup means given in Table 7.3. This is:

$$\overline{Y} = \frac{13(-3.33) + 22(1.89) + 16(-2.29) + 12\,(-3.47)}{13 + 22 + 16 + 12}$$
$$= -1.27$$

To compute SS_{TOTAL}, we now subtract the various observations in Table 7.3 from -1.27, square them, and sum them. That is:

$$SS_{\text{TOTAL}} = [-1.0 - (-1.27)]^2 + [-1.8 - (-1.27)]^2 + \ldots + [-10.3 - (-1.27)]^2$$
$$= 2{,}474.71$$

Now $SS_{BETWEEN}$ is computed using the formula given in Section 7.4.1:

$$\begin{aligned} SS_{BETWEEN} &= 13[-3.33 - (-1.27)]^2 + 22[1.89 - (-1.27)]^2 \\ &\quad + 16[-2.29 - (-1.27)]^2 \\ &\quad + 12[-3.47 - (-1.27)]^2 \\ &= 349.58 \end{aligned}$$

We can now compute SS_{WITHIN} as the difference between SS_{TOTAL} and $SS_{BETWEEN}$. This is:

$$SS_{WITHIN} = 2{,}474.71 - 349.58 = 2{,}125.13$$

As an exercise, you should calculate SS_{WITHIN} directly from the data.

To compute the mean squares, we need to calculate the degrees of freedom. For $SS_{BETWEEN}$ there are $4 - 1 = 3$ *df*, while for SS_{WITHIN} there are $63 - 4 = 59$ *df*. Therefore,

$$MS_{BETWEEN} = \frac{349.58}{3} = 116.53$$

$$MS_{WITHIN} = \frac{2{,}125.13}{59} = 36.02$$

With this information we can now calculate the test statistic. This is:

$$F_{3,59} = \frac{116.53}{36.02} = 3.24$$

Now for $\alpha = .05$, as Appendix E shows, the critical value is 2.76. Hence we can reject the null hypothesis of no difference in population loss as a function of crime. Table 7.4 summarizes the results of the ANOVA.

TABLE 7.4

Summary Table for One-Way Analysis of Variance of Population Change in Cities Classified by Crime Rate

Source	*Sum of Square*	*df*	*Mean Square*	*F*
Between groups	349.58	3	116.53	3.24*
Within groups (error)	2,125.13	59	36.02	
Total	2,474.71	62		

*$p < .05$.

Finally, to determine the strength of the relationship, we find eta squared:

$$\hat{\eta}^2 = 349.58/2{,}474.71 = .14$$

This finding, that $\hat{\eta}^2 = .14$, reveals that little of the variation in population change can be explained by crime rate. In spite of statistical significance there is no support for the research hypothesis, since percent population loss did not uniformly increase as a function of serious crime rate (as seen in Figure 7.2).

> As this example makes clear, one-way analysis of variance is not a test of the linear relationship between variables. The test is sensitive only to differences in the sample means, regardless of the order among the categories of the independent variable. Since the pattern of observed means does not follow the one we hypothesized, carrying out post hoc comparisons makes no sense.

Chapter 8 describes how to estimate *linear effects* for two continuous variables.

Review of Key Concepts

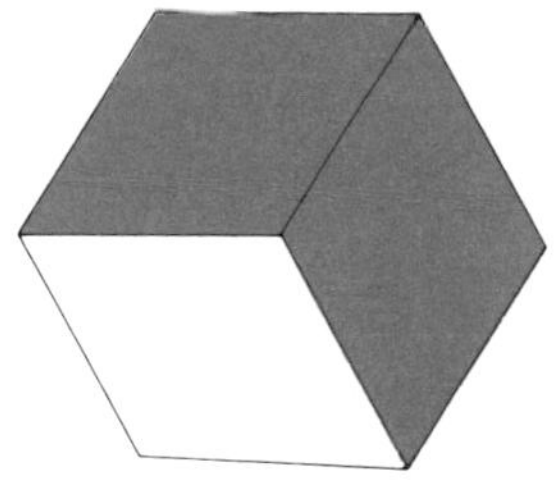

This list of the key concepts introduced in Chapter 7 is in the order of appearance in the text. Combined with the definitions in the margins, it will help you review the material and can serve as a self-test for mastery of the concepts.

analysis of variance
grand mean
effect
error (residual) term
total sum of squares
between sum of squares
within sum of squares
mean square
mean square between
mean square within
F ratio
homoscedasticity
F distribution
ANOVA summary table
eta squared
planned comparison
post hoc comparison
contrast

PROBLEMS

General Problems

1. A social psychologist hypothesizes that people who have high levels of introspection and low levels of self-esteem have higher rates of alcoholism than those with any other combination of levels of introspection and self-esteem. Write the null and alternative hypotheses in symbolic form.

2. A demographer hypothesizes that the rate of illegitimate births is greatest among teenage girls but does not differ between women in their 20s and those in their 30s. Write the null and alternative hypotheses in symbolic form.

3. On a measure of traditional religiosity, a sociologist finds that the overall mean is 10.6, and the mean of Catholics is 13.7, of Protestants is 11.1, of Jews is 10.1, and of those with no religious identification is 8.2. What are the effects on religious belief of being a member of each group (α_j)?

4. A psychologist administers differing dosages of THC to college students to observe the effects of the drug on a memory task. The results for these groups of $N = 10$ subjects are $\overline{Y}_{\text{control}} = 12.1$, $\overline{Y}_{\text{lo dose}} = 11.9$, and $\overline{Y}_{\text{hi dose}} = 6.0$. What are the effects associated with the dosage levels?

5. Find the degrees of freedom and the critical values of F for the following:
   a. $\alpha = .05$, 4 groups, 60 subjects.
   b. $\alpha = .05$, 6 groups, 24 subjects.
   c. $\alpha = .01$, 3 groups, 30 subjects.
   d. $\alpha = .01$, 7 groups, 40 subjects.
   e. $\alpha = .01$, 12 groups, 200 subjects.

6. Find the degrees of freedom and the critical values of F for the following:
   a. $\alpha = .01$, $N_1 = 8$, $N_2 = 6$, $N_3 = 8$.
   b. $\alpha = .05$, $N_1 = 20$, $N_2 = 20$, $N_3 = 20$, $N_4 = 20$.
   c. $\alpha = .001$, $N_1 = 220$, $N_2 = 150$.

7. If three groups are of equal size and two of the groups have means of 3.4 and 4.6, respectively, what must the mean of the third group equal if the overall or grand means equals 5.0?

8. Suppose you observe the following means and sample sizes: $\overline{Y}_1 = 10.0, N_1 = 10$; and $\overline{Y}_2 = 12.0, N_2 = 20$. What does the grand mean equal?

9. Suppose you observe the following means and sample sizes: $\overline{Y}_1 = 16.3, N_1 = 25$; $\overline{Y}_2 = 14.3, N_2 = 42$; and $\overline{Y}_3 = 17.6$, $N_3 = 12$. What does the grand mean equal?

10. Find the total sum of squares (SS_{TOTAL}) for the following scores: 5, 7, 6, 6, 4, 5, 8, 4, 5, 2, 6, 7.

11. If $s_Y^2 = 27.4$ and $N = 100$, what does SS_{TOTAL} equal?

12. Below are the reaction times for 15 subjects under three conditions in an experiment designed to study the effects of marijuana smoking on driving ability:

Control	*Light Smoking*	*Heavy Smoking*
4	5	7
2	7	9
2	7	10
3	8	10
5	8	11

Find the total sum of squares, the between sum of squares, and the within sum of squares. Find the mean square between and the mean square within. Find the F ratio and evaluate it against the critical value for $\alpha = .05$. Can you reject the null hypothesis that the three population means are equal?

13. Compute $\hat{\eta}^2$ for the relationship between marijuana smoking and driving ability in Problem 12.

14. Four groups of students are taught social statistics by four different methods. A test is given to random samples of five students from each group, with these results:

Group A	*Group B*	*Group C*	*Group D*
20	15	22	19
22	18	21	23
21	20	24	20
20	18	25	18
19	19	24	15

Perform a one-way ANOVA on this data, construct the ANOVA summary table, and decide whether to reject the null hypothesis that teaching method made no difference in the performance. Set $\alpha = .05$.

15. For the four groups in Problem 14, write a post hoc contrast which would test whether group C has a significantly better performance than the other three. Carry out the test with $\alpha = .05$.

16. An educational psychologist is interested in whether three types of reinforcers improve verbal learning. The results for the three methods, as applied to three different samples of children, are:

A	*B*	*C*
7	6	5
5	3	2
6	2	1
9	2	3
10	4	4
4	5	4
5	3	6
5	5	
7		

Present the results in an ANOVA summary table and test the null hypothesis of no difference, setting $\alpha = .05$.

17. Test the post hoc hypothesis in Problem 16 that reinforcement method A is superior to methods B and C. Set $\alpha = .05$. Show $\hat{\psi}$ and $\hat{\sigma}^2_{\hat{\psi}}$.

18. How much of the variance in verbal learning scores in Problem 16 can be explained as a function of type of reinforcement?

19. Here are some results from an ANOVA: $SS_{\text{BETWEEN}} = 315.68$; $df_{\text{BETWEEN}} = 6$; $SS_{\text{WITHIN}} = 1{,}427.32$; and $df_{\text{WITHIN}} = 60$.
    a. Construct the ANOVA summary table.
    b. Should the null hypothesis be rejected for $\alpha = .01$?
    c. What is $\hat{\eta}^2$ in this example?

20. In order to evaluate the effects of inmate self-government on recidivism, two groups of eight prisons were matched on background characteristics and type of offense. The prisons in each group either allowed prisoners to determine for themselves how to spend their free time, or the staffs made all decisions for them. Below are the rates of recidivism (in per-

cent) two years after release for inmates in the two groups of prisons.

Prison No.	*Self-Government*	*No Self-Government*
1	12%	21%
2	20	18
3	18	17
4	16	22
5	14	16
6	21	20
7	16	19
8	19	23

Do an ANOVA to determine whether the program affected the recidivism rate. Set $\alpha = .05$. Show the ANOVA summary table.

21. Do a t test for the difference between two means (Chapter 6), using the data in Problem 20 above, and show that the t^2_{14} equals the $F_{1,14}$ obtained in Problem 20.

22. Construct a box-and-whisker diagram (see Chapter 6) for the data in Problem 20.

Problems Requiring the 1980 General Social Survey

23. *a.* Do an analysis of variance examining the relationship between marital status (MARITAL) and health (HEALTH). Set $\alpha = .05$.
    b. Construct contrasts to test if (1) those currently married are healthier than all other groups combined, and (2) those never married are healthier as a group than all others. Set $\alpha = .05$.
    c. Can you think of an alternative explanation for these findings?

24. Given the history of discrimination against blacks in the United States, you might wonder whether blacks are less satisfied with where they live than whites are. To test this hypothesis ($\alpha = .01$), do an ANOVA with SATCITY as the dependent variable and RACE as the independent variable. (Eliminate the other racial groups from the analysis.)

25. Some sociologists hypothesize that interracial contact in-

creases positive attitudes toward minority group members. Test this hypothesis by first selecting only whites (RACE) and then doing an ANOVA using attitude towards racial segregation (RACSEG) as the dependent variable and the propinquity of blacks to one's home (RACDIS). Construct a contrast to determine whether whites living within three blocks of blacks have a more positive attitude than those living farther away. Set $\alpha = .01$ for all analyses.

26. Sociologists and economists hypothesize that current earnings are related to the size of the income of the family in which one is raised (along with several other variables).
    a. Test this hypothesis by doing an ANOVA using current family income (INCOME) as the dependent variable and family income when the respondent was 16 years old (INCOM16) as the independent variable. Set $\alpha = .01$.
    b. Interpret the pattern of means.
    c. Compute $\hat{\eta}^2$.

27. Perform an analysis similar to Problem 26 using the respondent's highest educational degree (DEGREE) as the independent variable. Set $\alpha = .01$. Do a contrast comparing those with at least a college degree to those with a high school degree or less. Set $\alpha = .01$.

Problems Requiring the 63-Cities Data

28. Does the percent of labor force in nonmanual occupations (WCPCT) vary by major region of the country (REGION)? Set $\alpha = .01$ and:
    a. Do an ANOVA to test this hypothesis.
    b. Construct a contrast to determine whether the West has a larger percentage of the labor force in nonmanual occupations than other regions do.

29. Cities in the Northeast and Midwest are said to be decaying. Set $\alpha = .05$ and:
    a. Do an ANOVA using CITYAGE as the dependent variable and REGION as the independent variable to test the null hypothesis that the mean age of cities is the same in all regions.
    b. Construct a contrast to see whether the Northeast

and Midwest do indeed have older cities than the South and the Far West.

30. Does per capita municipal expenditures (EXPENDIT) differ significantly with the number of municipal functions performed (FUNCTION)? Perform an analysis of variance to find out. Set $\alpha = .01$.

31. Compute per capita retail sales for the cities. Divide the percentage in white-collar occupations (WCPCT) into four equal-size categories. Perform an ANOVA to determine if per capita retail sales differ significantly with the occupational composition of cities. Set $\alpha = .05$. (To construct per capita retail sales, divide RETAIL by POPULAT and multiply the result by 1,000, using a COMPUTE statement.)

32. *a.* Is there a positive relationship between municipal revenues (REVENUE) and expenditures (EXPENDIT) in the 63 cities if revenues are recoded into 10 equal-sized categories? Set $\alpha = .05$ and do an ANOVA to find out.
    b. What does $\hat{\eta}^2$ equal?

8

Estimating Relations between Two Continuous Variables: Bivariate Regression and Correlation

Continuous measures, with their assumption of a meaningful distance between numbers along the scale, place at the researcher's disposal a powerful set of tools for analyzing the covariation of social variables. All you need to assume to apply the methods presented in this chapter, which continues our treatment of the analysis of relationships between pairs of variables, is first, that the variables are continuous, and second, that the distribution of the dependent variable follows a normal curve for every outcome of the independent variable. Even when these assumptions are violated, the results of our analyses in this chapter are quite *robust,* that is, we will not often decide that a finding is statistically significant when in fact it is not, and vice versa.[1]

8.1 An Example Using Regression and Correlation Techniques: Municipal Scope

When American cities began to experience fiscal crises in the 1970s, as happened in New York City and Cleveland, interest was stimulated in city government operations. Roland Liebert investi-

1. For more discussion of these issues, see J. Gaito, "Measurement Scales and Statistics: Resurgence of an Old Misconception," *Psychological Bulletin,* Vol. 87 (1980) pp. 564–67, and G. W. Bohrnstedt and T. M. Carter, "Robustness in Regression Analysis," in H. L. Costner (Ed.), *Sociological Methodology: 1971* (San Francisco: Jossey-Bass, 1971).

gated the changing functions of city governments and presented a comprehensive explanation of *municipal scope,* which he defined as the range of activities or services undertaken by a city government.[2] The concept is concerned with the number of distinctive functions a city government performs, usually in specialized departments and bureaus. In various cities this scope varies from a few to many tasks, and it has an important influence on the government's ability to avoid financial problems.

Liebert cited three major trends to account for the various levels of municipal scope to be found in American cities today. The first is that a city government is shaped by the forces of local tradition, political culture, and institutional forms. Cities emerging at different periods are more prone to adopt a certain mode of operation and persist in it. Second, growing national complexity and interdependence have weakened local autonomy and control over programs. Increasingly, various functions are being dispersed among federal, county, and special-district governments. Third, growing urban populations and increased pressures on local government to provide more services have splintered metropolitan governments into many specialized jurisdictions, particularly as suburbanization has taken hold.

These hypotheses led Liebert to specific empirical indicators in his research. Municipal scope was measured as the number of functions performed by a city government from a list of nine: welfare (federal aid), judicial, hospitals, education, welfare (general assistance), health, parks and recreation, sanitation, and sewers. He found that in the 668 cities of 25,000 or more population in 1960, the average number of functions performed was 3.7 (median and mode were both 4), with the range from zero (Joliet, Ill.) to 9 (New York City). Among the hypothesized historical causes of municipal scope were the city's age (the time since it first reached 10,000 population), city population size, and suburban status. Both older and larger cities were found to perform comparatively more functions, while suburbs performed fewer functions than central cities did.

Liebert's analyses of municipal scope used regression and correlation techniques similar to those described in this chapter.

---

2. Roland Liebert, *Disintegration and Political Action: The Changing Function of City Governments in America* (New York: Academic Press, 1976).

Our main illustrations replicate a portion of his analysis using the 63-cities data. We will use the 1960 measures provided by Liebert,[3] although a more realistic approach would be to remeasure them in 1970 to match the period during which population and age were measured for the 63 cities.

Following Liebert, we can state two propositions about the relationship of municipal government to city characteristics:

P1: The older a city, the greater its municipal scope.

P2: The larger a city, the greater its municipal scope.

By substituting the operational measures for these concepts, we arrive at two testable hypotheses:

H1: The more years since a city has exceeded 10,000 population, the larger the number of functions it performs.

H2: The larger a city's 1970 population, the more functions it performs.

8.1.1 Descriptive Statistics for Municipal Scope

For the 63 cities the mean age as dated by U.S. census year prior to 1976 was 88.1 years, with a standard deviation of 39.6 years. Mean population in 1970 was 58,754, with a standard deviation of 111,449. For the 62 cities with municipal function scores (Washington, D.C., was omitted, since it was not an autonomous city in 1960), the mean was 4.66, and standard deviation was 1.83. This higher average for municipal scope among the 62 cities compared to the 668 cities Liebert studied is probably due to the inclusion of many smaller cities in his study.

A new notational convention for designating variables is now necessary. The dependent variable in an analysis such as the number of city functions is usually designated with a Y, while the independent variable is usually labeled X. The reason will become apparent in the next section, where joint frequency plots on Cartesian coordinates are discussed. If more than one independent variable is investigated, the variables are distinguished by numerical subscripts. Thus, in this example, city age is X_1 and city

3. Liebert, *Disintegration and Political Action,* pp. 189–203.

population size is X_2. These letter designations are useful in writing shorthand formulas for the mathematical manipulation of the data. If an individual observation—a person or a city—is to be designated, a second subscript may be added to stand for the case number. In general formulas, *the subscript i is used to reference an individual case.* Thus X_{1i} stands for the scores of the ith case on variable X_1, and X_{2i} stands for the score of the ith case on variable X_2.

When we estimate the relationship between the dependent variable, Y, and an independent variable, X, we speak of *regressing* Y on X (i.e., we regress the dependent variable on the independent variables).

8.2 Scatterplots and Regression Lines

When both variables are continuous measures, displaying their relationship requires different techniques than those that have already been described. The box-and-whisker diagram is unsuitable because often very few observations have the same score on the independent variable. Combining scores of the independent variable in order to create broader categories for display purposes is one solution, but this would destroy much of the continuous character of the measures.

scatterplot—a type of diagram that displays the covariation of two continuous variables as a set of points on a Cartesian coordinate system

An alternative way to present the data visually is with a scatterplot. To construct a **scatterplot,** a set of Cartesian axes is drawn with the independent variable on the horizontal *(X)* axis and the dependent variable on the vertical *(Y)* axis. Then the position of each observation in this two-dimensional space is designated by a point corresponding to its X and Y values. If a strong relationship is present between the pair of variables, it should be evident from the scatter of points on the figure.

Figure 8.1 graphs the joint distribution of the outcomes for city age and number of functions (data for this figure are given in Table 8.1 in the next section). A clear trend is evident from lower left to upper right, suggesting that low values, like high values, tend to cluster together. A few outlying points are labeled. This figure is thus a visual indication that age and function covary positively.

As useful as a scatterplot is, we need a more precise statement

FIGURE 8.1
Scatterplot of City Age and Functions Variables

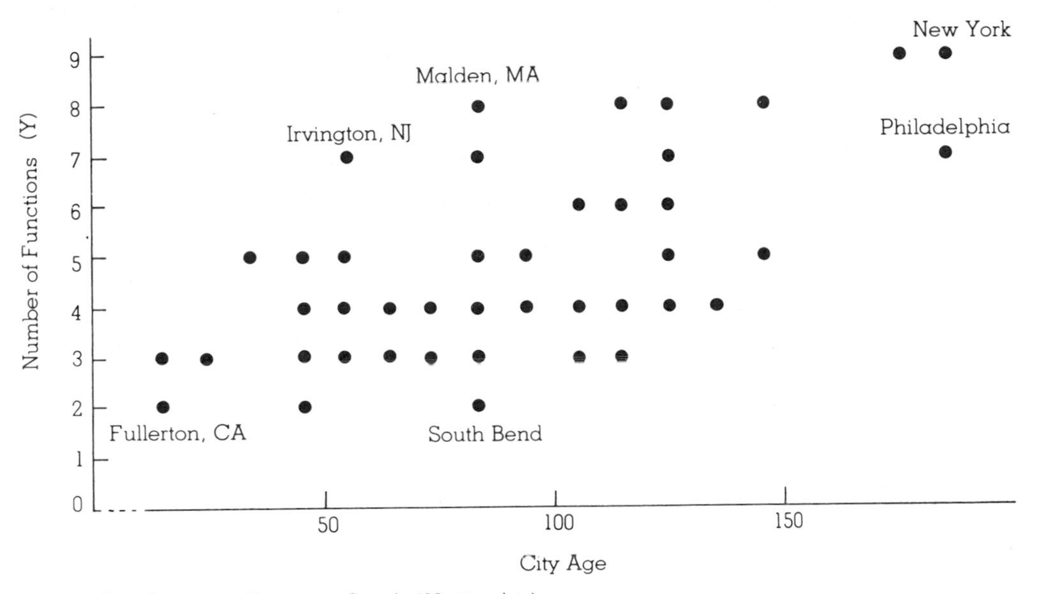

Source: NORC Permanent Community Sample (63-cities data).

of the relationship between the variables. One possibility is to measure the degree to which the data approximate a **linear relationship**. That is, if the relationship between city age and function is perfectly linear, all the data points would fall onto a straight line, with a greater or lesser slope to the axes. We assume a linear relationship rather than an exponential or more general polynomial function simply because a linear relationship is among the most elemental of all functions. Since *parsimony,* or economy in the use of means to an end, is one of the major goals of science, asking whether a linear function can account for the relationship makes sense. If a straight line is not adequate to account for the data, then more complex functional forms will have to be considered.

linear relationship—covariation in which the value of the dependent variable is proportional to the value of the independent variable

In this example, the empirical relationship falls considerably short of perfect linearity, but it does not depart so far from linearity as to form a random scatter pattern. The scatterplot does *not* suggest that a curvilinear function would fit the data better. Therefore we will describe a technique for measuring how well a straight line describes the tendency for the two continuous variables to covary.

In high school algebra you learned how a straight line is written for Cartesian coordinates:

$$Y = a + bX$$

The ordinate, or Y value, is the sum of a constant, a, which is the point at which the line crosses the Y axis, plus the product of the slope of the line, b, times the X value. For a line to be drawn through the scatter of points, the line must come closer, on the average, to all 62 points than any other line that could be drawn.

The equation $Y = a + bX$ assumes that the Ys are exact functions of the Xs. But the model should allow for errors in the prediction of Y from X. Hence the **regression model** employed is:

regression model—an equation for the linear relationship between a continuous dependent variable and one or more independent variables, plus an error term

$$Y_i = a + bX_i + e_i$$

for the ith observation. The term e_i takes into account that the ith observation is not accounted for perfectly by the **prediction equation,** which is:

prediction equation—a regression equation without the error term, useful for predicting the score on the dependent variable from the independent variable(s)

$$\hat{Y}_i = a + bX_i$$

For this reason, e_i here is called the *error* (or *residual*) *term,* a concept introduced in Chapter 7.

This equation, in which the values of a and b are estimated from the scores on Y and X for all observations, is also called the **regression line.** It has the unique property that, given any value for X—whether or not such a value occurs for any actual case—we can predict Y_i, on the assumption that the relationship between X and Y is linear. This predicted score of Y_i, written $\hat{Y}_i$, can be compared with the actual score, Y_i, for any individual i to see how well the relationship holds and to determine the error term:

regression line—a line that is the best fit to the points in a scatterplot, computed by ordinary least squares regression

$$e_i = Y_i - \hat{Y}_i$$

The next section develops the necessary tools to construct a regression equation for fitting a line through the scatterplot.

8.3 Linear Regression Equations

In the municipal scope example, the procedure for estimating the regression line must make use of all the information on both variables for each city (the Y value for the X value observed for each community). The estimates of a and b should have the property that the sum of the squared differences between the observed Y_i and the score predicted by the regression equation $\hat{Y}_i$ is a minimum sum compared to the quantity obtained using any other line. That is:

$$\sum_{i=1}^{N}(Y_i - \hat{Y}_i)^2 = \sum_{i=1}^{N} e_i^2$$

should be a minimum.

This estimation criterion is called the least-squares error sum, or, more simply, a and b are said to be estimated using **ordinary least squares** (OLS). Note the similarity between the OLS regression criterion and the mean. As Chapter 3 showed, the mean has the desirable property of minimizing the sum of the squared differences for a set of scores. Similarly, the regression line equation minimizes the sum of squared errors of prediction.

ordinary least squares—a method for obtaining estimates of regression equation coefficients that minimizes the error sum of squares

One useful interpretation of linear regression is that it estimates a **conditional mean** for Y. That is, *regression shows the expected value of Y* (see Section 5.2.1 of Chapter 5) *for a given value of the independent variable, X_i, on the assumption that the relationship between the variables is linear.* If no relationship exists between the variables, the regression slope, b, will be zero, and the expected value, $\hat{Y}_i$, for any level of X_i will simply be the intercept, a. And the value of a will be the mean score of Y for all observations (i.e., $\bar{Y}$). If Y and X are linearly related to each other, we will find that the conditional mean of Y differs systematically according to the level of X. Furthermore, the conditional means will fall on a straight line when plotted.

conditional mean—the expected average score on the dependent variable, Y, for a given value of the independent variable, X

BOX 8.1 Deriving Estimators of *a* and *b*

The basic principle behind ordinary least squares estimation (OLS) calls for finding values for *a* and *b* which make the sum of squared differences between observed and predicted *Y* scores, $(Y_i - \hat{Y}_i)^2$, as small as possible. If we replace $\hat{Y}_i$ with the regression estimate, we are trying to minimize $\Sigma(a + bX_i - Y_i)^2$. According to fundamental rules of calculus, any quadratic equation attains a minimal value at the point where the derivative is equal to zero. Hence, by taking partial derivatives of the expressions for *a* and for *b* and setting them equal to zero, we have:

$$2\Sigma(a + bX_i - Y_i) = 0$$

$$2\Sigma X_i(a + bX_i - Y_i) = 0$$

Simplifying and rearranging gives the usual form of two normal equations for a straight line:

$$Na + b\Sigma X_i = \Sigma Y_i$$

$$a\Sigma X_i + b\Sigma X_i^2 = \Sigma X_i Y_i$$

Two equations in two unknown quantities can always be solved for unique estimates of the unknowns. Dividing through the first normal equation by *N* and rearranging gives:

$$a = \overline{Y} - b\overline{X}$$

This shows that *the least-squares line passes through the point of the means,* $\overline{X}$ and $\overline{Y}$.

To find the formula for *b,* we begin with the definition of the error in prediction, using the regression equation:

$$e_i = Y_i - \hat{Y}_i$$

Since $\hat{Y}_i = a + bX_i$ and $a = \overline{Y} - b\overline{X}$, substituting $\overline{Y} - b\overline{X}$ for *a* gives:

$$\hat{Y}_i = \overline{Y} + b(X_i - \overline{X})$$

It then follows that:

Box 8.1 (continued)

$$e_i = (Y_i - \hat{Y}_i) = Y_i - [\overline{Y} + b(X_i - \overline{X})]$$
$$= (Y_i - \overline{Y}) - b(X_i - \overline{X})$$

Since the OLS estimator for b must minimize the sum of these squared error terms, we first create the sums and then take the derivative of them with respect to b and set it equal to zero. That is, we form the sums of squared values:

$$\Sigma e_i^2 = \Sigma(Y_i - \hat{Y}_i)^2 = \Sigma[(Y_i - \overline{Y}) - b(X_i - \overline{X})]^2$$

and then take the derivative of the expression on the right with respect to b:

$$-2\Sigma[(X_i - \overline{X})[(Y_i - \overline{Y}) - b(X_i - \overline{X})]] = 0$$

This can be rearranged:

$$b = \frac{\Sigma(X_i - \overline{X})(Y_i - \overline{Y})}{\Sigma(X_i - \overline{X})^2}$$

which is the value shown in the text.

Empirical estimates of b, the **bivariate regression coefficient,** or the slope of the regression line, are obtained from the observed pairs of X and Y scores by applying the following formula:

$$b = \frac{\Sigma(X_i - \overline{X})(Y_i - \overline{Y})}{\Sigma(X_i - \overline{X})^2}$$

(Box 8.1 shows how this OLS estimate is derived using elementary calculus and **normal equations.**) The numerator is the sum of the product of the deviations of the Xs and Ys about their respective means. When divided by $N - 1$ in the sample (or N in the population), this term is the **covariance**, labeled s_{XY}. That is,

bivariate regression coefficient—a parameter estimate of a bivariate regression equation that measures the amount of increase or decrease in the dependent variable for a one-unit difference in the independent variable

normal equations—algebraic equations used in the estimation of linear regression coefficient values

covariance—the sum of the product of deviations of the Xs and Ys about their respective means, divided by $N - 1$ in the sample and N in the population

$$s_{XY} = \frac{\Sigma(X_i - \overline{X})(Y_i - \overline{Y})}{N - 1}$$

The denominator of the regression coefficient is the sum of the squared deviations of the independent variable, X, about its mean. If divided by the $N - 1$, the term would be the *variance,* a statistic you first encountered in Chapter 3:

$$s_X^2 = \frac{\Sigma(X_i - \overline{X})^2}{N - 1}$$

Note that in the formula for b, the $N - 1$ in the numerator has canceled with the $N - 1$ in the denominator. Hence, the slope of the regression line (i.e., the regression coefficient) is estimated by the ratio of the covariance between Y and X to the variance of X. This is:

$$b = \frac{s_{YX}}{s_X^2}$$

intercept—a constant value in a regression equation showing the point at which the regression line crosses the Y axis when values of X equal zero

Although the **intercept** *(a)* is not derived here, it can be easily estimated by:

$$a = \overline{Y} - b\overline{X}$$

Note that if the regression slope, b, is zero, the second term drops out. Then our estimate of the intercept is the mean of Y. Thus when Y and X are unrelated, the best-fitting regression line will be parallel to the X axis, passing through the Y axis at the sample mean. Knowing a specific value of X will not yield a predicted value of Y different from the mean.

8.3.1. Linear Regression Applied to Municipal Scope

To show how the linear regression formulas are applied to data, Table 8.1 gives the estimation of the covariance between city age (X_1) and government function (Y), along with the calculation of

TABLE 8.1

Calculation of Covariance, Means, and Variances for City Age (X_1) and Government Function (Y)

(1) Observation No.	(2) Y	(3) X_1	(4) $(Y - \bar{Y})$	(5) $(X_1 - \bar{X}_1)$	(6) $(Y - \bar{Y})(X_1 - \bar{X}_1)$	(7) $(Y - \bar{Y})^2$	(8) $(X_1 - \bar{X}_1)^2$
1	4	86	−0.66	−1.13	0.75	0.44	1.28
2	5	146	0.34	58.87	20.02	0.12	8,465.68
3	4	46	−0.66	−41.13	27.15	0.44	1,691.68
4	4	106	−0.66	18.87	−12.45	0.44	356.08
5	4	66	−0.66	−21.13	13.95	0.44	446.48
6	4	86	−0.66	−1.13	0.75	0.44	1.28
7	3	16	−1.66	−71.13	118.08	2.76	5,059.48
8	9	176	4.34	88.87	385.70	18.84	7,897.88
9	4	126	−0.66	38.87	−25.65	0.44	1,510.88
10	8	116	3.34	28.87	96.43	11.16	833.48
11	3	66	−1.66	−21.13	35.08	2.76	446.48
12	5	56	0.34	−31.13	−10.58	0.12	969.08
13	3	86	−1.66	−1.13	1.88	2.76	1.28
14	3	26	−1.66	−61.13	101.48	2.76	3,736.88
15	4	86	−0.66	−1.13	0.75	0.44	1.28
16	2	16	−2.66	−71.13	189.21	7.08	5,059.48
17	3	56	−1.66	−31.13	51.68	2.76	969.08
18	3	76	1.66	−11.13	18.48	2.76	123.88
19	3	66	−1.66	−21.13	35.08	2.76	446.48
20	3	106	−1.66	18.87	−31.32	2.76	356.08
21	7	56	2.34	−31.13	−72.84	5.48	969.08
22	4	76	−0.66	−11.13	7.35	0.44	123.88
23	4	56	−0.66	−31.13	20.55	0.44	969.08
24	8	86	3.34	−1.13	−3.77	11.16	1.28
25	6	116	1.34	28.87	38.69	1.80	833.48
26	7	126	2.34	28.87	67.56	5.48	833.48
27	5	126	0.34	38.87	13.22	0.12	1,510.88
28	5	96	0.34	8.87	3.02	0.12	78.68
29	7	126	2.34	38.87	90.96	5.48	1,510.88
30	3	26	−1.66	−61.13	101.48	2.76	3,736.88
31	4	66	−0.66	−21.13	13.95	0.44	446.48
32	3	56	−1.66	−31.13	51.68	2.76	969.08
33	4	136	−0.66	48.87	−32.25	0.44	2,388.28
34	8	126	3.34	38.87	129.83	11.16	1,510.88
35	6	106	1.34	18.87	25.29	1.80	356.08
36	5	46	0.34	−41.13	−13.98	0.12	1,691.68
37	4	96	−0.66	8.87	−5.85	0.44	78.68
38	8	116	3.34	28.87	96.43	11.16	833.48
39	2	46	−2.66	−41.13	109.41	7.08	1,691.68
40	4	76	−0.66	−11.13	7.35	0.34	123.88
41	3	46	−1.66	−41.13	68.28	2.76	1,691.68
42	3	76	−1.66	−11.13	18.48	2.76	123.88
43	4	86	−0.66	−1.13	0.75	0.44	1.28
44	2	86	−2.66	−1.13	3.01	7.08	1.28
45	4	66	−0.66	−21.13	13.95	0.34	446.48
46	5	36	0.34	−51.13	−17.38	0.12	2,614.28
47	4	116	−0.66	28.87	−19.05	0.44	833.48
48	4	76	−0.66	−11.13	7.35	0.44	123.88
49	3	16	−1.66	−71.13	118.08	2.76	5,059.48
50	7	86	2.34	−1.13	−2.64	5.48	1.28

(Continued next page)

TABLE 8.1 (continued)

(1) Observation No.	(2) Y	(3) X_1	(4) $(Y - \bar{Y})$	(5) $(X_1 - \bar{X}_1)$	(6) $(Y - \bar{Y})(X_1 - \bar{X}_1)$	(7) $(Y - \bar{Y})^2$	(8) $(X_1 - \bar{X}_1)^2$
51	3	46	−1.66	−41.13	68.28	2.76	1,691.68
52	9	186	4.34	98.87	429.10	18.84	9,775.28
53	6	126	1.34	38.87	52.09	1.80	1,510.88
54	5	86	0.34	−1.13	−0.38	0.12	1.28
55	7	186	2.34	98.87	231.36	5.48	9,775.28
56	7	126	2.34	38.87	90.96	5.48	1,510.88
57	4	86	−0.66	−1.13	0.75	0.44	1.28
58	8	146	3.34	58.87	196.63	11.16	3,465.68
59	4	86	−0.66	−1.13	0.75	0.44	1.28
60	4	66	−0.66	−21.13	13.95	0.44	446.48
61	5	96	0.34	8.87	3.02	0.12	98.68
62	3	116	−1.66	28.87	−47.92	2.76	833.48
Sums	289	5,402			2,893.97	203.89	94,021.13

$$\bar{Y} = 289/62 = 4.6613$$
$$\bar{X}_1 = 5{,}402/62 = 87.1290$$
$$s_Y^2 = 203.89/61 = 3.3425$$
$$s_{X_1}^2 = 94{,}021.13/61 = 1{,}541.3300$$
$$s_{XY} = 2{,}893.97/61 = 47.4421$$

their means and variances. Columns 2 and 3 show the pairs of observed values for each of the 62 cities. The sums, means, and variances for the two variables are shown at the bottom of the table. In columns 4 and 5, the means ($\bar{Y} = 4.66$ and $\bar{X}_1 = 87.13$) have been subtracted from each observed score. In column 6, these deviations have been multiplied together. Note the preponderance of positive products, indicating the tendency for *pairs* of scores to be either *both* below their means or *both* above their means. Only 14 cities have scores where one value is above the mean and the other below.

This condition, of course, reflects the previously noted tendency for the data points in the scatterplot to vary systematically in a positive direction. The sums entry in column 6 shows that the sum of the products is 2,893.97. Dividing by $N - 1 = 61$ gives a covariance, $s_{YX} = 47.44$. The variance of the 62 scores on city age is 1,541.33. Dividing this value into the covariance gives the ordinary least-squares estimate of the regression coefficient, $b = (47.44/1{,}541.33) = 0.0308$.

This coefficient has a simple interpretation: The *bivariate re-*

gression coefficient measures the amount of increase or decrease in the dependent variable for a one-unit difference in the independent variable. In this example, therefore, every year of city age is associated with a three-hundredths increase in the number of functions performed on average. Two cities which differ in age by 32.47 years are thus likely to differ in the number of municipal tasks they undertake, with the older city on average performing one more function than the younger. A hundred-year difference in city ages would yield a prediction of 3.08 more functions performed by the older city.

To complete the calculation of the bivariate OLS regression equation, we need to estimate the intercept term. Since mean city age is $\overline{X}_1 = 87.13$ and mean government functions is $\overline{Y} = 4.66$, the estimated OLS value is $a = 4.66 - (0.0308)87.13 = 1.976$. Thus, the prediction equation may be written:

$$\hat{Y}_i = 1.976 + 0.308X_{1_i}$$

We will soon show how to test for whether this relationship is statistically significant or not. First, however, we will show the regression line for the equation graphed onto its scatterplot, as in Figure 8.2.

In this figure, the regression line through the scatterplot intercepts the Y axis at 1.976, and it has a positive 0.0308 slope. While the line passes through or close to a few data points, most observations fall off the line, about as many above as below. Note that the regression line passes exactly through the point which is the mean of both Y and X. This relationship follows from the formula for calculation of the intercept, a.

Figure 8.2 can also be used to illustrate the *residuals,* or *errors in prediction.* As we noted in Section 8.2, a residual, or error, is the difference between an observed value and the value predicted by the regression line; that is, $e_i = Y_i - \hat{Y}_i$. Two such residuals, for Irvington, N.J., and Albany, N.Y., are calculated and shown in the figure. With an age of 56 years, Irvington is expected to perform 3.70 functions, but it actually performs seven. Thus the regression relationship underestimates this particular value of Y by 3.30. On the other hand, the equation overestimates 146-year-old Albany's number of functions at 6.47, when it actually undertakes only 5.0 tasks. Our ability to calculate the error of prediction in each case provides a measure of the degree of linear association between continuous variables, as explained in the next section.

FIGURE 8.2
Scatterplot of City Age and Functions Variables, with Regression Line

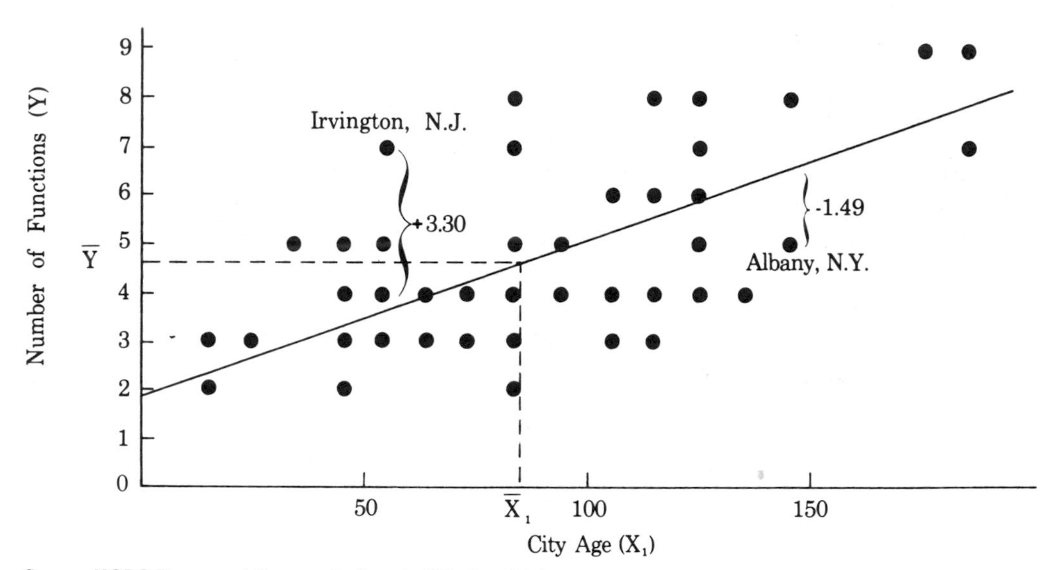

Source: NORC Permanent Community Sample (63-cities data).

8.4 Measures of Association: The Coefficient of Determination and Correlation Coefficient

Linear regression analysis is not limited to estimating an equation showing the precise quantitative relationship of two continuous variables. It can also be used to determine the strength of association between the pair. One measure of strength of association is the answer to the question: How close do the observed data points lie to the regression line? As pointed out several times above, in the ideal case all observations would fall exactly on the regression line, and we would have the ability to predict the Y scores without error on the basis of our information about the X values. But usually the observed scores will be due to a *systematic component* (predicted by regression) and to an *error component* (random, and hence not predicted from regression). That is, we can partition the total sum of squares much as we did in ANOVA.

FIGURE 8.3
Method of Accounting for an Observation by the Regression Line and an Error Component

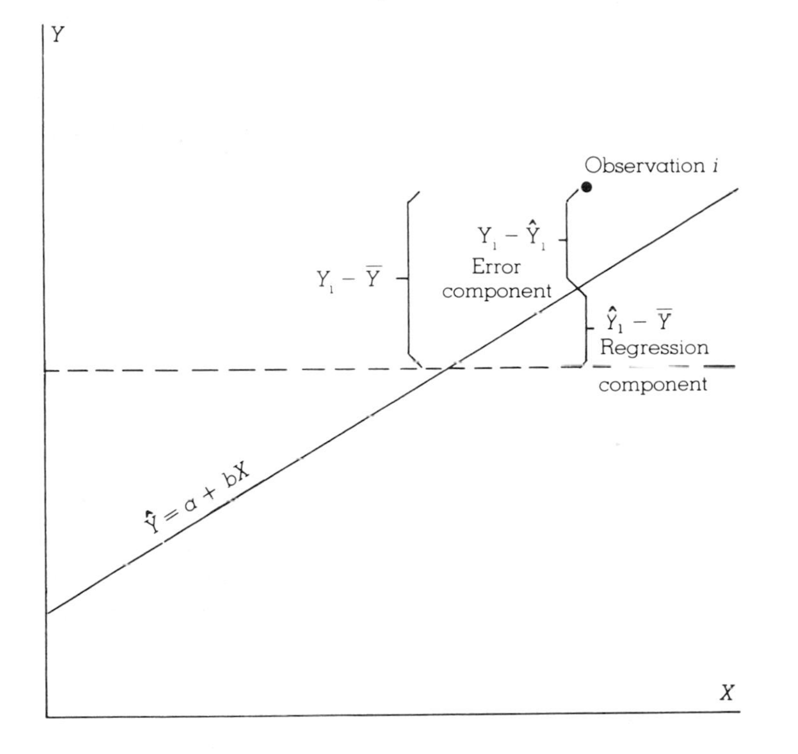

If we begin with the observation Y_i, which has been centered about its mean (i.e., $Y_i - \overline{Y}$), the following identity can be constructed:

$$(Y_i - \overline{Y}) = (Y_i - \hat{Y}_i) + (\hat{Y}_i - \overline{Y})$$

Thus, an observation can be seen to be a function of two components. The first component, $Y_i - \hat{Y}_i$, indicates the discrepancy between an observation and the predicted value for that observation. This discrepancy is e_i. The second component, or

$\hat{Y}_i - \overline{Y}$, is that part of the observed score that can be accounted for by the regression line. This can be seen by examining Figure 8.3, where a single observation deviated from the mean is shown as a function of both a regression component and an error component.

regression sum of squares—a number obtained in linear regression by subtracting the mean of a set of scores from the value predicted by linear regression, squaring, and summing these values

error sum of squares—a numerical value obtained in linear regression by subtracting the regression sum of squares from the total sum of squares

If we square and sum the components shown above, they are called the **regression sum of squares** and the **error sum of squares.** If we rearrange terms, it turns out that:

$$\Sigma(Y_i - \overline{Y})^2 = \Sigma(\hat{Y}_i - \overline{Y})^2 + \Sigma(Y_i - \hat{Y}_i)^2$$

or

$$SS_{\text{TOTAL}} = SS_{\text{REGRESSION}} + SS_{\text{ERROR}}$$

This formulation is obviously very similar to that used in ANOVA, in which, as described in Chapter 7, $SS_{\text{TOTAL}} = SS_{\text{BETWEEN}} + SS_{\text{WITHIN}}$. (The term SS_{ERROR} used in regression is analagous to SS_{WITHIN} used in ANOVA.) The difference, however, is an important one. In ANOVA we make no assumption about the form of the relationship between the independent and dependent variable. We only seek to know how much variation in SS_{TOTAL} can be explained as a function of the various outcomes of the independent variable, that is, $Y = f(X)$, where the nature of $f(X)$ is not specified. But in regression analysis we assume that the relationship between the independent and dependent variables can be described by a straight line, $Y = f(X)$, where $f(X) = a + bX$. Therefore ANOVA can be thought of as a more general model than regression analysis (not better, just more general). When we use regression analysis we make the assumption that Y and X are related by a straight line.

What is the possible *maximum error of prediction* we might make using a regression model? If we had no information about the regression equation, the best we could do is to predict that $\hat{Y}_i$ equals the sample mean of $\overline{Y}$ for every score. That is, the sum of squared errors,

$$\Sigma(Y_i - \hat{Y}_i)^2$$

becomes

$$\Sigma(Y_i - \overline{Y})^2$$

on substituting the mean for $\hat{Y}_i$. As we noted in Section 3.2.5 in Chapter 3, $\sum_{i=1}^{N}(Y_i - \overline{Y})^2$ is the numerator of the variance of Y, or s_Y^2. These two items, the sum of squared errors for regression ($SS_{REGRESSION}$) and the sum of squared errors about the mean (SS_{TOTAL}), give us all the information we need to construct a **proportional reduction in error,** or PRE measure of association for linear regression. All PRE measures take on this general form:

proportional reduction in error—a characteristic of some measures of association that allows the calculation of reduction in errors in predicting the dependent variable, given knowledge of its relationship to an independent variable

$$\text{PRE statistic} = \frac{\text{Error without decision rule} - \text{Error with decision rule}}{\text{Error without decision rule}}$$

The linear regression equation can be considered one rule for making decisions (predictions) about expected Y values (i.e., means conditional on the X values). If any relationship exists between Y and X, the linear prediction will result in a smaller error of prediction than relying only on guessing the mean. Therefore, as the degree of predictability from X to Y is greater, the error is proportionately smaller.

8.4.1. The Coefficient of Determination

The specific PRE statistic for linear regression is called the **coefficient of determination,** because it is the proportion of total variation in Y "determined" by X. Its symbol is $R^2_{Y \cdot X}$ (read "R squared"). Substituting the two prediction components into the general PRE formula yields:

coefficient of determination—a PRE statistic for linear regression that expresses the amount of variation in the dependent variable explained or accounted for by the independent variable(s) in a regression equation

$$R^2_{Y \cdot X} = \frac{\Sigma(Y_i - \overline{Y})^2 - \Sigma(Y_i - \hat{Y}_i)^2}{\Sigma(Y_i - \overline{Y})^2}$$

or

$$R^2_{Y \cdot X} = \frac{SS_{TOTAL} - SS_{ERROR}}{SS_{TOTAL}}$$

$$= 1 - \frac{SS_{ERROR}}{SS_{TOTAL}}$$

That is, the coefficient of determination, $R^2_{Y \cdot X}$, is 1 minus the ratio of error sum squares to the total sum of squares. Note that when SS_{ERROR} is zero, $R^2_{Y \cdot X} = 1.0$. But when $SS_{\text{ERROR}} = SS_{\text{TOTAL}}$ (i.e., all the variation is error variance), $R^2_{Y \cdot X} = 0.0$. Also note that since $SS_{\text{REGRESSION}} = SS_{\text{TOTAL}} - SS_{\text{ERROR}}$, it follows that:

$$R^2_{Y \cdot X} = \frac{SS_{\text{REGRESSION}}}{SS_{\text{TOTAL}}}$$

(Section 7.9 established that $\eta^2 = SS_{\text{BETWEEN}}/SS_{\text{TOTAL}}$.) The coefficient of determination, therefore, equals the proportion of the total sum of squares accounted for by fitting a least squares regression line to the data.

To use these formulas to calculate the empirical values of $R^2_{Y \cdot X}$ involves a lot of tedious subtracting, squaring, summing, and dividing. An alternative approach which gives the same result makes use of values that have already been estimated in arriving at the regression coefficients. This formula states that the coefficient of determination equals the ratio of the square of the covariance between X and Y to the product of the variances of each variable:

$$R^2_{Y \cdot X} = \frac{s^2_{XY}}{s^2_X s^2_Y}$$

For the city age–government function example, Table 8.1 above gives the values for the covariance (47.4421), variance of Y (3.3425), and the variance of X (1,541.3300). In this case:

$$R^2_{Y \cdot X} = \frac{(47.4421)^2}{(1{,}541.3300)\ (3.3425)} = 0.4369$$

Since the largest value $R^2_{Y \cdot X}$ can reach is 1.00 and the minimum value is zero, *the coefficient of determination can be interpreted as the proportion or percentage of total variation in Y which can be attributed to its linear relationship with X.* In the example above, 43.7% of the variation in city government functions can be "explained" (in a statistical rather than a causal sense) by the age of the city. A bivariate association of this magnitude is quite large by nonexperimental social science standards. Indeed, the association among the 62 cities is larger than Liebert's finding for all 668 cities, where $R^2_{Y.X}$ was 0.31.

The quantity $1 - R^2_{Y.X}$ is called the **coefficient of nondetermination.** In this example this coefficient is $1 - 0.4369 = 0.5631$, or 56.3% of the variance left *un*explained.

coefficient of nondetermination—a statistic that expresses the amount of variation in a dependent variable that is left *un*explained by the independent variable(s) in a regression equation

8.4.2. The Correlation Coefficient

In social science research the linear relationship between two continuous measures often is presented not as $R^2_{Y.X}$ but as its square root. This is called the Pearson product-moment **correlation coefficient,** after the famous statistician Karl Pearson. It is given by:

correlation coefficient—a measure of association between two continuous variables that estimates the direction and strength of linear relationship

$$r_{XY} = \sqrt{R^2_{Y.X}}$$

Pearson's r_{XY} has a positive or negative sign attached to it to indicate the direction of the covariation. This sign must agree with the sign of the regression coefficient *(b).* In the municipal function example, the correlation is +0.66 between city function and city age. Unlike $R^2_{Y.X}$, r_{XY} can range between −1.00 for a perfect inverse association to +1.00 for a perfect positive covariation, with $r_{XY} = 0$ indicating no relationship. The usefulness of the correlation coefficient lies in its communication of directionality as well as magnitude of association, unlike $R^2_{Y.X}$, which conceals whether the variables are directly or inversely related.

From the definition of $R^2_{Y.X}$ in Section 8.4.1, it follows that $r_{XY} = r_{YX}$, since:

$$r_{XY} = \frac{s_{XY}}{s_X s_Y}$$

$$= \frac{\Sigma(X_i - \overline{X})(Y_i - \overline{Y})/(N - 1)}{s_X s_Y}$$

$$= \frac{\Sigma(Y_i - \overline{Y})(X_i - \overline{X})/(N - 1)}{s_Y s_X}$$

$$= \frac{s_{YX}}{s_Y s_X}$$

$$= r_{YX}$$

To get this result you only have to recall that $s_{XY} = \Sigma(X_i - \overline{X})(Y_i - \overline{Y})/(N - 1)$ and recall that the products $(X_i - \overline{X})(Y_i - \overline{Y})$ and $s_X s_Y$ equal $(Y_i - \overline{Y})(X_i - \overline{X})$ and $s_Y s_X$, respectively. (This is the same as $ab = ba$ in general.)

Although $r_{XY} = r_{YX}$, *it is not the case that the b from regressing Y on X is the same as the b from regressing X on Y.* Can you prove this assertion?

8.4.3. Correlating Z Scores

When we correlate two variables that have first been converted to Z scores, some interesting results emerge. The standard deviation of a Z score is always 1, as we noted in Section 3.3 in Chapter 3. Since, in general, $r_{XY} = s_{XY}/(s_X s_Y)$, for two Z scores it follows that:

$$r_{Z_X Z_Y} = \frac{s_{Z_X Z_Y}}{s_{Z_X} s_{Z_Y}} = \frac{s_{Z_X Z_Y}}{(1)(1)} = s_{Z_X Z_Y}$$

That is, the correlation of two Z scores equals the covariance of two Z scores.

Furthermore, since $s_{Z_X Z_Y} = \Sigma(Z_X - \overline{Z}_X)(Z_Y - \overline{Z}_Y)/(N - 1)$ and $\overline{Z}_X = \overline{Z}_Y = 0$ (i.e., the mean of a Z score is zero), it follows that:

$$r_{Z_X Z_Y} = s_{Z_X Z_Y} = \frac{\Sigma Z_X Z_Y}{N - 1}$$

The correlation between two standardized variables, Z_X and Z_Y, is the sum of the product of the two variables divided by $N - 1$. (In the population, this correlation is the sum of products divided by N instead of $N - 1$.) This result will be especially useful in Chapter 12.

Since $Z_X = (X - \overline{X})/s_X$ and $Z_Y = (Y - \overline{Y})/s_Y$, it follows that:

$$
\begin{aligned}
r_{Z_X Z_Y} &= \frac{\Sigma Z_X Z_Y}{N - 1} \\
&= \frac{\Sigma[(X - \overline{X})/s_X]\,[(Y - \overline{Y})/s_Y]}{N - 1} \\
&= \frac{\Sigma(X - \overline{X})\,(Y - \overline{Y})/(N - 1)}{s_X s_Y} \\
&= \frac{s_{XY}}{s_X s_Y} \\
&= r_{XY}
\end{aligned}
$$

The correlation between two Z-transformed variables, Z_X and Z_Y, is the same as the correlation between the original variables, X and Y. That is, the correlation coefficient is unaffected by whether the variables are Z scores or not.

8.4.4. The Relation of Regression and Correlation Coefficients

The relationship between a regression coefficient and a correlation coefficient is easy to show for the two-variable case. First, we identify b as the ratio of covariance to variance:

$$b = \frac{s_{XY}}{s_X^2}$$

Next, we note that $r_{XY} = s_{XY}/(s_X s_Y)$. Now we multiply r_{XY} by the ratio of the standard deviation of Y to the standard deviation of X. We obtain:

$$r_{XY}\left(\frac{s_Y}{s_X}\right) = \left(\frac{s_{XY}}{s_X s_Y}\right)\left(\frac{s_Y}{s_X}\right) = \frac{s_{XY}}{s_X^2}$$

But, since this last value is equal to b:

$$b = r_{XY}\left(\frac{s_Y}{s_X}\right)$$

Conversely,

$$r_{XY} = b\left(\frac{s_X}{s_Y}\right)$$

Therefore, if we know b, s_X, and s_Y, we can easily determine r_{XY}. Or if we know r_{XY}, s_X, and s_Y, we can easily determine b.

8.5 Standardized Regression, or Beta, Coefficients

Regression coefficients provide a clear interpretation of the linear relationship between two variables with unambiguous units of measurement. However, they can be difficult to interpret when the metric of the variables is not clear. Many variables in the social sciences that researchers regard as continuous have no natural or agreed-on unit of measure. Examples of such variables include industrialization, religiosity, alienation, and socioeconomic status. Because of this lack of a natural unit of measure, many researchers prefer to convert their variables to Z scores prior to carrying out their analyses.

Z scores always have a mean of zero as well as a standard deviation of 1, and, as we showed in the preceding section, in general $r_{XY} = b(s_X/s_Y)$. But if we standardize both X and Y by converting them to Z scores, s_X and s_Y both equal unity (1.0), and the regression coefficient for the standardized scores equals the correlation coefficient, r_{XY}.

beta coefficient—a standardized regression coefficient indicating the amount of net change, in standard deviation units, of the dependent variable for an independent variable change of 1 standard deviation

The regression coefficient for standardized variables is commonly called the **beta coefficient,** or *beta weight,* for which β^* is the symbol.[4] As the preceding section suggests, for bivariate regression analysis:

---

4. We have starred (*) the coefficient beta (β) to differentiate it from the population parameter β, which will be introduced in the next section.

$$\beta^* = r_{XY}$$

Since Z scores convert variables to standard deviation units, the interpretation of β^* is as follows: *For a standard deviation difference in X, the predicted difference in Y is β^* standard deviations.* In the municipal function example, the correlation coefficient of 0.66 between function and city age indicates that, whereas each year an average of $b = 0.031$ more functions were performed, a 1 standard deviation difference on age (which is 39.26 years) produces an expected difference of 0.66 standard deviation in city function (or (.66)(1.828) = 1.21 functions).

Since $a = \overline{Y} - b\overline{X}$ for raw, unstandardized data, and since the mean of a standardized variable is zero, the intercept of the regression equation when both X and Y are converted to Z scores is zero. This is true because $a = 0 - b(0) = 0$. The prediction equation for two standardized variables, Z_X and Z_Y, is:

$$\begin{aligned}\hat{Z}_Y &= \beta^* Z_X \\ &= r_{XY} Z_X\end{aligned}$$

8.5.1. Regression toward the Mean

The standardized form of the prediction equation is useful for understanding how the term *regression* came to be applied to this form of analysis. Suppose we are interested in the relationship between mothers' and daughters' weights. We sample 500 mothers, match them to their oldest daughters, and weigh both pairs. We find that $\overline{X} = \overline{Y} = 135$; $s_X = s_Y = 15$; and $r_{XY} = 0.70$, where X and Y refer to mothers' and daughters' weights, respectively.

Now if we standardize the two variables, our prediction is:

$$\hat{Z}_Y = 0.70 Z_X$$

Notice that mothers whose weight is 1 standard deviation above the mean (i.e., $Z_X = 1$) are predicted to have daughters whose weights on the average are only 0.70 standard deviation above the mean. In contrast, mothers whose weights place them 1 standard deviation below the mean ($Z_X = -1$) are predicted on average to have daughters who are only -0.70 standard deviations below the mean. That is, a daughter's weight demonstrates **regression toward the mean;** or, on the average, daughters are predicted to be closer to the mean in weight than their mothers were.

regression toward the mean—a condition demonstrated when the predicted scores on the dependent variable show less variability about the mean than the observed scores do, due to the imperfect correlation between two variables

This regression can be seen clearly by examining Table 8.2, which gives selected, observed and predicted values for mothers' and daughters' weights (both standardized and unstandardized). On average, mothers who are above the mean have daughters who weigh less than they do, and mothers who are below the mean have daughters who are heavier than they are. This seems to imply that the population is becoming more homogeneous with respect to weight in the daughters' generation. Indeed there is clearly less variability in the *predicted* weight of the daughters than in the observed weight of the mothers, as the second and fourth columns in Table 8.2 show. But the fact that the predicted

TABLE 8.2

Illustration of Regression Toward the Mean, Using the Prediction of Daughter's Weight from Mother's Weight

Standardized Weight		*Unstandardized Weight*	
Observed Mother's Weight	*Predicted Daughter's Weight*	*Observed Mother's Weight*	*Predicted Daughter's Weight*
−3	−2.1	180	166.5
−2	−1.4	165	156.0
−1	−0.7	150	145.5
0	0	135	135.0
+1	+.7	120	124.5
+2	+1.4	105	114.0
+3	+2.1	90	103.5
Prediction Equation		*Prediction Equation*	
$\hat{Z}_Y = 0.7Z_Y$		$\hat{Y} = 12.5 + 0.7X$	

scores are closer to the mean and have less variance should *not* lead you to an erroneous conclusion. Since we found in our sample that s_X and s_Y were identical, there is as much variance in the *observed* weight of the daughters as in the observed weight of the mothers.

Imperfect correlation causes regression toward the mean. Notice that if two variables are perfectly correlated, $\hat{Z}_Y = (1.00)Z_X$, then no regression toward the mean will occur. But if $r_{XY} = 0.2$, the amount of regression will be substantial.

These examples reaffirm that the weaker the relationship between two variables, the more the mean becomes the best predicted outcome of Y for any outcome of X. And the smaller the correlation between X and Y, the smaller the variance of $\hat{Y}$ (the *predicted* dependent variable).

We state without proof that:

$$s^2_{\hat{Z}_y} = r^2_{XY}$$

As the relationship between X and Y approaches zero, the variance of $\hat{Y}$ approaches zero as well, thereby accounting for the *apparent* homogenization of observations.

The ability to standardize regression coefficients is especially useful when trying to compare magnitudes of relationships among variables measured on different scales. The utility of Pearson's correlation coefficient as a workhorse in social statistics will become apparent when we consider multivariate regression in Chapter 11.

8.6 Significance Tests for Regression and Correlation

Like other descriptive statistics, regression and correlation coefficients are estimated on data sampled from a larger population of potential observations. The inferential question to be answered is: What is the probability that, given the observed sample values for the descriptive statistics, the population parameter being estimated is zero? In other words, what inferential statistics can be calculated to test the statistical significance of the regression and correlation coefficients?

The t test, introduced in Chapters 5 and 6, and the F test, introduced in Chapter 7, are the basic inferential statistics for testing the significance of relationships among continuous data. In this section we will show how these tests can be applied to $R^2_{Y.X}$, b, a, and r_{XY}.

8.6.1. Testing the Significance of the Coefficient of Determination

To test $R^2_{Y.X}$ for statistical significance, we will use an F test very similar to that used in the analysis of variance in Chapter 7.

We know from Section 7.5 in Chapter 7 that SS_{TOTAL} has $N - 1$ degrees of freedom associated with it. And we know that $SS_{\text{REGRESSION}}$, defined in Section 8.4, is estimated from a single function of the X_is, namely b, and hence it has *1 degree of freedom associated with it.* Now, since:

$$df_{\text{TOTAL}} = df_{\text{REGRESSION}} + df_{\text{ERROR}}$$

and

$$N - 1 = 1 + df_{\text{ERROR}}$$

It straightforwardly follows that $df_{\text{ERROR}} = N - 2$.

mean square regression—a value in linear regression obtained by dividing the regression sum of squares by its degrees of freedom

mean square error—a value in linear regression obtained by dividing the error sum of squares by its degrees of freedom

We can construct the **mean square regression** and **mean square error** by dividing the appropriate sums of squares by their associated degrees of freedom, in much the same way as we did for mean square within and mean square between in Chapter 7. That is:

$$MS_{\text{REGRESSION}} = \frac{SS_{\text{REGRESSION}}}{1}$$

$$MS_{\text{ERROR}} = \frac{SS_{\text{ERROR}}}{N - 2}$$

Where $\rho^2_{Y.X}$ (Greek letter *rho*) refers to the population coefficient of determination, it can be proven that if the null hypothesis, $H_0: \rho^2_{Y.X} = 0$, is true, then both $MS_{\text{REGRESSION}}$ and MS_{ERROR} are unbiased estimates of σ^2_e, the population variance of the errors of prediction (i.e., the e_i). If, however, $\rho^2_{Y.X} > 0$ in the population, we would also expect $MS_{\text{REGRESSION}} > MS_{\text{ERROR}}$ in the sample. Since an F ratio is simply the ratio of two estimates of the same variance, σ^2_e in this case, it follows that we can test the null hypothesis that $\rho^2_{Y.X} = 0$ in the population by choosing an α level and calculating:

$$F_{1,N-2} = \frac{MS_{\text{REGRESSION}}}{MS_{\text{ERROR}}}$$

If the obtained value of F is as large or larger than the critical value for a given α level, the obtained $R^2_{Y.X}$ is significantly greater than zero. If the obtained F is not larger, we cannot reject the hypothesis that $\rho^2_{Y.X} = 0$ in the population.

There are several ways to compute $SS_{\text{REGRESSION}}$. The most straightforward one follows from knowing that $R^2_{Y.X} = SS_{\text{REGRESSION}} / SS_{\text{TOTAL}}$, as we showed in Section 8.4.1. From this it follows directly that:

$$SS_{\text{REGRESSION}} = R^2_{Y.X} SS_{\text{TOTAL}}$$

Note that, from the definition given in Chapter 3:

$$s^2_Y = \Sigma \frac{(Y_i - \overline{Y})^2}{N - 1}$$

and the expression on page 246:

$$SS_{\text{TOTAL}} = \Sigma(Y_i - \overline{Y})^2$$

Therefore it follows that:

$$SS_{\text{TOTAL}} = s^2_Y (N - 1)$$

We can easily use this information to deduce SS_{ERROR}. Since $SS_{\text{TOTAL}} = SS_{\text{REGRESSION}} + SS_{\text{ERROR}}$, it follows that:

$$SS_{\text{ERROR}} = SS_{\text{TOTAL}} - SS_{\text{REGRESSION}}$$

In the municipal scope example, as Table 8.1 shows, $s^2_Y = 3.3425$. We know that $R^2_{Y.X} = 0.4369$ and $N = 62$. Therefore:

$$SS_{\text{TOTAL}} = (3.3425)(61) = 203.893$$

and

$$SS_{\text{REGRESSION}} = (0.4369)(203.89) = 89.080$$

Therefore:

$$SS_{\text{ERROR}} = 203.890 - 89.080 = 114.810$$

Furthermore,

$$MS_{\text{REGRESSION}} = \frac{89.080}{1} = 89.080$$

and

$$MS_{\text{ERROR}} = \frac{114.810}{62 - 2} = 1.914$$

If we fix $\alpha = .05$, we see from Appendix E that the critical value for an F with 1 and 60 degrees of freedom is 4.00. But our test statistic is $F_{1,60} = 89.080/1.914 = 46.541$. Hence we can easily reject the null hypothesis that $\rho^2_{Y.X} = 0$ in the population.

Because $SS_{\text{REGRESSION}}$ involves the multiplication of $R^2_{Y.X}$ and s^2_Y, it is very important that these quantities not be rounded to fewer than three significant (nonzero) digits. Even greater accuracy is desirable. This need for accuracy is a problem only when doing examples by hand, since the computer stores these quantities many places beyond those that can be done practically with hand computation.

Once we have finished these computations we can summarize them in an ANOVA table, such as Table 8.3.

TABLE 8.3
ANOVA Summary Table for Municipal Scope

Source	*df*	*SS*	*MS*	*F*
Regression	1	89.080	89.080	46.541*
Error	60	114.810	1.914	
Total	61	203.890		

*Significant for $\alpha = .05$.

8.6.2. Testing the Significance of b and a

The conventional representation of a sample statistic is to use italic English letters, while Greek letters stand for the population parameters. Hence, the bivariate regression equation for sample data is:

$$\hat{Y} = a + bX$$

while the **population regression equation** is written:

population regression equation—a regression equation for a population rather than a sample

$$\hat{Y} = \alpha + \beta X$$

where α and β are population parameters to be estimated by a and b in the sample data.

Do not confuse β with the beta coefficient (β^*) described above. Also do not confuse these alphas and betas with the same symbols used to designate probability levels. This need for many symbols to play double and triple duty is an unfortunate aspect of the standard statistical repertoire. You will have to accommodate to the practice and keep alert to the context in which each symbol is used.

The null hypothesis asserts that the population parameter is zero: $H_0: \beta = 0$. As with many other statistics drawn from large populations, a sample regression coefficient, b, has a known sampling distribution. To determine whether b is significantly different from zero, we will construct a t test. This is the same procedure we used for testing whether a single mean equals some population value. In testing the null hypothesis in bivariate regression:

$$t = \frac{b - \beta}{s_b} = \frac{b - 0}{s_b}$$

Where:

s_b = The sample estimate of σ_b.

To be able to construct this ratio we need the sample estimate of β, namely b, and an estimate of the standard error of b, or s_b. The *standard error,* a concept introduced in Chapter 5, is the standard deviation of the sampling distribution of b.

If we assume that, in the population being sampled, Y is normally distributed for every outcome of X, and we further assume that the variance of the errors in prediction is the same for every outcome of X (i.e., the variance exhibits homoscedasticity), then the sampling distribution of b will be normally distributed as N gets large. The mean of the sampling distribution will equal β, the true population regression coefficient, with:

$$\sigma_b^2 = \frac{\sigma_e^2}{\Sigma(X_i - \overline{X})^2}$$

It may not be obvious, but these results are due to the central limit theorem, which was used in Chapter 5 to derive the sampling distribution for the mean.

As an estimate of the variance of errors in prediction, σ_e^2, we can simply use MS_{ERROR} from the significance test for $R^2_{Y \cdot X}$:

$$\hat{\sigma}_e = \sqrt{MS_{\text{ERROR}}}$$

There are $N - 2$ degrees of freedom associated with this t test, since those are the df associated with MS_{ERROR}. With these facts established we now construct a t ratio:

$$t_{N-2} = \frac{b - \beta}{\sqrt{\dfrac{MS_{\text{ERROR}}}{\Sigma(X_i - \overline{X})^2}}}$$

$$= \frac{b - 0}{\sqrt{\dfrac{MS_{\text{ERROR}}}{s_X^2(N - 1)}}}$$

In the municipal scope example,

$$t_{(62-2)} = \frac{0.0308}{\sqrt{\dfrac{1.914}{(1{,}541.330)(61)}}}$$

$$= \frac{0.0308}{0.0045} = 6.826$$

For $\alpha = .05$ and $df = 60$, the critical value given in Appendix D is 2.00. Hence we reject the hypothesis that $\beta = 0$ in the population. We can conclude, with little chance of being wrong, that in the population of cities from which our sample was taken, the age of the city is positively related to the number of functions the government performs. In similar fashion, you can show yourself that the regression of function on city size is significant.

The intercept, *a*, can also be tested for statistical significance with a *t* test. It is:

$$t_{N-2} = \frac{a - \alpha}{\sqrt{\dfrac{MS_{\text{ERROR}}}{N}}}$$

In this example, for $\alpha = .05$, $a = 1.976$, and $df = 60$, we can test the hypothesis that $\alpha = 0$ in the population as follows:

$$t_{60} = \frac{1.976-0}{\sqrt{\dfrac{1.914}{62}}} = 11.246$$

This result is highly significant, since the critical value is 2.00, as it was for testing *b* for significance.

Researchers do not ordinarily test for the significance of the intercept. But many computer programs, including the Statistical Package in the Social Sciences (SPSS), routinely compute the test, so presenting it here is important.

8.6.3. The Relationship between F and t^2

In Chapter 7 we showed that an F with 1 and ν degrees of freedom equals a t^2 with ν degrees of freedom. In the bivariate case, the F test associated with testing whether $\rho^2_{Y \cdot X} = 0$ in the population is equal to the *square of the t test* for testing whether $\beta^2_{Y \cdot X} = 0$ in the population. Recall that $F_{1,60} = 46.541$ (see Table 8.3) in testing whether $\rho^2_{Y \cdot X} = 0$. This is identical (save for rounding error) to $t^2_{60} = 6.826^2 = 46.594$ for testing whether β equals zero. Therefore, in bivariate regression, one of the two tests will be redundant. As we will show in Chapter 11, however, *the tests of significance for the coefficient of determination and the regression coefficients differ when Y is regressed on two or more independent variables.*

8.6.4. Confidence Intervals

The standard error of b, s_b, can be used to construct a confidence interval around the sample estimate, in a fashion similar to the confidence interval constructed around the sample mean in Chapters 5 and 6. Again, we decide on a probability level, this time selecting the corresponding t value for a two-tailed test (since the interval will be symmetrical around the observed value). If we pick $\alpha = .05$, the critical value for $df = 60$ is exactly 2.00. Then the upper and lower limits for the 95% confidence interval are $b + s_b(2.00)$ and $b - s_b(2.00)$.

In the municipal scope example, $0.0308 + 0.0045(2.00) = 0.040$ and $0.031 - 0.0045(2.00) = 0.022$. Therefore the confidence interval ranges from .022 to .040. Remember that the correct interpretation of a confidence interval is *not* that the population parameter has a 95% chance of being inside the interval (it either is in the interval or it is not). Rather, *with repeated samples, the confidence interval will contain the population parameter, β, 95% of the time.*

If we had a much larger sample of cases, the t ratio for the $\alpha = .05$ interval would be 1.96 (i.e., the Z score for a two-tailed test using the normal distribution). Since this value is very close to 2.00, a statistical rule of thumb has been developed which asserts that if a regression coefficient is twice its standard error, the b is significant at the $\alpha = .05$ level. In general, the upper and lower confidence intervals for a given α-level are $b \pm s_b(\text{c.v.})$.

8.6.5. Testing the Significance of the Correlation Coefficient

Since the correlation coefficient, r_{XY}, is the square root of the coefficient of determination, $R^2_{Y \cdot X}$, in the bivariate case, it is a foregone conclusion that the observed correlation of 0.66 between government function and city age will be statistically significant. The test of significance is presented here in anticipation of a later development in which the coefficient of determination will *not* be the same as r_{XY}.

To test the null hypothesis, $H_0: \rho_{XY} = 0$, we use a transformation of r to Z which is approximately normal. This allows us to use the Z-score table in Appendix C to determine the probability of observing a given r_{XY} under H_0. The ***r*-to-*Z* transformation,** which was developed by the celebrated English statistician R. A. Fisher, is a function of the natural logarithm:

r-to-Z transformation—a natural logarithm transformation in the value of the correlation coefficient to a Z score, to test the probability of observing r under the null hypothesis

$$Z = \left(\frac{1}{2}\right) \ln \left(\frac{1 + r_{XY}}{1 - r_{XY}}\right)$$

Modern pocket calculators with a natural log key make such transformations effortless. In case you do not have this function on your calculator, an r-to-Z table in Appendix F gives values for all possible correlation coefficients.

The variance of Z is a function of the sample size:

$$\hat{\sigma}^2_Z = \frac{1}{N - 3}$$

The test statistic is then:

$$Z = \frac{Z_r - Z_{\rho_0}}{\hat{\sigma}_Z}$$

If the hypothesized population value of the correlation is zero, then $Z_{\rho_0} = 0$. Since we already know that $r_{XY} = 0.66$ is significantly different from zero, we can test a different null hypothesis. What is the probability that $\rho_{XY} = 0.70$, given that the sample correlation is 0.66 and $N = 62$? The r-to-Z transformations are .7928 and .8673, respectively, while the $\hat{\sigma}^2_Z$ is .0169. Hence the ratio, distributed approximately as a Z score following the normal distribution, is:

$$\frac{.7928 - .8673}{\sqrt{.0169}} = \frac{-.0745}{.1300} = -.573$$

This Z value falls considerably short of the critical value necessary to reject the hypothesis, even at $\alpha = .05$ using a one-tailed test. Therefore, we conclude that the population correlation could easily be 0.70.

8.7. A Problem with an Outlier: Testing the Second Hypothesis on Municipal Scope

According to the second hypothesis about municipal scope presented in Section 8.1, city size and number of municipal functions covary in a positive direction. The summary data to test this hypothesis are shown in Table 8.4 and in scatterplot form in Figure 8.4. We use the data for all 62 cities and regress the number of city functions on city size. When New York is included, this yields:

$$\hat{Y} = 4.271 + 0.0006X_2$$

We use the data from Table 8.4 to calculate:

$$R^2_{Y \cdot X_2} = \frac{s^2_{YX_2}}{s^2_Y s^2_{X_2}} = \frac{(824.663)^2}{(3.342)(1{,}242{,}090.858)} = 0.1638$$

Then we take the square root and see that $r_{YX_2} = 0.4047$.

To test the regression coefficient for statistical significance, we use the variance of the number of city functions, $s^2_Y = 3.342$, along with the coefficient of determination, $R^2_{Y \cdot X_2} = 0.1638$. Now we find that:

$$SS_{\text{TOTAL}} = s^2_Y(N - 1) = 3.342(61) = 203.862$$

$$SS_{\text{REGRESSION}} = R^2_{YX_2} SS_{\text{TOTAL}} = (.1638)(203.862) = 33.393$$

Therefore:

$$SS_{\text{ERROR}} = 203.862 - 33.393 = 170.469$$

$$MS_{\text{REGRESSION}} = \frac{33.393}{1} = 33.393$$

$$MS_{\text{ERROR}} = \frac{170.469}{60} = 2.841$$

Now we set $\alpha = .05$, and from Appendix E we find that the critical value is 4.00 for 1 and 60 degrees of freedom. The test statistic is:

$$F_{1,60} = \frac{33.393}{2.841} = 11.754$$

Thus the observed R^2 is statistically significant.

The computation of this example has been straightforward, but we have been leading you down the primrose path in order to show the dangers of a mechanical approach to data analysis. Notice New York in Figure 8.4. It appears to almost be a "deviant" observation, since it is far from any other. This distance is due to the fact that it is far larger than any other city in the data list, and it performs more city functions than any other. *The covariance, correlation, and regression coefficient are all sensitive to such* **outliers**.

outliers—extreme values of observed variables in a scatterplot that can distort estimates of regression coefficients

This sensitivity can be illustrated by recalling that the covariance is a weighted sum of products. These products are formed by deviating joint observations from their respective means and then multiplying these deviations together. This product contributes substantially to the covariance, since both city size and number of functions in New York City are far above their respective means. *Unfortunately, however, outliers of this sort can give a false sense of ability to predict a dependent variable.*

TABLE 8.4

Regression of Municipal Function on City Size, Including and Excluding New York City

A. New York City Included (N = 62)

Variable	*Mean*	*Standard Deviation*	*Variance*
Function	4.661	1.828	3.342
City size (000)	587.548	1,114.491	1,242,090.858

$r_{YX_2} = 0.40474 \quad R^2_{Y \cdot X_2} = 0.1638 \quad s_{YX_2} = 824.663$

$\hat{Y} = 4.271 + 0.0006X_2$

B. New York City Not Included (N = 61)

Variable	*Mean*	*Standard Deviation*	*Variance*
Function	4.590	1.755	3.056
City size (000)	469.667	610.577	372,804.151

$r_{YX_2} = 0.2865 \quad R^2_{Y \cdot X_2} = 0.0821 \quad s_{YX_2} = 305.821$

$\hat{Y} = 4.205 + 0.00082X_2$

FIGURE 8.4
Scatterplot of Number of Functions and City Size, with Regression Lines Including and Excluding New York City

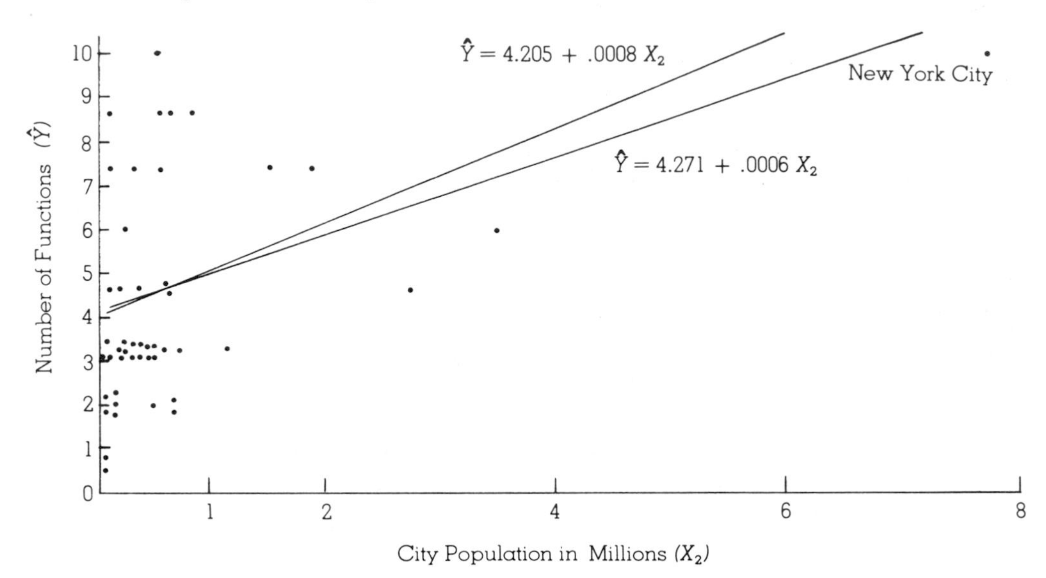

Source: NORC Permanent Community Sample (63-cities data).

To show the influence of outliers, Table 8.4 also gives the summary statistics for the other 61 cities when New York City is excluded. Notice that the covariance between municipal function and city size has been substantially reduced. Furthermore, the variance of city size is much smaller. What is more important,

$$R^2_{\hat{Y}\cdot X_2} = \frac{(305.821)^2}{(3.056)(372{,}804.151)} = 0.0821$$

and therefore

$$r_{YX_2} = \sqrt{0.0821} = 0.2865$$

While both r_{YX_2} and $R^2_{\hat{Y}\cdot X_2}$ are significant for $\alpha = .05$, when New York City is omitted the latter, or the coefficient of determination (0.0821) is almost *half* the size it is when New York is included (0.1638). That is, *half of the explained variance using all 62 cities is due to the inclusion of a single observation!* For this reason when the results of the test of H2 are reported New York City should be eliminated from the analyses, noting why. While the hypothesis of a positive relationship between city size and municipal function remains supported, the strength of the relationship is substantially less than we originally calculated.

As an exercise, you can estimate the regression equation and test it for statistical significance without New York included in the data set.

While it is difficult to provide a precise definition of an outlier, the effects of doing analyses without first examining the scatterplot are substantial, as we just saw.

> In doing a regression or correlation analysis, it is very important to examine the scattergram for outliers. If you see one or more observations which appear substantially deviant from the others, eliminate it or them from the analyses. If the results then are substantially different, the analyses based on the reduced data set should be reported, and a note explaining why the observations were deleted should be provided. If in doubt, both sets of analyses can be reported.

8.8. Nonlinear Regression

Regression and correlation refer to *linear* relationships between continuous variables. On the basis of theory or empirical evidence, a nonlinear relationship might be expected, however. Basic algebra suggests a number of nonlinear functional relationships between Y and X values which might be found: quadratic (Y is a function of X^2), logarithmic (Y is a function of the log of X), power (Y is a function of X to the b power), reciprocal (Y is a function of $1/X$), and various polynominal relationships (Y is a function of X^2, X^3, X^4, etc.). Some examples of the various **nonlinear regression** lines are illustrated in Figure 8.5.

nonlinear regression—regression analysis performed on variables that do not bear a linear relationship to one another

In principle, the solutions to nonlinear regression problems are simple. We merely transform or "recode" the raw scores to their new values before performing the regression or correlation. These estimated coefficients then must be interpreted in terms of the effects of *transformed* independent variables on the dependent measures. Computer packages such as SPSS frequently have transformation functions to make possible rapid change of observed values to logs, square roots, powers, and the like.

FIGURE 8.5
Examples of Nonlinear Regressions

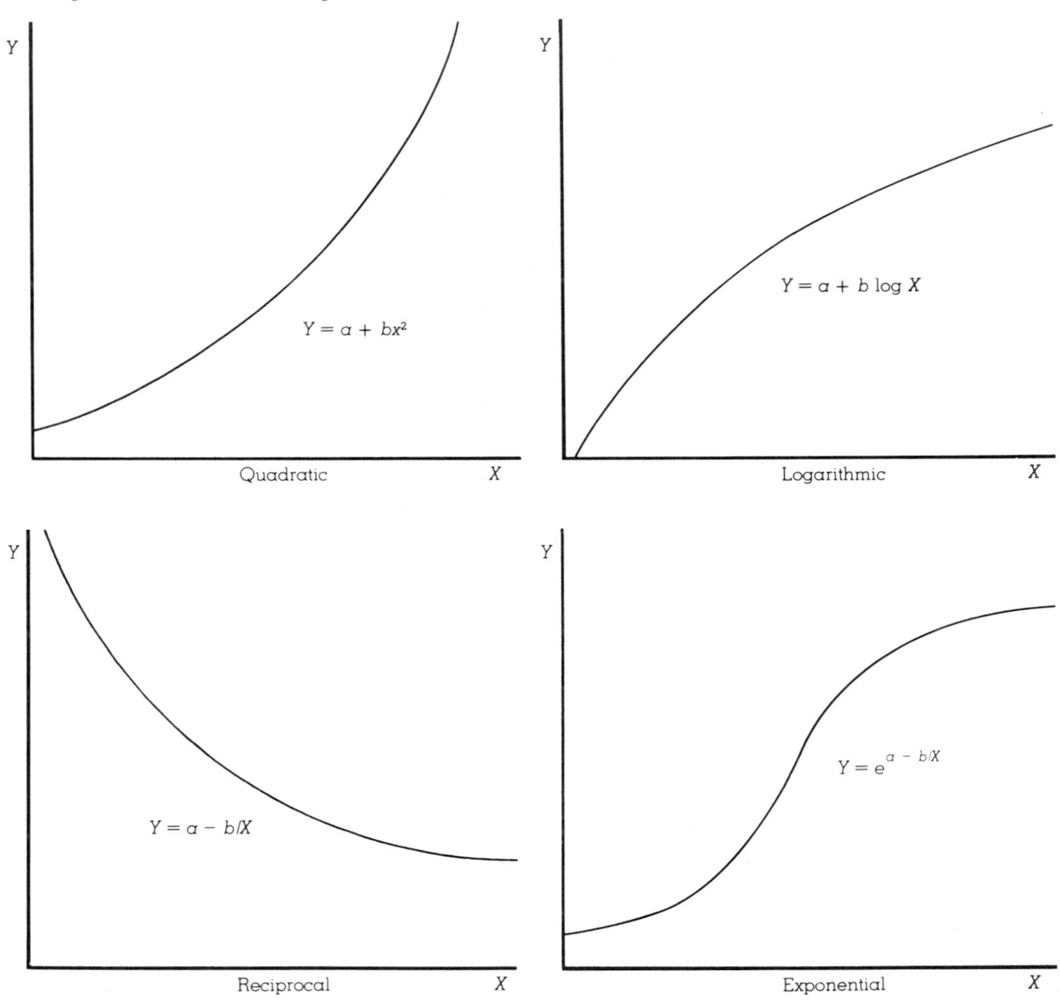

8.8.1. Testing for Curvilinearity

In bivariate regression analysis, the researcher assumes that two variables are linearly related, but this assumption may or may not be met in practice. There is a statistical procedure available for testing whether or not two variables are linearly related. This test uses information with which you should be already be familiar. It involves the following steps:

1. Collapse the independent variable into a set of categories which do not seriously distort its original distribution.
2. Do a one-way analysis of variance (ANOVA), using the categories established in step 1 as treatment conditions.
3. Compute $\hat{\eta}^2$—the measure of variance in the dependent variable explained by the independent variable.
4. Regress the dependent variable on the original independent variable, treating it as an continuous variable. Then compute $R^2_{Y.X}$—the measure of vari-

FIGURE 8.6
Scatterplot of Antiabortion Attitude and Education Variables

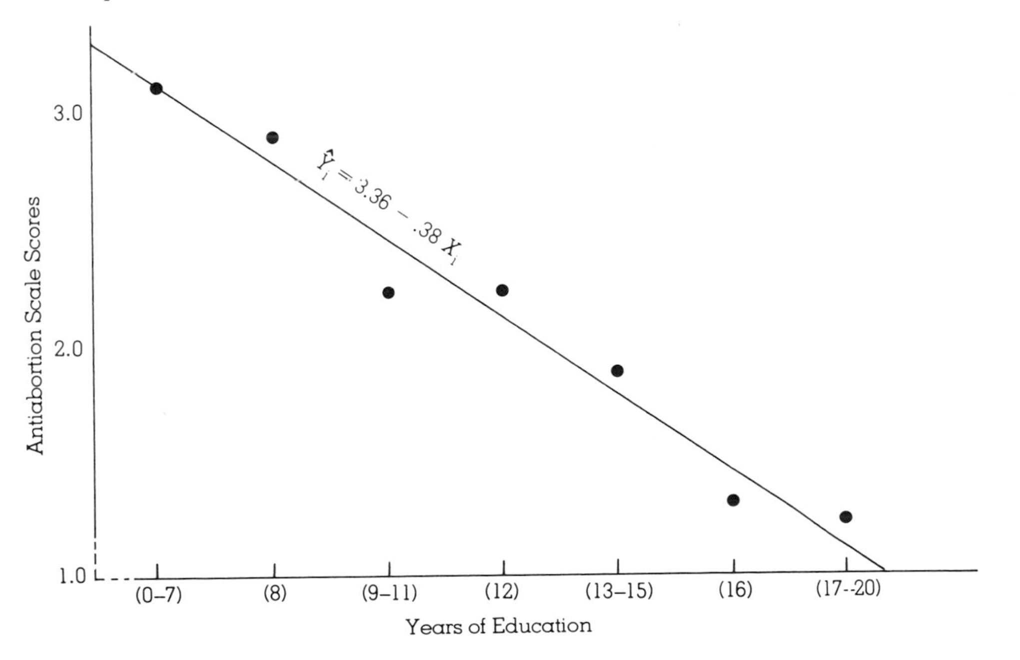

Source: 1978 General Social Survey.

ance explained by assuming the variables are *linearly* related.

5. Test whether $\hat{\eta}^2$ is significantly larger than $R^2_{Y.X}$. If it is, reject the hypothesis that the variables are linearly related.

(Subprogram BREAKDOWN in SPSS will do all of these steps for you in one pass.)

To illustrate, in Figure 8.6 we use the 1978 GSS data and plot the means on an antiabortion scale for seven educational categories, treating education as a discrete rather than continuous variable. The antiabortion scale provides a count of the number of times respondents say they disagree that a woman should be able to have a legal abortion under six different conditions: if the baby would be born defective; if the woman were poor; if the woman had been raped; if the woman wanted no more children; if the woman were single; and if the woman's health were in danger.

The scatterplot in Figure 8.6 shows the means on this scale for each educational level. Notice that the means decline with educational level, indicating that better educated respondents are on average less likely to be against abortion, under many conditions. The category means fall very close to the regression line, $\overline{Y}_i = 3.36 - 0.32X_i$. While the means fall close to a linear pattern, considerable within-category variation about these conditional means also exists. The coefficient of determination is 0.0547, revealing that just 5.47% of the variation in the abortion scale can be accounted for by assuming a linear relationship with education. If we relax the linear restriction and compute a one-way ANOVA, we find that η^2 increases to only 5.85% variance explained, while requiring an additional five degrees of freedom.

A statistical test for determining whether the null hypothesis, H_0: $\eta^2 = \rho^2_{Y.X}$, can be rejected is given by:

$$F_{k-2,N-k} = \frac{(\eta^2 - R^2_{Y.X})/(k-2)}{(1-\eta^2)/(N-k)}$$

Where:

k = The number of categories associated with the independent variable on which the ANOVA is performed.

FIGURE 8.7
Scatterplot of Television Viewing and Education Variables

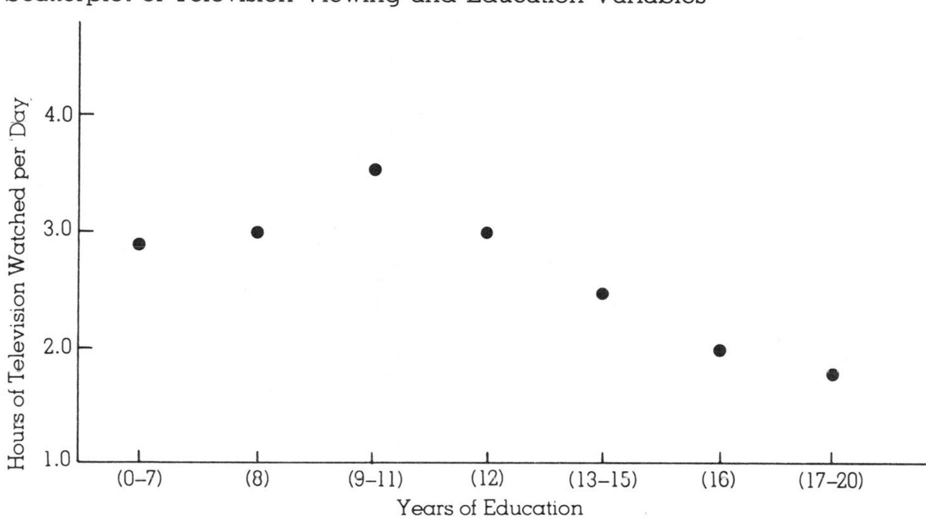

Source: 1978 General Social Survey.

The F test is distributed with $(k - 2)$ and $(N - k)$ *df.* Setting $\alpha = .05$ and carrying out this test for the data in Figure 8.6, we find:

$$F_{(7-2),(1356-7)} = \frac{(0.0585 - 0.547)/(7 - 2)}{(1 - 0.0585)/(1{,}356 - 7)} = 1.09$$

Since the critical value of F necessary to reject the hypothesis of linearity in favor of nonlinearity is 2.21 (see Appendix E), we conclude that the covariation of antiabortion attitude with education is linear.

Another example is illustrated in Figure 8.7, where we have plotted the number of hours of television watched each day by 1978 GSS respondents in the seven educational levels. Several departures from linearity are obvious, most notably the high level of TV viewing by those with some high school education. Linear regression yields $R^2_{Y.X} = 0.0368$, while $\hat{\eta}^2 = 0.0548$. The test for increase in variance explained is:

$$F_{(7-2),\,(1519-7)} = \frac{(0.0548 - 0.0368)/(7 - 2)}{(1 - 0.0548)/(1{,}519 - 7)} = 5.76$$

This is sufficient to reject a linear relationship at the $\alpha = .01$ level of significance. Why TV viewing is highest among middle-educated groups is a topic for additional research.

8.9. Special Cases of Inference Involving Correlations

Correlations can be used in several special cases to generalize or draw conclusions about the attributes of a population on the basis of the evidence provided by a sample. Two of the tests in which they are used—to find the difference between two means in nonindependent populations and to find the difference between two correlations in independent populations—are described in this section.

8.9.1. Testing the Difference between Two Means in Nonindependent Populations

The procedures for testing the difference between two means described in Chapter 6 were based on means drawn from independent populations. Sometimes, however, the assumption that the observations are independent is unreasonable. The researcher may be interested, for example, in change in some variable across time in the same population, as we noted in the last section of Chapter 6. In this case it is not possible to assume independent observations, since the *same* persons are measured at two points in time.

With **nonindependent samples**, the appropriate test statistic is as follows:

nonindependent samples—samples drawn according to procedures in which the choice of one observation for a sample determines the probability of another observation being chosen for a different sample

$$t_{(N-1)} = \frac{\bar{Y}_2 - \bar{Y}_1}{\sqrt{\dfrac{s^2_{Y_1}}{N} + \dfrac{s^2_{Y_2}}{N} - \dfrac{2r_{Y_1Y_2}s_{Y_1}s_{Y_2}}{N}}}$$

where Y_1 and Y_2 are the same variables measured on the same subjects at times 1 and 2, respectively.

We could not present this material in Chapter 6, since the standard error of the sampling distribution assumes knowledge of the correlation coefficient, $r_{Y_1 Y_2}$, as well as the variances of the two variables.

In 1964 one of the authors collected data on incoming freshmen at the University of Wisconsin. We can use these data to determine whether changes occurred between the freshman and senior years on several variables, including religiosity. In this survey, religiosity was an index (see Section 11.1 for more on indexes) composed of six items to which respondents agreed or disagreed, on a four-point scale. Examples of items were "There is an almighty God who watches over us," and "There is a life after death." For a sample of $N = 2{,}506$ (both males and females), for the freshman year the mean ($\overline{Y}_1$) was 11.30 and standard deviation (s_{Y_1}) was 4.38, whereas for the senior year the values were $\overline{Y}_2 = 8.39$ and $s_{Y_2} = 4.66$. Furthermore, the correlation of religiosity from freshman to senior years was $r_{Y_1 Y_2} = 0.645$.

Since the sample was so large we set $\alpha = .01$. Our two hypotheses are H_0: $\mu_{Y_2} \geq \mu_{Y_1}$ and H_1: $\mu_{Y_2} < \mu_{Y_1}$. Thus with the alternative hypothesis we assume that college will have a liberalizing effect and that students on average would be less religious in their senior year than they were in their freshman year.

For a one-tailed t test with $df = 2{,}505$, the critical value is -2.58 (this can be verified in Appendix D). The test statistic is:

$$t_{2505} = \frac{8.39 - 11.30}{\sqrt{\dfrac{(4.38)^2}{2{,}506} + \dfrac{(4.66)^2}{2{,}506} - \dfrac{2(.645)(4.38)(4.66)}{2{,}506}}}$$
$$= -38.16$$

Therefore we reject the null hypothesis of no change and instead conclude that college students at the University of Wisconsin were less religious by their senior year. Before concluding that college caused the change, however, we would need a control group of students who did not attend college but who were like the students who did enter the university in other aspects. Finding such a group of controls would be difficult at best.

8.9.2. Testing the Difference between Two Correlations in Independent Populations

correlation difference test—a statistical test to determine whether two correlation coefficients differ in the population

Sometimes the researcher is interested in whether the correlations between two variables, X and Y, are the same in two populations, 1 and 2. To find out, a **correlation difference test** is done. If ρ_1 and ρ_2 are used to designate the correlations between X and Y in populations 1 and 2, the null and alternative hypotheses are H_0: $\rho_1 = \rho_2$ and H_1: $\rho_1 \neq \rho_2$.

Now assuming that we have independent random samples of size N_1 and N_2 from the two populations, we test the null hypothesis by first transforming the two observed *sample* correlations, r_1 and r_2, to Z_{r_1} and Z_{r_2}, using Fisher's r-to-Z transformation table in Appendix F. Following these transformations, an α level is chosen and the test statistic is computed as follows:

$$Z = \frac{Z_{r_1} - Z_{r_2}}{\sqrt{\dfrac{1}{N_1 - 3} + \dfrac{1}{N_2 - 3}}}$$

We would expect that education is positively correlated with tolerance toward the marriage of a close relative to someone of a different race. But is the relationship between education and tolerance of interracial marriage the same for blacks and whites? We can answer this question using the 1977 GSS data, since in this survey respondents were asked: "How would it make you feel if a close relative of yours were planning to marry a (Negro/black)/white? Would you be very uneasy (coded 1), somewhat uneasy (coded 2), or not uneasy at all (coded 3)?" Number of years of education was coded directly. For a two-tailed test the critical value is ± 1.96. The two zero-order correlations are 0.230 and 0.193 for whites and blacks, respectively, and the two sample sizes are $N_1 = 1{,}299$ and $N_2 = 163$.

Appendix F indicates that $Z_{r_1} = .234$ and $Z_{r_2} = .195$. Therefore the test statistic is

$$Z = \frac{.234 - .195}{\sqrt{\dfrac{1}{1{,}299 - 3} + \dfrac{1}{163 - 3}}}$$
$$= .465$$

Thus we *cannot* reject the null hypothesis. The relationship between education and tolerance of interracial marriage appears to be the same for both blacks and whites.

This chapter concludes our treatment of topics related to bivariate regression and correlation. In Chapter 9 we present statistics for examining bivariate relations with discrete variables. In Chapter 10 we introduce the logic of multivariate analysis in general, and in Chapter 11 we generalize the materials on bivariate regression to the case of several independent variables. The materials in Chapters 7–11 form the core of data analysis. You should learn them thoroughly.

Review of Key Concepts

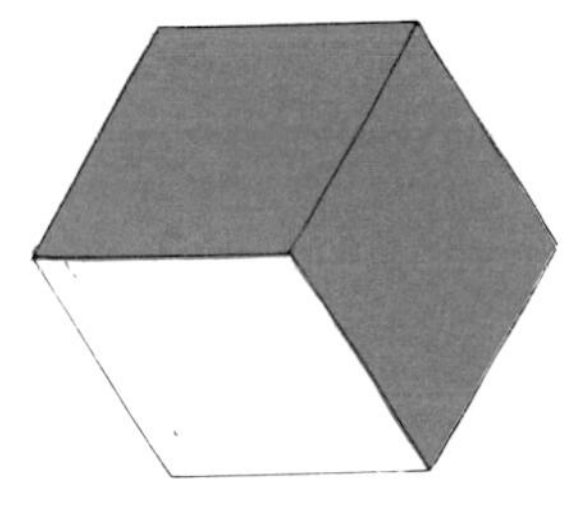

This list of the key concepts introduced in Chapter 8 is in the order of appearance in the text. Combined with the definitions in the margins, it will help you review the material and can serve as a self-test for mastery of the concepts.

scatterplot
linear relationship
regression model
prediction equation
regression line
ordinary least squares (OLS)
conditional mean
bivariate regression coefficient
normal equations
covariance
intercept
regression sum of squares
error sum of squares
proportional reduction in error (PRE)
coefficient of determination
coefficient of nondetermination
correlation coefficient
beta coefficient (beta weight)
regression toward the mean
mean square regression
mean square error
population regression equation
r to Z transformation
outliers
nonlinear regression
nonindependent samples
correlation difference test

PROBLEMS

General Problems

1. For the following data on 10 persons, construct a scatterplot showing the relationship between income and number of children, and describe the relationship in verbal terms:

Person (i)	*Annual Income (in Thousands X_i)*	No. of Children (Y_i)
1	$25	0
2	17	0
3	20	1
4	14	2
5	11	2
6	10	3
7	6	4
8	8	5
9	6	6
10	4	7

2. For the data in Problem 1:
   a. Calculate the predicted regression equation.
   b. Calculate the expected number of children for a family with $15,000 a year income.
   c. Give the covariance of income and number of children.

3. For the data in Problem 1:
   a. Report the residuals for each person.
   b. What is the least-squared error sum?

4. In a study of social change and anomie, a social scientist regresses the annual suicide rate (number per 100,000 population) among white adult men (Y) on the annual unemployment percentage (X), yielding the following equation: $\hat{Y}_t = 5.40 + 3.5X_t$, where t refers to the year, from 1950 to 1980.
   a. For every increase of 1% in the unemployment level, how many more suicides can be expected?
   b. Given that the unemployment level fluctuated between 3.5% and 8.5% over the three decades, what is the expected range of suicide rates?
   c. What would the unemployment level have to be to expect less than 15 suicides per 100,000?

5. A college admissions officer has the following regression equation for predicting grade point average (Y) from Scholastic Aptitude Test (SAT) scores (X): $\hat{Y}_i = 0.60 + 0.002\ X_i$. If

the college has a rule against admitting students likely to average C− or lower (1.5 or lower GPA), what is the minimum SAT score a prospective student must have to gain admittance?

6. Given that $\bar{X} = 20$, $\bar{Y} = 40$, $N = 60$, $s_{XY} = 1{,}600$, $s_X = 20$, and $s_Y = 100$, what is the regression equation for predicting Y from X?

7. Problem 1 of Chapter 3 gives some data on 16 noncommunist European nations. Compute the regression equation for the prediction of birth rate from percentage employed in agriculture.

8. Fill in the missing values in the table below:

	SS_{TOTAL}	$SS_{REGRESSION}$	SS_{ERROR}
a.		450	278
b.	12.46		8.93
c.	4,673.48	2,718.74	
d.	138.2		63.7

9. For the data in Problem 1, find *(a)* SS_{TOTAL}, *(b)* $SS_{REGRESSION}$, *(c)* SS_{ERROR}, and *(d)* $R^2_{Y \cdot X}$.

10. In a regression of presidential vote on party preference, a political scientist finds that the coefficient of nondetermination is 0.634. Given that the total sum of squares is 13,950, find *(a)* $SS_{REGRESSION}$, *(b)* SS_{ERROR}, and *(c)* $R^2_{Y \cdot X}$.

11. For the data given in Problem 6, find the *(a)* coefficient of determination, *(b)* coefficient of nondetermination, and *(c)* correlation coefficient.

12. What is the correlation between birth rate and percentage employed in agriculture for the data given in Problem 1 of Chapter 3?

13. Prove that, in general, the b obtained by regressing Y on X is not the same as the b obtained by regressing X on Y.

14. Fill in the missing blanks in the table below:

	b	s_X	s_Y	r_{XY}
a.		40	60	+0.25
b.	0.45	40	60	
c.	−4.5	50	300	
d.		75	75	−0.60
e.	−0.67	75	300	
f.		150	50	+0.80

15. Standardize the regression coefficient in Problem 4 if $s_X = 10$ and $s_Y = 50$.

16. Find β^* for the following:

	b	s_Y	s_X
a.	19.5	100	3
b.	0.375	35	50
c.	−3.00	90	10
d.	50	500	8
e.	−4.30	240	15

17. In a regression of number of delinquent acts on closeness of supervision by parents for a sample of 50 male juveniles, a criminologist obtains the following statistics: $MS_{\text{REGRESSION}} =$ 1,100 and $MS_{\text{ERROR}} = 250$. Test the null hypothesis that parental supervision is unrelated to delinquent behavior, using $\alpha = .05$. Summarize in an ANOVA summary table.

18. Test the data in Problem 1 for the significance of $R^2_{Y \cdot X}$. Give the F ratio and state whether the null hypothesis that $\rho^2_{Y \cdot X} =$ 0 may be rejected at $\alpha = .001$.

19. Test the following for significance at $\alpha = .01$:

	$R^2_{Y \cdot X}$	N	SS_{TOTAL}
a.	0.10	40	436.0
b.	0.10	400	436.0
c.	0.55	13	52.5
d.	0.30	16	134.3
e.	0.26	24	28.4

Report the F ratio and your decision.

20. Test whether $\beta = 0$ in the population from which the data in Problem 1 were selected. Give *(a)* s_b, *(b)* t_{N-2}, and *(c)* the decision at $\alpha = .01$.

21. Construct the 95% and 99% confidence intervals around a regression coefficient of 4.78 which has a standard error of 1.46. Can you be 99% confident that the population β is not 8.00, if N = 500?

22. In predicting the ownership of handguns from a question about whether respondents like to hunt, a sociologist finds the following statistics in a sample of 27 respondents: $SS_{\text{REGRESSION}} = 11.40$ and $SS_{\text{ERROR}} = 153.20$. Compute the coefficient of determination for this relationship and test whether it is significantly different from zero at $\alpha = .05$.

23. Where H_0: $\beta = 0$, calculate t ratios and test hypotheses for the following:

	b	MS_{ERROR}	s_X	N	α
a.	1.9	2,150	8	25	.05
b.	0.55	750	50	37	.01
c.	0.43	630	14	100	.01
d.	6.2	1,800	3	48	.01

24. Construct the 95% confidence intervals around the b's given in Problem 23.

25. Calculate the test statistics for the hypothesis that $\rho = 0$ for the following:

	r	N	α
a.	0.20	12	.05
b.	0.20	120	.05
c.	0.45	33	.05
d.	−0.45	333	.001
e.	−0.88	28	.01

Give Z and the decision.

26. Using the values given in Panel B of Table 8.4, test the statistical significance of the $R^2_{Y \cdot X_2}$ between number of functions and city size when New York City is not included. Report the F ratio and your decision for $\alpha = .05$.

27. In a linear regression of frequency of sexual intercourse on age, a sex researcher finds $R^2_{Y \cdot X} = .36$. Upon breaking the age measure into six decade-wide categories, the researcher finds that $\hat{\eta}^2 = 0.44$. Is there a significant nonlinear relationship at $\alpha = .05$, if the sample size is 84?

28. In a panel study of 80 respondents, a researcher found that 55% thought President Reagan was doing a good or excellent job before the assassination attempt early in 1981. After the attempt, the approval rating jumped to 68%. The correlation between the two measures was 0.82. Was the amount of increase in the panel sample statistically significant at $\alpha =$ .05?

29. Find the t ratios for the following nonindependent samples:

	$\overline{Y}_1$	$\overline{Y}_2$	s_{Y_1}	s_{Y_2}	$r_{Y_1 Y_2}$	N
a.	16	14	3	2	0.30	35
b.	56	66	12	8	0.50	47
c.	25	28	4	3	−0.60	22
d.	36	28	6	5	0.20	70

30. In a sample of 165 women, a researcher finds a correlation of +0.57 between measures of religiosity and frequency of church attendance. The correlation coefficient among 137 males is +0.49. Is there a significant difference in covariation between the sexes, at $\alpha = .01$?

Problems Requiring the 1980 General Social Survey

31. Differential fertility refers to the rate of reproduction as it varies across social class. Perform and interpret three regression equations for women aged 40 or older (AGE):
    a. The regression of number of children ever born (CHILDS) on years of formal schooling (EDUC),
    b. The regression of number of children on spouse's occupational prestige (SPPRES).
    c. The regression of number of children on family income (INCOME).
32. Some urban sociologists hypothesize that large places have experienced a loss of the sense of community that smaller places may provide. Determine whether a significant positive linear relationship occurs between SATCITY and SIZE of place.
33. One of the factors that may affect people's satisfaction with their jobs is the amount of money they earn. But the amount of satisfaction per dollar may differ for men and for women. Test this hypothesis by performing separate regressions by SEX of SATJOB on RINCOME (recoded to interval midpoints). Interpret the regression coefficients and test for differences in the size of the correlations.
34. Sexual attitudes and religiosity are sometimes believed to covary as parts of an underlying "morality" orientation. Treating attitudes toward extramarital sex (XMARSEX) and homosexuality (HOMOSEX) as the dependent variables, obtain two regression equations with church attendance (ATTEND) as the independent variable. Determine whether significant covariation occurs as predicted.
35. Test whether a significant nonlinear relationship exists between the number of voluntary organizations to which respondents belong (MEMNUM) and their age measured in decades (i.e., recode AGE as 10–19, 20–29, etc.).

Problems Requiring the 63-Cities Data

36. Construct a scatterplot of the relationship between population change (POPCHANG) and municipal revenue (REVENUE). What is the regression equation, and how much variance in revenue is explained by population change?

37. Older cities presumably have older houses, but what is the precise relationship? For each 10 years of age (CITYAGE), how much larger is the percentage of housing built prior to 1950 (HOUSEPCT)?

38. What is the magnitude of the correlation between birth rate (BIRTHS) and average income (INCOMEPC) at the city level of analysis?

39. Determine whether the correlation between size of the middle class (WCPCT) and volume of retail sales (RETAIL) differs significantly between Southern and non-Southern cities.

40. A criminologist hypothesizes that the more women there are in the labor force (FEMLABOR), the greater the crime rate (CRIMRATE) because more homes are left unoccupied during working hours, thus increasing the opportunities for burglaries. Is there any evidence to support this supposition?

Statistics for Social Data Analysis

9

Measuring Association with Discrete Variables

In this text we show how the most powerful statistical tools available can be used to study social behavior. As a general proposition, the descriptive and inferential statistics applicable to continuous measures introduced in Chapter 8 can be used whenever the data meet or approximate the underlying assumptions. In some instances, however, researchers may have variables that can be measured only at the discrete level, or they may consider an assumption of continuous properties to be unwarranted. A large variety of statistics for measuring association between pairs of discrete variables has been developed. In this chapter, we introduce several of the most important of these measures of association, as well as tests of their statistical significance when they are available.

9.1. Measures of Association for Nonorderable Discrete Variables

As we noted in Chapter 1, the most elementary level of variable measurement classifies cases into two or more discrete categories, distinguishable only by the presence or absence of some attribute—such as skin color for the race variable, country for nationality, or identification with a denomination for religion. No intrinsic order or quantity can be meaningfully placed on a nonorderable discrete measure.

9.1.1. An Example: Religion and School Prayer Attitude

An issue of current concern in the United States is whether state or local governments may permit or require the reading of prayers in public schools. This practice was banned by a decision of the Supreme Court in the 1960s. The recent conservative revival, as evidenced by Ronald Reagan's victory in the 1980 presidential election and the prominence of such groups as the Moral Majority, has given impetus to the call for its reinstatement. Since the major support for reinstating prayers in public schools has come from members of fundamentalist churches, we advance the following hypothesis:

H1: Protestants are more likely than Catholics and others to disapprove of the Supreme Court decision banning prayer in public schools.

Table 9.1 shows a crosstabulation of the 1977 General Social Survey respondents by religious preference and attitude toward

TABLE 9.1
Crosstabulation of Religion and School Prayer Decision Variables

School Prayer Decision	*Religion*			
	Protestant	*Catholic*	*Others*	*Total*
Approve	27.9%	39.7%	64.2%	34.2%
(N)	(274)	(146)	(88)	(508)
Disapprove	72.1	60.3	35.8	65.8
(N)	(708)	(222)	(49)	(979)
Total	100.0%	100.0%	100.0%	100.0%
(N)	(982)	(368)	(137)	(1,487)

Source: 1977 General Social Survey.

prayer in public schools. Attitude toward prayer in public schools was assessed by the item: "The United States Supreme Court has ruled that no state or local government may *require* the reading of the Lord's Prayer or Bible verses in public schools. What are your views on this—do you approve or disapprove of the court ruling?"

Since there is no intrinsic order among the three broad religious categories (Protestants, Catholics, and others), Table 9.1 displays a relationship among nonorderable discrete measures.

The percentages show that Protestants favor school prayers more often than Catholics do, while those with other religious affiliations or no religion overwhelmingly approve the Court's decision on school prayer. There is clearly a statistically significant relationship between these two variables, since $\chi_2^2 = 77.16$ (see Sections 4.4 and 4.5 for a discussion on the computation of chi square). But a test for statistical significance does not indicate the strength of covariation.

Like the standardized measure of association for continuous measures discussed in Chapter 8, an ideal measure of association for two nonorderable discrete variables should take on a value of zero when no covariation occurs and a maximum value of 1.00 (or −1.00) when perfect (error-free) prediction from independent to dependent variable is possible. A measure of association for nonorderable discrete variables which has such properties is lambda, which is discussed and applied to the data in Table 9.1 in the following section.

9.1.2. Lambda

All good measures of association make use of the *proportionate reduction in error (PRE)* approach described in Section 8.4 in Chapter 8. Such statistics are based on comparisons of the errors in classification or prediction when no information about the independent variable is available, to the errors made when information about the independent variable helps improve prediction of the dependent measure. Thus, *a PRE statistic reflects how well knowledge of one variable improves prediction of the second variable.*

The general formula for any PRE statistic was given in Chapter 8 in terms of decision rules about expected Y values, conditioned on X values. In terms of first and second variables, the formula is:

$$\text{PRE} = \frac{\text{Errors without 2nd variable} - \text{Errors with 2nd variable}}{\text{Errors without 2nd variable}}$$

When variables Y and X are unrelated, the capacity of the second variable to reduce errors in estimating values of the first variable is nil, and the PRE value is zero. When a perfect prediction from one variable to the other is possible, no errors exist, and the PRE statistic takes its maximum value of 1.00. Intermediate values of the PRE measure show greater or lesser degrees of predictability.

lambda—an asymmetric measure of association for nonorderable discrete variables based on prediction from modes

Lambda (λ), a PRE statistic applicable to nonorderable discrete measures, is based on the ability to predict *modes,* the value of the response category for the dependent variable that has the largest number or percentage of cases (see Section 3.1.1 in Chapter 3). The first type of error is found by guessing that all respondents will be found in the modal category of Y, as would be shown in the total marginal of a crosstabulation (see Chapter 4). For the religion-prayer relationship, the mode of the prayer variable's marginal distribution is "disapprove." If X is the independent variable, we next determine the second type of errors of prediction by assuming that *within* each category of X, all respondents are located at the mode for Y. Thus we would predict a "disapprove" response for Protestants and Catholics and an "approve" response for others. The value for lambda is then calculated:

$$\lambda = \frac{\text{Errors in assuming mode of } Y - \text{Sum of errors in assuming modes on } Y \text{ within levels of } X}{\text{Errors in assuming mode of } Y}$$

To illustrate, by assuming all respondents favor school prayer, we would misclassify 508 respondents (those who in fact say they are opposed). By predicting that all persons of a given religion fall into the mode on prayer for that religion, we would make 469 errors (274 for Protestants, 146 for Catholics, 49 for others), a reduction in error of only 39 cases. The ratio of this difference to the initial amount of error is:

$$\lambda = \frac{508 - 469}{508} = 0.077$$

This is hardly a stunning improvement, so we conclude that knowledge of religious preference gives only a modest ability to predict a respondent's attitude toward school prayer.

Like all good measures of nonorderable discrete association, lambda varies in magnitude between zero and 1.00. *Zero occurs whenever the same mode on the dependent variable is predicted for each category of the independent variable.* Note that lambda is an *asymmetric* coefficient: *The value obtained by predicting Y from X is usually not the value found when predicting X from Y.* In the example, when religion is the dependent variable, lambda is zero: Regardless of attitude, there are more Protestants than Catholics or others. Therefore, in reporting your calculations of lambda, you should be careful to indicate clearly which variable is dependent and which is independent.

Since the sampling distribution for lambda is complex, it is not presented in this text.[1] However, when lambda equals zero in the population, chi square also will equal zero. Therefore the null hypothesis that lambda equals zero can be evaluated by a simple chi-square test with $R - 1$ and $C - 1$ degrees of freedom. (To refresh your memory, reread Sections 4.4 and 4.5 in Chapter 4, where the chi-square statistic was introduced.)

For the data in Table 9.1, $\chi^2_2 = 77.16$, as we noted in Section 9.1. For $\alpha = .05$ the critical value is 5.99. Therefore we reject the null hypothesis that lambda equals zero in the population. However, since $\lambda = 0.077$, the strength of support for the research hypothesis that Protestants disapprove of the Supreme Court ruling more than Catholics and others do is weak. The apparent inconsistency between the magnitudes of the chi-square and lambda statistics results from the chi-square statistic's sensitivity to sample size. In this example, as Table 9.1 showed, $N = 1{,}487$.

Box 9.1 introduces two other measures of association for nonorderable discrete variables—**Cramer's V** and the **contingency coefficient** (C). These are commonly encountered in research instruments such as SPSS outputs. Because neither V nor C is a PRE measure, however, in our opinion neither is as useful as lambda for measuring association between nonorderable discrete measures.

Cramer's V—a measure of association for discrete variables based on chi-square

contingency coefficient—a measure of association for nonorderable discrete variables based on chi-square

9.2. Measures of Association for Orderable Discrete Variables

In variables measured at the orderable discrete level, as we noted in Section 2.3 in Chapter 2, response categories are arranged in an ascending or descending sequence but lack continuous mea-

1. This sampling distribution is given in Graham J. G. Upton, *The Analysis of Cross-Tabulated Data* (Chicester: Wiley, 1978), pp. 31–32.

BOX 9.1 Cramer's *V* and the Contingency Coefficient, *C*

Cramer's *V* adjusts chi-square for sample size. Its formula is:

$$V = \sqrt{\frac{\chi^2}{(N)(\text{Min}(r-1, c-1))}}$$

The term $\text{Min}(r-1, c-1)$ in the denominator directs us to choose the smaller of the two values, $r-1$ or $c-1$, where $r-1$ is the number of rows less 1 and $c-1$ is the number of columns less 1.

Cramer's V is zero when the two nominal measures are independent, since χ^2 is also zero. *Its upper limit is 1.00;* this value can occur even if the number of rows and columns is not the same.

The contingency coefficient, *C,* is also based on chi-square. Like Cramer's *V, it takes the value of zero when no relationship exists between two variables.* But unlike *V, its upper limit depends on the number of rows and columns in the table. This makes Cramer's V easier to interpret.*

The formula for *C* shows why it can never attain the maximum value of 1.00:

$$C = \sqrt{\frac{\chi^2}{\chi^2 + N}}$$

In a 2×2 table, the largest value χ^2 can obtain is *N,* when only one cell in a given row and column has nonzero entries. Hence, the maximum value for *C* is the square root of one half, or 0.707. Similarly, the largest χ^2 possible in a 3×3 table is *2N.* In this case *C* can never exceed 0.816. As the dimensions of the crosstabulation increase, so does the upper limit on *C,* but it can never reach 1.00.

In the example in the text, which proposes a correlation

Box 9.1 (continued)

between religious preference and attitudes toward school prayer, the observed values for V and C are:

$$V = \sqrt{\frac{77.16}{(1,487)(2 - 1)}} = 0.228$$

$$C = \sqrt{\frac{77.16}{77.16 + 1,487}} = 0.222$$

surement properties. Although lambda may be applied to cross-tabulations of orderable discrete variables, more appropriate measures of association are available. These measures of association have the useful property of indicating the *direction* of association between the two orderable discrete variables, ranging between $+1.00$ and -1.00 and with zero indicating no relationship. This section presents four measures of association for orderable discrete variables: gamma, tau b, tau c, and Somers's d_{yx}.

9.2.1. An Example: ERA Support and Sex Role Attitude

One of the important current social and political issues is the status of women. We expect that in the general population, persons who support sex role equality will also favor legal changes to bring about greater equality in the positions of men and women. Our general proposition is:

P1: The greater the belief in sex role equality, the stronger the support for legislation to grant equal rights to women.

Using the 1977 General Social Survey, belief in sex role equality is operationalized (see Section 1.5 in Chapter 1) by an item about woman's role in the household: "It is much better for everyone involved if the man is the achiever outside the home and the woman takes care of the house and family." The response categories range from "strongly agree" to "strongly disagree," with disagreement indicating greater support for sex role equality. Similarly, legislative support for women's rights is operationalized by an item measuring respondents' support for the Equal Rights Amendment, also ranging from "strongly favor" to "strongly oppose'' (but omitting persons with no opinion or no understanding of the amendment's meaning). Using these indicators, this proposition can be rewritten:

> H1: The greater the disagreement that a woman's place is in the home, the stronger the support for passage of the Equal Rights Amendment.

Table 9.2 displays the crosstabulation of these two measures, with sex role attitude treated as the independent variable. The two variables are plainly related, as, taken together, 40% of those holding the most traditional sex role attitude oppose the ERA (16.1% + 23.9%), while 85.7% of those who strongly disagree with the traditional stereotype also favor passage of the ERA. For this 4 × 4 table, $\chi^2_9 = 128.36$, leading to the conclusion that the 1977 GSS sample probably reflects a nonzero covariation in the population.

If the tabular data were displayed as a scatterplot, as in bivariate regression (see Section 8.2 in Chapter 8), we would expect to find the largest number of cell frequencies along the *main diagonal* if the relationship is one of positive covariation. (In a crosstabulation the main diagonal is the four cells from lower left to upper right; in Table 9.2, the frequencies are 49, 116, 200, and 39.) In contrast, if the ERA–sex role relation were inverse or negative, we would expect to observe a concentration of frequencies in the diagonal cells running from upper left to lower right. Inspection of the table reveals the tendency toward a positive relationship,

TABLE 9.2
Crosstabulation of ERA Support and Sex Role Attitude Variables

Equal Rights Amendment Support	*Sex Role Attitude: A Woman's Place Is in the Home*				
	Strongly Agree	***Agree***	***Disagree***	***Strongly Disagree***	***Total***
Strongly favor	16.6%	16.8%	29.2%	50.6%	22.7%
(N)	(34)	(91)	(104)	(39)	(268)
Favor somewhat	43.4	51.7	56.2	35.1	50.6
(N)	(89)	(281)	(200)	(27)	(597)
Oppose somewhat	16.1	21.4	11.5	11.7	16.9
(N)	(33)	(116)	(41)	(9)	(199)
Strongly oppose	23.9	10.1	3.1	2.6	9.9
(N)	(49)	(55)	(11)	(2)	(117)
Total	100.0%	100.0%	100.0%	100.0%	100.1%
(N)	(205)	(543)	(356)	(77)	(1,181)

Source: 1977 General Social Survey.

although the relationship does not appear to be particularly strong.

Among the desirable attributes of orderable discrete measures of association are that the direction of the relationship should be indicated by the *sign* of the coefficient, with a plus indicating positive covariation and a minus indicating an inverse relationship. Like nonorderable discrete measures of association, *orderable measures of association should also be normed to vary between zero (for no relationship) and 1,* to provide maximum predictability of the dependent variable values from the independent variable. Thus, like the correlation coefficient, these measures of association should take on values ranging between −1.00 and +1.00. As a standard of comparison, for the four measures of association discussed in this section, if the two variables in Table 9.2 are assigned values of 1 through 4, the Pearson $r = 0.274$. This indicates at best a modest covariation (e.g., only 7.5% of the variance in one variable can be "explained" by the other).

9.2.2 Gamma

Gamma *is the most frequently used measure of ordered crosstabular association.* Like lambda, it is a proportionate-reduction-in-error (PRE) statistic. But, unlike lambda, gamma is *symmetric,* so that predicting the second variable from the first yields the same gamma value as the opposite order of prediction, even for nonsquare tables. It has the desired characteristic of ranging between +1.00 and −1.00, with zero indicating no relationship. Gamma is also a "margin-free" measure of association; its value does not depend on the row or column marginal totals.

gamma—a symmetric measure of association for orderable discrete variables that takes into account only the number of untied pairs

The calculation of gamma uses the observed cell frequencies (n_{ij}) in a crosstabulation. This statistic systematically compares *every* pair of observations eliminating those that are identical on at least the categories of one variable (i.e., *tied* cases). *Only the untied pairs of cases are used to calculate gamma.*

concordant pairs—in a crosstabulation of two orderable discrete variables, the number of pairs having the same rank order of inequality on both variables

discordant pairs—in a crosstabulation of two orderable discrete variables, the number of pairs having reverse rank order of inequality on both variables

The total number of untied pairs is sorted into two groups: (1) **concordant pairs**: the number of pairs having the same rank order of inequality on both variables (n_s), and (2) **discordant pairs**: the number of pairs with the reverse rank order of inequality on both variables (n_d).

Using these two sets of untied pairs, the formula for the sample statistic for gamma (G) is:

$$G = \frac{n_s - n_d}{n_s + n_d}$$

The PRE nature of gamma can be clearly seen in this formula. If any pair of cases is drawn at random from the crosstabulation table and we try to predict whether the same or reverse order occurs, our chances of being correct depend on the relative preponderance of concordant and discordant pairs in the table. If $n_s = n_d$, we will be unable to predict at better than chance, and gamma will equal zero. But if n_s is substantially larger than n_d, gamma will be positive, and we will be more successful in predicting that the respondent with the higher value on one variable will also have the higher value on the second variable, compared to the other respondent in the pair. Note especially that when $n_d = 0$ (i.e., there are no discordant pairs), gamma equals unity (1.00). The rate of prediction error is reduced, however, when n_d is substantially larger than n_s. In this case, gamma will be negative. For a negative gamma we predict that if person A is higher than person B on variable X, the reverse order holds for the pair on variable Y. The maximum negative value of gamma (−1.00) occurs when n_s is zero, and there are no concordant pairs.

Although the calculation of gamma is simple once we have the numbers of concordant and discordant pairs, obtaining these values from a crosstabulation requires a lot of arithmetic. To illustrate the general procedure, think of three respondents—A, B, and C. Suppose that person A strongly favors ERA and strongly disagrees that a woman's place is in the home; person B somewhat opposes ERA and agrees that women should stay home; and person C strongly favors ERA but also strongly agrees that a woman's place is in the home. If we examine all three pairs of respondents, we find first that A is higher than B, both in favoring ERA and in disagreeing with sex role stereotypes. Hence this pair should be placed in the concordant, or n_s, group. Since there are 39 people like A and 116 like B in Table 9.2, there is a total of (39)(116) = 4,524 concordant pairs exactly like this one. Next, although A and C have different attitudes about traditional sex roles, they agree on ERA; hence, this pair is tied on one of the variables and must be dropped from further consideration. Altogether, 1,326 pairs of this kind will be eliminated (i.e., the 34 respondents in the upper-left cell and the 39 in the upper-right cell form (34)(39) = 1,326 pairs). The B-C pair is "untied," but the direction is not consistent for both variables: B is less favorable to ERA than C, but B shows greater disagreement, compared to C, on the sex role measure. Hence, this pair should be placed in the n_d (discordant) sum. There are (34)(116) = 3,944 pairs identical to this one.

All of the $(N)(N - 1)/2 = (1{,}181)(1{,}180)/2 = 696{,}790$ unique pairs of respondents in Table 9.2 must be sorted into the n_s, n_d, or eliminated groups. Such a task would be formidable if every pair had to be laboriously inspected, one at a time, as described in the previous paragraph. Fortunately, a relatively easy procedure exists for making these comparisons rapidly, even when a computer is not handy. This *algorithm,* or procedure, for determining n_s and n_d is as follows:

1. With the ordered table laid out in standard fashion, begin at the *upper-right* cell. Ignoring all entries in the same row and column, add the remaining cell frequencies, that is, those n_{ij}s in the cells below and to the left. (Example: in Table 9.2 the sum of all entries below and to the left of the 39 respondents who strongly favor ERA and also strongly disagree that a woman's place is in the home is 89 + 281 + 200 + 33 + 116 + 41 + 49 + 55 + 11 = 875.) Multiply this sum by the cell frequency in the upper-right cell (e.g., 39), and add the product to the n_s total. All these pairs have the same order of inequality on both variables (concordant), and thus (875)(39) = 34,125 pairs forms the first count to be added to the n_s total.
2. Still within the top row of the table, move over one column to the *left* and again add up all cell frequencies below and to the left. (Example: The sum of all entries below and to the left of the 104 respondents who strongly favor ERA but merely disagree that a

woman's place is in the home is 89 + 281 + 33 + 116 + 49 + 55 = 623.) Once again obtain the product of this sum and the frequency in the initial cell, and add it to the n_s total. (Example: (623)(104) = 64,792; thus far we have 64,792 + 34,125 = 98,917 in n_s.)

3. Proceed in this fashion, moving to the left across the first row, multiplying the cell frequency by the sum of all frequencies below and to the left, and cumulating the total count of same-order pairs. When any row has been completed, move to the row below, always starting with the right-most column. When finished, the total number of same-order untied pairs, n_s, will have been found. (There are 207,338 n_s pairs in this example.)
4. To calculate the number of *discordant untied pairs,* n_d, follow the same process but begin with the *upper-left* cell and multiply it by the sum of all cell frequencies below and to the *right.* (Example: target the upper-left cell—34 cases. Find the sum of cell frequencies in cells below and to the right of the first row and first column: 281 + 200 + 27 + 116 + 41 + 9 + 55 + 11 + 2 = 742 cases. Multiply this sum by the number of cases in the targeted cell to obtain the number of discordant pairs associated with that cell: (742)(34) = 25,228.)
5. Still within the top row, move one column to the *right* and again add up all cell frequencies below and to the right. (Example: The target cell has 91 cases, and the sum of cases below and to the right is 290. Their product is (91)(290) = 26,390 discordant pairs. Add this to the number obtained in the preceding step: 25,228 + 26,390 = 51,618.)
6. Continue in this fashion until all discordant pairs have been calculated. (Convince yourself that Table 9.2 contains 100,133 n_d pairs.)

Using the formula for gamma, we can now find the association between favoring ERA and rejecting traditional sex role attitudes:

$$G = \frac{(207{,}338) - (100{,}133)}{(207{,}338) + (100{,}133)} = 0.349$$

Since the maximum positive value that gamma can attain is 1.00, we interpret this result as indicating a moderate positive association between the two attitudes. Note that gamma is slightly larger than the correlation coefficient obtained by assuming continuous variables ($r = 0.274$, as shown in Section 9.2.1). Squaring gamma to obtain an indicator of "variance explained" is *not* a legitimate operation, as it is with the correlation coefficient, however.

The population parameter that G estimates is labeled γ (Greek lowercase *gamma*). If we have a simple random sample of cases, the sampling distribution of G approaches normality as N becomes large (50 or more). The test statistic is roughly approximated by:

$$Z = (G - \gamma)\sqrt{\frac{n_s + n_d}{N(1 - G^2)}}$$

Where:

γ = The population value of gamma under the null hypothesis.

Since this formula gives a rather conservative estimate of Z, there may be circumstances where the absolute value of Z is in fact larger than that computed this way. The exact calculations are given elsewhere.[2]

In this example the null and alternative hypotheses are H_0: $\gamma \leq 0$ and H_1: $\gamma > 0$. To test the null hypothesis we choose $\alpha = .01$. Appendix C shows that the critical value is 2.33. Now, using the data calculated above, the standard score, (that is, the test statistic), is

$$Z = (0.349 - 0)\sqrt{\frac{207{,}338 + 100{,}133}{1{,}181\ (1 - 0.349^2)}}$$

$$= 6.01$$

2. See H. T. Reynolds, *The Analysis of Cross-Classification* (New York: Free Press, 1977), pp. 85–88.

Since 6.01 greatly exceeds 2.33, we reject the null hypothesis and instead conclude that there is a positive relationship between a liberal attitude toward woman's role and support for the ERA.

9.2.3. Tau b

tau b—a symmetric measure of association for two orderable discrete variables with the same number of categories that takes into account the number of both tied cases and untied cases.

Like gamma, **tau b** (τ_b) uses information about two orderable discrete variables by considering every possible pair of observations in the crosstabulation table. Unlike gamma, τ_b *takes into account the number of ties on the independent variable, the number of ties on the dependent variable, but not the number of ties on both variables.* It too ranges in value from -1.0 to $+1.0$ if the table being analyzed is square (i.e., $R = C$). And τ_b equals zero if the two variables being analyzed are unrelated. (If the table being analyzed is not square, τ_c, presented in the next section, is the more appropriate statistic to use.)

The computing formula for τ_b defines n_s and n_d the same as for gamma and is given by:

$$\tau_b = \frac{n_s - n_d}{\sqrt{n_s + n_d + T_r}\sqrt{n_s + n_d + T_c}}$$

Where:

T_r = The number of ties associated with the row variables.
T_c = The number of ties associated with the column variables.

Since the numerator of τ_b is identical to the numerator of G (see Section 9.2.2), it need not be recalculated.

The steps to be followed in computing T_c are:

1. With the table laid out in standard fashion, *begin in the upper-left corner.* Add all of the values in the column *below* the upper-left cell. (For example, in Table 9.2, 89 + 33 + 49 = 171.) Multiply this value times the cell frequency. (Example: (34)(171) = 5,814.) Call this term C_{11}.
2. Move to the cell immediately below the upper-left cell. Add the values in the two cells in the same column just *below* this cell (e.g., 33 + 49 = 82),

and multiply this sum times the targeted cell frequency (e.g., (82)(89) = 7,298). Call this term C_{21}.

3. In the same column, move to the cell immediately below, and multiply this cell frequency times the lower-left cell (e.g., (33)(49) = 1,617). Call this term C_{31}.
4. Move to the top cell of the next column to the right. Multiply this cell frequency times the sum of cell frequencies in the column below the cell (e.g., (91) (281 + 116 + 55) = 41,132). Call this term C_{12}.
5. Move to the cell immediately below and multiply this cell frequency times the sum of cell frequencies below the cell (e.g., (281)(116 + 55) = 48,051). Call this term C_{22}.
6. Continue this process of targeting a given cell and multiplying its frequency times the sum of all cell frequencies beneath the cell, within the same column, until all cells have been targeted except for those in the last row of the table. When you have finished the process, there will be $(R \times C) - R = R(C - 1)$ terms that comprise T_c. (In this example, there will be $4(4 - 1) = 12$ terms to add together to obtain T_c.)
7. Add the $R(C - 1)$ terms together to form T_c. If the term associated with a given cell is labeled C_{ij}, where i is the row and j the column for the cell, then:

$$T_c = \sum_{j=1}^{R-1} \sum_{i=1}^{C} C_{ij}$$

In the example using Table 9.2, verify that the terms associated with each cell are:

$$\begin{array}{llll} C_{11} = 5{,}814 & C_{12} = 41{,}132 & C_{13} = 26{,}208 & C_{14} = 1{,}482 \\ C_{21} = 7{,}298 & C_{22} = 48{,}051 & C_{23} = 10{,}400 & C_{24} = 297 \\ C_{31} = 1{,}617 & C_{32} = 6{,}380 & C_{33} = 451 & C_{34} = 18 \end{array}$$

and that:

$$T_c = \sum_{j=1}^{3} \sum_{i=1}^{4} C = 149{,}148$$

The value T_c is the pairs tied on the column variable (the sex-role item investigating attitudes toward woman's "place" in the home).

To compute T_r, the pairs tied, as the column variable, the same general algorithm is followed:

1. *Target the upper-left cell.* Multiply this cell frequency times the sum of all the other cell frequencies in the same *row* and to the right of the cell (e.g., 34(91 + 104 + 39) = 7,956). Call this term R_{11}.
2. Target the next cell to the right and in the same row. Multiply this cell frequency times the sum of the other cell frequencies to the right and in the same row (e.g., 91(104 + 39) = 13,013). Call this term R_{12}.
3. Continue this process of targeting a given cell and multiplying its frequency times the sum of all cell frequencies to the right of it and in the same row until all cells have been targeted, except for those in the last column. There will be $(C \times R) - C = C(R - 1)$ terms in all.
4. Add all of the $C(R - 1)$ terms together to form T_r. If the term associated with a given cell is labeled R_{ij}, then:

$$T_r = \sum_{i=1}^{C-1} \sum_{j=1}^{R} R_{ij}$$

In the example, verify that:

$$\begin{array}{lll} R_{11} = 7{,}956 & R_{12} = 13{,}013 & R_{13} = 4{,}056 \\ R_{21} = 45{,}212 & R_{22} = 63{,}787 & R_{23} = 5{,}400 \\ R_{31} = 5{,}478 & R_{32} = 5{,}800 & R_{33} = 369 \\ R_{41} = 3{,}332 & R_{42} = 715 & R_{43} = 22 \end{array}$$

Then:

$$T_r = \sum_{i=1}^{3} \sum_{j=1}^{4} R_{ij} = 155{,}140$$

Now we have all the information we need to compute t_b—the sample estimate of the population parameter, τ_b:

$$t_b = \frac{207{,}338 - 100{,}133}{\sqrt{207{,}338 + 100{,}133 + 155{,}140}\,\sqrt{207{,}338 + 100{,}133 + 149{,}148}}$$

$$= 0.233$$

The standard error of the sampling distribution is very complex and will not be presented here. However a quick and not-too-dirty estimate has been developed by Somers.[3] This is:

$$\hat{\sigma}_{t_b} = \sqrt{\frac{4(R+1)(C+1)}{9\,NRC}}$$

This formula can only be used as an approximation when there is simple random sampling, and the null hypothesis states that $\tau_b = 0$ in the population.

To test whether $t_b = 0.233$ is significantly different from zero we choose $\alpha = .01$. The critical value is then 2.33 for a one-tailed test (see Appendix C). The test statistic is:

$$Z = \frac{t_b}{\hat{\sigma}_{t_b}}$$

In this example, the test statistic is:

$$Z = \frac{0.233}{\sqrt{\dfrac{4(4+1)\,(4+1)}{(9)\,(1{,}181)\,(4)\,(4)}}}$$

$$= \frac{0.233}{0.024}$$

$$= 9.61$$

3. Robert Somers, "Simple Approximations to Null Sampling Variances: Goodman and Kruskal's Gamma, Kendall's Tau, and Somers's d_{yx}," *Sociological Methods and Research*, Vol. 9 (August 1980), pp. 115–26.

Therefore we reject the hypothesis that attitudes towards sex roles and support for the ERA are unrelated, using τ_b as well as γ.

9.2.4. Tau *c*

Tau *b*, described above, is used when the number of rows and columns in a crosstabulation is the same. When the number of rows does not equal the number of columns for two discrete orderable variables, a better statistic to use is **tau *c*** (τ_c). The formula for computing this statistic in the sample is:

tau *c*—a symmetric measure of association for two orderable discrete variables with unequal numbers of categories that takes into account only the number of untied pairs

$$t_c = \frac{2m(n_s - n_d)}{N^2(m - 1)}$$

Where:

m = The smaller of R or C.

To illustrate, we will use a different example than the ERA support—sex role attitude data used elsewhere in this section. Suppose we are interested in whether men or women in the United States are happier. Because we have no basis to assume that either sex is happier, the null hypothesis is simply that the two do not differ in level of happiness. Although a variable with only two outcomes may not be intrinsically orderable, we can nevertheless proceed to analyze the data as though it were. One outcome is arbitrarily treated as the positive outcome, and the other is treated as the negative outcome. The General Social Survey asks, "Taken all together, how would you say things are these days—would you say you are very happy, pretty happy, or not too happy?"

In Table 9.3 we have crosstabulated sex by happiness for the 1977 GSS data. The distributions suggest that there is very little difference in happiness by sex. This is verified by the fact that $\chi^2_2 = 0.712$, which, as Appendix B reveals, is far from significant for $\alpha = .05$.

To compute t_c, the sample estimate, we need to compute n_s and n_d. Verify for yourself that in Table 9.3 these values are $n_s = 157{,}300$ and $n_d = 156{,}260$. Now you can see that:

TABLE 9.3
Crosstabulation of Sex and Happiness Variables

	Sex		
Happiness	***Male***	***Female***	***Total***
Very happy	34.7%	34.1%	34.3%
(N)	(221)	(300)	(521)
Pretty happy	55.1	56.8	56.1
(N)	(351)	(500)	(851)
Not too happy	10.2	9.1	9.6
(N)	(65)	(80)	(145)
Total	100.0%	100.0%	100.0%
(N)	(637)	(880)	(1,517)

Source: 1977 General Social Survey.

$$t_c = \frac{(2)(2)(157{,}300 - 156{,}260)}{1{,}517^2(2-1)}$$
$$= 0.0018$$

We can use the approximation for the standard error introduced for t_b to help decide whether this value is statistically significant. Since we have no reason to hypothesize whether men or women are happier, we use a two-tailed test with $\alpha = .05$. The critical value is ± 1.96, as can be verified from Appendix C. Now:

$$Z = \frac{0.0018}{\sqrt{\dfrac{4(3+1)(2+1)}{9(1{,}517)(3)(2)}}}$$
$$= 0.075$$

Since this is nowhere near 1.96, we cannot reject the hypothesis that the sexes do not differ on happiness.

9.2.5. Somers's d_{yx}

Unlike gamma and τ_b, which yield the same value regardless of which variable is considered independent and which is considered dependent, **Somers's d_{yx}** is an *asymmetric* measure of ordinal association whose value depends on which variable plays which role. Suppose we are trying to predict the value of variable Y from our knowledge of variable X. We can take into account those pairs of observations that are tied on variable Y, the depen-

Somers's d_{yx}—an asymmetric measure of association for two orderable discrete variables that takes into account the numbers of united pairs and of pairs tied only on the dependent variable

dent variable. But we ignore any pairs on which both observations are tied in the independent variable, X. Somers's d for predicting Y from X, assuming Y is the row variable and X is the column variable in a crosstabulation, is given by:

$$d_{yx} = \frac{n_s - n_d}{n_s + n_d + T_r}$$

This is a PRE-type measure of association in which we predict the ranking on the dependent variable from a prediction rule which includes one type of tie but not the other type. Like lambda for nonorderable discrete variables, the asymmetric nature of Somers's d means that two values which are unlikely to be identical can be calculated from one table. (For d_{xy} instead of d_{yx}, T_r is replaced with T_c in the denominator.) When reporting this statistic, therefore, you must specify which variable is assumed to be the dependent measure.

When support for ERA is treated as a dependent variable in Table 9.2, and the sample estimate of the population parameter, d_{yx}, is indicated by $\hat{d}_{yx}$,

$$\hat{d}_{yx} = \frac{207{,}338 - 100{,}133}{207{,}338 + 100{,}133 + 155{,}140}$$
$$= 0.232$$

If the sex role item is taken as the dependent variable and support for ERA is considered the independent variable,

$$\hat{d}_{yx} = \frac{207{,}338 - 100{,}133}{207{,}338 + 100{,}133 + 149{,}148}$$
$$= 0.235$$

Somers presents an approximation to the standard error of the sampling distribution that can be used to test for the significance of $\hat{d}_{yx}$.[4] This is:

4. Somers, "Simple Approximations to Null Sampling Variances."

$$\hat{\sigma}_{\hat{d}_{yx}} = \frac{2}{3R}\sqrt{\frac{(R^2 - 1)(C + 1)}{N(C - 1)}}$$

The test statistic for $\hat{d}_{yx}$ is:

$$Z = \frac{\hat{d}_{yx}}{\hat{\sigma}_{\hat{d}_{yx}}}$$

If we again choose $\alpha = .01$, Appendix C reveals that the critical value is 2.33 for a one-tailed test. We compute:

$$Z = \frac{0.232}{\frac{2}{(3)(4)}\sqrt{\frac{(4^2 - 1)(4 + 1)}{1{,}181\,(4 - 1)}}}$$
$$= 9.57$$

Since 9.57 is clearly larger than 2.33, we reject the null hypothesis that sex role orientation and support for ERA are unrelated.

9.2.6. Comparing Orderable Measures of Association

All of the statistics presented for the analysis of discrete orderable data are PRE measures (except for t_c). When deciding which statistic to use, you should first ask whether you are merely estimating association between the two variables or whether one of the variables is the dependent and the other the independent variable. *If one of the variables is clearly dependent, then d_{yx} should be used,* since it is the only asymmetric coefficient among the choices.

If you are merely interested in assessing association, you have three choices. Since gamma excludes all tied pairs in its computations, it is always larger than τ_b and τ_c. We do *not* see this as a positive feature, and for that reason we recommend one of the tau measures as more appropriate because they take tied pairs into account. Of these two measures, τ_b is better when the table being analyzed is square ($R = C$), and τ_c is more appropriate when the table is rectangular ($R \neq C$) rather than square.

Our own preference is to use the product-moment correlation coefficient (see Chapter 8) for data where there is clearly an underlying continuous variable, even if it is measured only at the discrete level. There is disagreement among statisticians and practitioners about this issue, however. Many published research reports employ the statistics presented in this chapter, and for these reasons it is important for you to be familiar with them.

9.3. The Association of Ranked Data: Spearman's Rho

The measures of association in the preceding sections all apply to data in crosstabulations where many cases are placed in the same category of each variable. There is another measurement procedure which results in each observation being scored uniquely. This involves **ranked data,** or orderable discrete measures, *in which each observation is assigned a number from 1 to N (the sample size) which reflects its standing relative to the other observations.* The position of a basketball team in the league standings is a ranking, as are seeds or standings for a tennis tournament. All students in a statistics class can be ranked according to their placement on the midterm exam if there are no tied scores. Or each state in the union can be ranked from 1 to 50 on its industrial

ranked data—orderable discrete measures in which each observation is assigned a number from 1 to *N*, to reflect its standing relative to the other observations

TABLE 9.4

Rank-Order Correlation for 10 Largest Cities on Population and Municipal Function

City	*Population (000)*	*Functions (N)*	*Population Rank*	*Functions Rank*	D_i	D_i^2
New York	7,895	9	1	1.0	0	0.00
Chicago	3,369	6	2	5.0	−3	9.00
Los Angeles	2,811	5	3	6.5	−3.5	12.25
Philadelphia	1,949	7	4	3.5	0.5	0.25
Detroit	1,514	7	5	3.5	1.5	2.25
Houston	1,234	4	6	8.5	−2.5	6.25
Baltimore	906	8	7	2.0	5	25.00
Dallas	844	4	8	8.5	−0.5	0.25
Cleveland	751	3	9	10.0	−1	1.00
Milwaukee	717	5	10	6.5	3.5	12.25
Total						68.50

Source: NORC Permanent Community Sample (63-cities data).

wealth, population size, number of traffic fatalities, and numerous other variables.

A descriptive statistic which measures the relationship between ranked data is **Spearman's rho** (ρ_s). The two sets of rankings of the same observations on the two variables are compared by: (1) taking the difference of ranks (D_i), (2) squaring the difference in ranks (D_i^2), and (3) adding up these squared differences $\left(\sum_{i=1}^{N} D_i^2\right)$. Then this value is placed in the formula below, with r_s referring to the sample estimate of the population parameter, that is, ρ_s:

Spearman's rho—an asymmetric measure of association for two ranked variables

$$r_s = 1 - \frac{6 \sum_{i=1}^{N} (D_i^2)}{(N)(N^2 - 1)}$$

9.3.1. An Example: City Population and Municipal Function Rank Order

A brief illustration of r_s uses data from the 63-cities study. Table 9.4 displays the 1970 populations of the 10 largest cities and the number of municipal functions performed by them, as well as the ranks of the cities on each variable. Notice that *whenever two or more observations are tied for the same place, the ranks they could hold are averaged, and this average rank is assigned to all the tied observations.* The last two columns in the table show the difference in ranks for the two variables and the square of these differences, whose sum is used in the calculation of r_s. The rank-order correlation is:

$$r_s = 1 - \frac{(6)(68.50)}{(10)(10^2 - 1)} = 1 - \frac{411}{990} = +\,0.585$$

In comparison, Pearson's product-moment correlation coefficient for the continuous scores (see Section 8.4.2 in Chapter 8) is $r = +0.613$, which is quite similar.

In this example we rank cities on the basis of continuous variables in order to illustrate the principles of rank correlation. The preferable method for showing covariation with continuous measures is, of course, the correlation coefficient.

If N is equal to or greater than 10 and you are testing the null hypothesis, that is, $H_0: \rho_s = 0$ in the population, you can use this formula:

$$t_{N-2} = r_s \sqrt{\frac{N-2}{1-r_s^2}}$$

When N is smaller than 10 this test should be performed with caution, since it may yield a poor approximation. In this example, suppose we set $\alpha = .05$. For a one-tailed test, the critical value is 1.86. To test the null hypothesis that $\rho_2 = 0$ in the population, we compute:

$$t = 0.585 \sqrt{\frac{10-2}{1-0.585^2}}$$
$$= 2.04$$

Hence we reject the null hypothesis and conclude that there is a positive relationship between city size and municipal functions performed.

9.4 The 2 × 2 Table

Two measures of association in popular use—Yule's Q and phi—are applicable only to 2 × 2 tables. This section, therefore, gives special attention to estimating relationships in crosstabulations of dichotomous variables.

In standard labeling for 2 × 2 tables, the first four italic letters are used to designate cell frequencies, as follows:

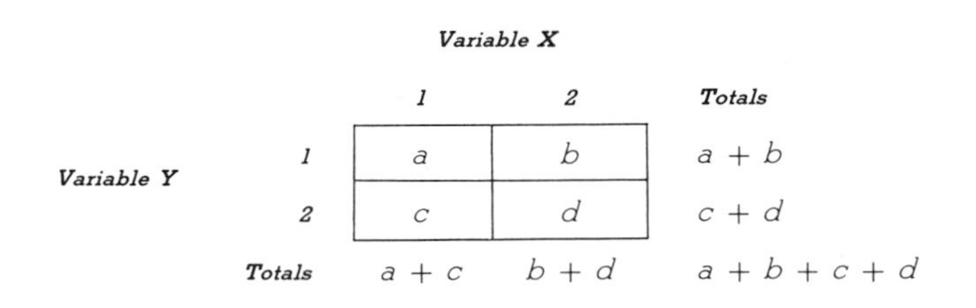

		Variable X		
		1	*2*	**Totals**
Variable Y	*1*	a	b	$a + b$
	2	c	d	$c + d$
	Totals	$a + c$	$b + d$	$a + b + c + d$

An equivalent notation uses f_{ij} to represent the frequencies in row i and column j; the dot in the subscript indicates the sum of all the values across the subscript. That is, $a = f_{11}$, $b = f_{12}$, $c = f_{21}$, $d = f_{22}$, $a + b = f_{1.}$, $c + d = f_{2.}$, $a + c = f_{.1}$, $b + d = f_{.2}$, and $a + b + c + d = f_{..} = N$. Therefore, we can write an approximation formula for χ^2 for a 2×2 table in two separate notations.[5] These are:

$$\chi_1^2 = \frac{N(bc - ad)^2}{(a + b)(a + c)(b + d)(c + d)}$$

$$\chi_1^2 = \frac{N(f_{11} f_{22} - f_{12} f_{21})^2}{f_{.1} f_{.2} f_{1.} f_{2.}}$$

In general, the degree of freedom for a χ^2 test of an $R \times C$ table is $(R - 1)(C - 1)$. Thus, for a 2×2 table, $df = (2 - 1)(2 - 1) = 1$.

9.4.1. An Example: Effects of Race on Busing Attitudes

In investigating measures of association for 2×2 tables, we will use an example from race relations. Racial and ethnic conflict is one of the most deeply ingrained social problems of contemporary American society. Whites and blacks, for example, express wide disagreement over solutions to end the harmful effects of racism on housing, employment, and educational opportunities. In general, we propose, blacks tend to favor integrationist policies more than whites do. Specifically, with regard to public education:

H1: Blacks favor busing for school integration more than whites do.

5. Many earlier statistics texts advised that something called Yates's correction for continuity should be applied when the expected frequency in a cell is equal to or less than 5.0. Recent studies indicate that both the uncorrected χ^2 formula and Yates's correction give biased estimates of true α levels. The formulas that give better estimates of the true α levels for 2×2 tables are too cumbersome to present here.

Unfortunately, SPSS CROSSTABS (see Preface) automatically applies Yates's correction for continuity to all 2×2 tables with $N \geq 20$. All we can do is to caution you that this problem exists; presumably, the SPSS package will be modified sometime in the future to deal with this problem.

To test this hypothesis using the 1977 General Social Survey, we measured a respondent's race by the interviewer's perception (or inquiry if race was not obvious). Then we dichotomized the measure into black and white classifications. Preference on school busing was operationalized by the item, "In general, do you favor or oppose the busing of black and white children from one district to another?"

TABLE 9.5
Crosstabulation of Race and Busing Attitude Variables

School Busing	*Race*		
	Black	*White*	*Total*
Favor	47.6%	12.6%	16.6%
(N)	(80)	(164)	(244)
Oppose	52.4	87.4	83.4
(N)	(88)	(1,136)	(1,224)
Total	100.0%	100.0%	100.0%
(N)	(168)	(1,300)	(1,468)

Source: 1977 General Social Survey.

Table 9.5 displays the crosstabulation of these two dichotomous measures. Note the sharp racial cleavage on this issue, which is one of the most persistent problems in race relations. Not surprisingly, the χ^2 for this table is very large:

$$\chi_1^2 = \frac{(1{,}468)\,[(164)(88) - (80)(1{,}136)]^2}{(244)(1{,}224)(168)(1{,}300)}$$

$$= 131.53$$

This is highly significant, even for $\alpha = .001$. This formula for χ^2 is only an approximation, and it may fit poorly when sample size is small. In such cases, Fisher's exact test, which is described in most more advanced statistics texts, should be used.[6]

9.4.2 Yule's Q

Yule's Q—a symmetric measure of association for 2×2 cross tabulations, equivalent to gamma

One measure of association which is suitable only for 2×2 tables is **Yule's Q,** which makes use of cross-products, that is, the two

6. When $N \leq 20$, SPSS CROSSTABS calculates Fisher's exact test automatically.

TABLE 9.6
Three Hypothetical Examples of Q

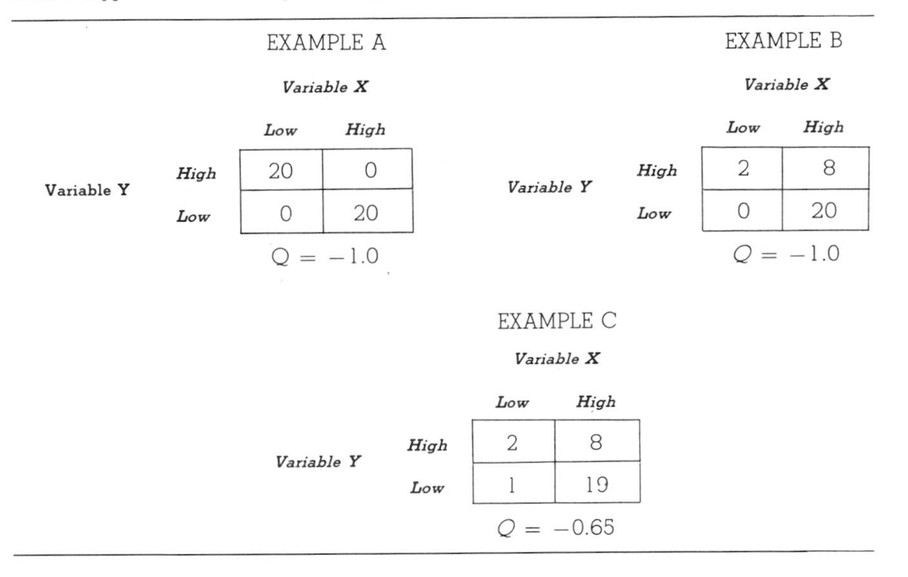

EXAMPLE A

Variable Y		Variable X: *Low*	Variable X: *High*
	High	20	0
	Low	0	20

$Q = -1.0$

EXAMPLE B

Variable Y		Variable X: *Low*	Variable X: *High*
	High	2	8
	Low	0	20

$Q = -1.0$

EXAMPLE C

Variable Y		Variable X: *Low*	Variable X: *High*
	High	2	8
	Low	1	19

$Q = -0.65$

products of the pair of cell frequencies diagonal to each other. By convention, as noted above, the four cells of a 2×2 table are identified by italic letters: starting at the upper left and ending at the lower right, these labels are *a, b, c,* and *d.* The two cross-products, (ad) and (bc), are then placed into the familiar PRE format:

$$Q = \frac{(bc) - (ad)}{(bc) + (ad)}$$

Note that if one of the four cell frequencies is zero, the value of Q must either be -1.00 or $+1.00$. Yet less than a "perfect" relationship may exist between the two dichotomous variables. In this case, *perfect* means that all cases fall either into the two main diagonal cells or into the two off-diagonal cells.

Clearly, Q can give misleading information when zero cells are present. This is shown in hypothetical illustrations in Table 9.6. In Example A, a maximum negative relationship is shown, with all 40 observations located in the two off-diagonal cells. Q attains its maximum value, -1.0. In Example B, Q also has a value of -1.0, but the cell entries show that the majority of the cases in the high level of variable Y are concentrated in the main diagonal cell. Example C shows that if just a single observation from example B were to be shifted from d to c, the value of Yule's Q would drop substantially, to -0.65.

These illustrations underscore *one of the major defects of Yule's Q—its sensitivity to distortion because of cell frequencies of zero.* The problem is most serious when the sample size is small (as in these hypothetical illustrations) or when the frequencies in some variable categories are rare. For these reasons, you should always carefully inspect the cell frequencies in a 2 × 2 table and examine it for the presence of zero cell frequencies.

Using the real data in Table 9.5, we can readily calculate:

$$Q = \frac{(164)(88) - (80)(1{,}136)}{(164)(88) + (80)(1{,}136)}$$
$$= 0.726$$

Notice that *this value is the same as would be obtained for gamma.* A close look at the formulas for both measures of association reveals that *Q is a special case of gamma* when $R = C = 2$. The difference is that while gamma can be calculated for crosstabulations of any size, Q can be calculated only for 2 × 2 tables.

To test the statistical significance of Q, we can use the following estimate of the variance of Q:

$$\hat{\sigma}^2_Q = \tfrac{1}{4}(1 - Q^2)^2 \left(\frac{1}{a} + \frac{1}{b} + \frac{1}{c} + \frac{1}{d} \right)$$

Using the example data:

$$\hat{\sigma}^2_Q = 1/4\ (1-(-0.726)^2)^2 \left(1/80 + 1/164 + 1/88 + 1/1{,}136 \right)$$
$$= 0.0017$$

Thus the standard error of the sampling distribution of Q is:

$$\hat{\sigma}_Q = \sqrt{0.0017} = 0.0415$$

In a Z test of the null hypothesis that Q equals zero in the population for $\alpha = .01$, the critical value for a one-tailed test is 2.33. Thus:

$$Z = \frac{0.726}{0.0415} = 17.5$$

We would therefore reject the null hypothesis and instead conclude that whites are more opposed to busing than blacks are.[7]

9.4.3. Phi

Phi (ϕ) is a correlation coefficient which is used to estimate association in a 2 × 2 table; in fact, it is identical to r, the correlation coefficient in a sample (see Section 8.6.5 in Chapter 8), but we do not show this. The range of phi lies between -1.00 and $+1.00$. In the case of nonorderable discrete variables, however, the sign of phi depends solely on how the outcomes of both variables are coded. For example, males may be coded 1 and females 0, or vice versa.

phi—a symmetric measure of association for 2 × 2 crosstabulations, equivalent to the correlation coefficient

Unlike Yule's Q, phi is sensitive to the distribution of cases in the row and column marginals of the table, as can be seen by the formula:

$$\phi = \frac{bc - ad}{\sqrt{(a + b)(c + d)(a + c)(b + d)}}$$

TABLE 9.7
Observed Value and Maximum Absolute Value of Phi, Hypothetical Data

A. OBSERVED VALUE

		Variable A		
		0	1	Total
Variable B	1	15	10	25
	0	5	20	25
	Total	20	30	50

$\phi = -0.41$

B. MAXIMUM VALUE

		Variable A		
		0	1	Total
Variable B	1	20	5	25
	0	0	25	25
	Total	20	30	50

$\phi_{max} = -0.82$

7. If the formula for computing Z for gamma is used instead of the formulas for Q, a somewhat smaller Z value (although also significant at the .01 level) is obtained. As pointed out in Section 9.2.2, the formula for estimating Z for gamma gives conservative estimates, which explains this discrepancy.

With Q, phi shares the cross-product difference in its numerator. But phi's denominator is the square root of the product of the four marginal totals. For any 2×2 table with a specific set of row and column totals, phi can attain a maximum or minimum value which may be considerably short of the hypothetical range between -1.00 and $+1.00$. Some researchers prefer to adjust phi to take this limitation into account. **Phi adjusted** (ϕ_{adj}) is found by dividing the observed value of phi by the maximum absolute value (ϕ_{max}) which might be obtained with the given set of marginals:

phi adjusted—a symmetric measure of association for 2×2 crosstabulation in which phi is divided by phi maximum to take into account the largest covariation possible, given the marginals

$$\phi_{adj} = \frac{\phi}{|\phi_{max}|}$$

In this formula, the parallel bars on the denominator indicate "absolute value."

To find **phi maximum** (ϕ_{max}), we simply reduce the entries in one of the four cells to zero and alter the frequencies in the other three cells. Table 9.7 gives an illustration with hypothetical data. The observed value is -0.41. The maximum value, obtained by removing all cases from cell c (any of the four cells could be used as the focal cell), is -0.82. Then $\phi_{adj} = -0.41/|-0.82| = -0.50$. This indicates a somewhat stronger inverse relationship than suggested in the observed data. In the race and busing data in Table 9.5, ϕ is -0.296, ϕ_{max} is -0.798, and therefore $\phi_{adj} = -0.371$.

phi maximum—the largest value that phi can attain for a given 2×2 crosstabulation; used in adjusting phi for its marginals.

While ϕ_{adj} is sometimes reported in the analysis of data from a 2×2 table, our recommendation is to do so with caution. Unlike ϕ, the square of ϕ_{max} does *not* equal the amount of variance in one variable explained by the other.

9.4.4. Relation to Chi-Square-Based Measures

In this section we have examined several measures of association for crosstabulations of discrete measures. You may wonder whether we can reduce these measures to a smaller set, or at least indicate how they are related to each other. To examine this possibility, we use the formula for χ^2 for a 2×2 table:

$$\chi^2 = \frac{N(bc - ad)^2}{(a + b)(a + c)(b + d)(c + d)}$$

We also use the formula for phi presented in Section 9.4.3:

$$\phi = \frac{bc - ad}{\sqrt{(a + b)(a + c)(b + d)(c + d)}}$$

Then we conclude that:

$$\phi^2 = \frac{\chi^2}{N}$$

$$\phi = \pm\sqrt{\frac{\chi^2}{N}}$$

We divide the χ^2 for a 2×2 table by the sample size and then take the square root (positive or negative root, depending on the direction of the association), which gives the value of ϕ. Since both Cramer's V and C, the contingency coefficient (see Box 9.1 above), are based on χ^2, we can state the relationship between these two measures of association and phi:

$$V = \frac{|\phi|}{\sqrt{\text{Min}(r - 1, c - 1)}} = |\phi|$$

$$C = \frac{|\phi|}{|\sqrt{\phi^2 + 1}|}$$

(Here again, the parallel lines mean "absolute value.")

As an exercise, you can calculate V and C using these formulas on the data in Table 9.5. Then compare the results to those found with the standard formulas for V and C.

Superficially, phi and Yule's Q appear to be simple transformations of each other, since both formulas contain the cross-product $(bc) - (ad)$ in their numerators. But this surface resemblance is misleading, since neither denominator contains common terms which can be canceled. Thus, the relationship of Q to ϕ is cumbersome and unrevealing:

$$\phi = Q\,\frac{(bc) + (ad)}{\sqrt{(a + b)(a + c)(b + d)(c + d)}}$$

This relationship tells us only that ϕ takes on smaller values than Q (except when zero and ± 1.00 occur without empty cells). This is true because the adjustment factor has a numerator that is always smaller than the denominator.

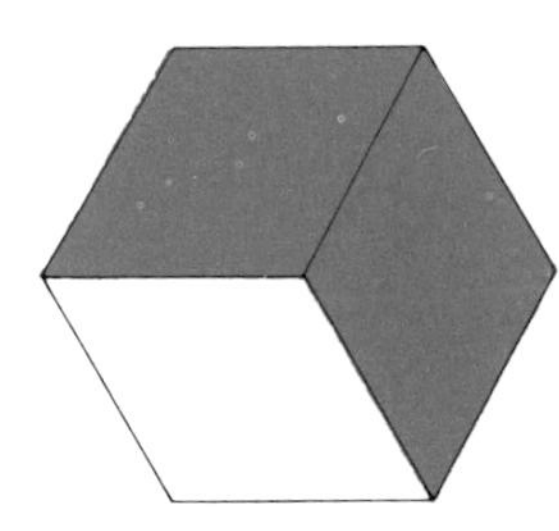

Review of Key Concepts

This list of the key concepts introduced in Chapter 9 is in the order of appearance in the text. Combined with the definitions in the margins, it will help you review the material and can serve as a self-test for mastery of the concepts.

lambda
Cramer's V
contingency coefficient, (C)
gamma
concordant pairs
discordant pairs
tau b
tau c
Somers's d_{yx}
ranked data
Spearman's rho
Yule's Q
phi
phi adjusted
phi maximum

PROBLEMS

General Problems

1. In the 1976 General Social Survey the crosstabulation between religious identification and approval of abortion if a woman has been raped is:

Religious Identification	*Approval of Abortion*		
	Yes	*No*	*Total*
Protestant	774	135	909
Catholic	286	89	375
Jewish	27	0	27
None	103	10	113
Other	13	1	14
Total	1,203	235	1,438

a. Compute the λ for predicting abortion attitude from religious identification. Your conclusion?

b. Compute χ^2 as a test of statistical significance with $\alpha = .01$. What do you conclude?

c. Compute V and C (see Box 9.1).

2. We might expect a substantial relationship between respondents' current religious identifications and those they grew up with and had at age 16. In the 1978 General Social Survey, the data are:

Current Religious Identification	*Religious Identification at Age 16*					
	Protestant	*Catholic*	*Jewish*	*None*	*Other*	*Total*
Protestant	918	30	1	28	1	978
Catholic	27	351	1	5	0	384
Jewish	1	0	28	0	0	29
None	65	34	1	15	1	116
Other	5	3	0	0	9	17
Total	1,016	418	31	48	11	1,524

a. Compute λ with current religion as the dependent variable. What do the results suggest?

b. $\chi^2 = 3{,}119.85$. Can the null hypothesis be rejected for $\alpha = .001$?

3. To examine the relationship between crime rate and region of the country, the number of serious crimes per 1,000 population was crosstabulated with region, using the 63-cities data:

Region	*Serious Crimes per 1,000 People*			
	0–35	*36–54*	*55 or more*	*Total*
Northeast	9	1	5	15
South	3	9	4	16
Midwest	8	6	4	18
Far West	3	4	7	14
Total	23	20	20	63

a. Relatively, which region has the highest crime rate?

b. Compute λ with crime rate as the dependent variable.

c. Compute χ^2 as a test of significance for λ. Set $\alpha = .05$.

d. Compute V and C.

4. An investigator is interested in whether a person's perception of life as exciting or dull is affected by his or her current marital status. The data from the 1977 GSS are:

Marital Status	*Perception of Life*			*Total*
	Exciting	*Routine*	*Dull*	
Married	433	482	47	962
Widowed	46	89	24	159
Divorced	37	55	8	100
Separated	26	25	8	59
Never married	122	80	14	216
Total	664	731	101	1,496

a. Relatively, which marital group most describes life as "exciting"?
b. What is λ with attitude about life as dependent?
c. Given $\chi^2_8 = 52.06$, is the relationship significant for $\alpha = .001$?
d. Compute V and C.

5. Some people seem to think that the grass is greener on the other side of the fence. Using the 1977 GSS data, the U.S. regions where respondents live were crosstabulated with their self-reported happiness:

Region	*Self-Reports of Happiness*			*Total*
	Very Happy	*Pretty Happy*	*Not Too Happy*	
New England	15	42	5	62
Mid-Atlantic	79	119	39	237
East North Central	111	193	38	342
West North Central	48	57	8	113
South Atlantic	112	160	41	313
East South Central	28	39	9	76
West South Central	46	57	16	119
Mountain	24	33	4	61
Pacific	69	118	22	204
Total	532	813	182	1,527

a. Compute λ.
b. $\chi^2_{16} = 20.08$. If $\alpha = .05$, what do the results of *a* and *b* suggest?

6. In the 1974 General Social Survey, respondents were asked whether the reading of the Lord's Prayer or other Bible verses

should be either required or not allowed in all U.S. public schools, or whether this decision should be left to each state or local community. The crosstabulation of opinions on this matter and religious identification is:

Religion	Reading of Lord's Prayer			
	Required	Forbidden	Local Option	Total
Protestant	169	23	266	458
Catholic	48	10	115	173
Jewish	0	12	11	23
None	4	15	25	44
Other	2	0	3	5
Total	223	60	420	703

a. Compute χ^2, V, and C.
b. Compute λ with opinion as the dependent variable.
c. Percentage the table with opinion as the dependent variable.
d. What do the results of *a*, *b*, and *c* above suggest about the relationship?

7. In the 1977 General Social Survey, white respondents were asked whether they would object if someone in their family brought a black person home for dinner. The crosstabulation of responses to this question with educational degree is:

Educational Degree Held	Black to Dinner?			
	Strongly Object	Mildly Object	Not Object	Total
Less than high school	82	93	264	440
High school	64	109	486	659
Associate/ Jr. college	2	2	25	29
Bachelor	3	12	115	130
Graduate	2	4	66	72
Total	153	220	957	1,330

a. Percentage the table with attitude as the dependent variable. What do the results suggest?
b. Compute γ.
c. Is γ significant for $\alpha = .01$?

8. Some sociologists of medicine claim that health care and education are positively related. The crosstabulation in the 1977 GSS of perceived health and educational degree held is:

Educational Degree Held	Health				Total
	Excellent	Good	Fair	Poor	
Less than high school	101	195	165	75	536
High school	260	336	121	23	740
Associate/ Jr. College	15	13	4	1	33
Bachelor	78	44	14	3	139
Graduate	30	33	8	2	73
Total	484	621	312	104	1,521

a. Compute G.

b. Compute λ with health dependent.

c. For a $\chi^2_{12} = 177.92$, is the relationship significant for $\alpha = .001$?

9. Compute t_c for Problem 7. Set $\alpha = .001$ and test the null hypothesis that $t_c = 0$.

10. Compute $\hat{d}_{yx}$ for Problem 7 with the attitude as the dependent variable. Set $\alpha = .01$ and test the null hypothesis that $d_{yx} = 0$ in the population.

11. In the table below are ranked data for the first 10 cities in the 63-cities data set:

City ID	Name	Age Rank	Population Size Rank	No. of Functions Rank
0	Akron, Ohio	6.5	5	6.5
1	Albany, N.Y.	2	8	3
2	Amarillo, Tex.	9	6	6.5
3	Atlanta, Ga.	5	2	6.5
4	Berkeley, Calif.	8	7	6.5
5	Birmingham, Ala.	6.5	4	6.5
6	Bloomington, Minn.	10	10	10
7	Boston, Mass.	1	1	1
8	Buffalo, N.Y.	3	3	6.5
9	Cambridge, Mass.	4	9	2

a. Compute the rank-order correlation between age rank and population size rank.

b. Test the null hypothesis that $\rho_s = 0$, with $\alpha = .05$.

12. *a.* Using the data in Problem 11, calculate the rank-order correlation between population size rank and number of functions rank.

b. Test the null hypothesis that $\rho_s = 0$, setting $\alpha = .05$.

c. Explain how r_s in this problem could differ so much from that based on the data in Table 9.4, where $r_s = .0.585$.

13. Consider the following 2 × 2 table:

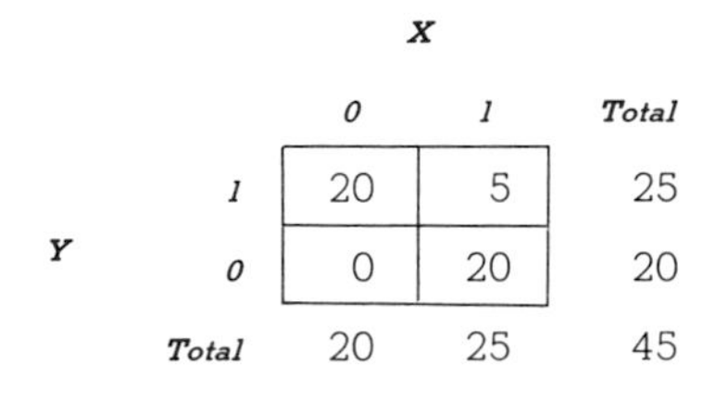

		X		
		0	*1*	*Total*
Y	*1*	20	5	25
	0	0	20	20
	Total	20	25	45

Compute *(a)* χ^2, *(b)* Q, and *(c)* ϕ.

14. In the 63-cities data, the following relationship between per capita income and region is found:

		Per Capita Income		
		$3,000 or less	*$3,000 or more*	*Total*
Region	*Other regions*	16	33	49
	Far West	0	14	14
	Total	16	47	63

Compute *(a)* χ^2, *(b)* Q, and *(c)* ϕ and ϕ_{max}.

15. The crosstabulation of population change between 1960 and 1970 and region in the 63-cities data is:

		Population Change		
		Negative or No Change	*Positive Change*	*Total*
Region	*Far West*	4	10	14
	Other regions	30	19	49
	Total	34	29	63

Determine whether the Far West is associated with population growth between 1960 and 1970 by *(a)* computing Q and testing it for statistical significance with $\alpha = .05$, and *(b)* computing ϕ.

16. Find what ϕ^2 equals if *(a)* $\chi_1^2 = 6.72$ and $N = 70$; *(b)* $\chi^2 = 0.200$ and $N = 100$; and *(c)* $\chi_1^2 = 9.24$ and $N = 50$.

17. If $\chi_1^2 = 21.2$ and $\phi^2 = 0.106$, what does N equal?

18. In a 3 × 4 table, if $\phi = -0.30$, *(a)* What does V equal? *(b)* What does C equal?

19. In a 2 × 2 table, if $\phi = 0.40$, *(a)* What does V equal? *(b)* What does C equal?

Problems Requiring the 1980 General Social Survey Data

20. There are some who believe that the astrological sign under which a person is born is related to such personal outcomes as health and happiness. Crosstabulate ZODIAC by HAPPY, computing χ^2, λ, V, and C. Set $\alpha = .01$. What do you conclude?

21. Perhaps the astrological sign under which one is born is related to one's health. Find out by crosstabulating ZODIAC and HEALTH, setting $\alpha = .01$. Again, calculate χ^2, λ, V, and C. Your conclusion?

22. Since the Catholic religion teaches that divorce is a sin, you might expect that Catholics are divorced less than members of other religions. To test this hypothesis, set $\alpha = .01$ and crosstabulate RELIG by MARITAL, after recoding MARITAL into two categories—divorced and not divorced. Compute χ^2, λ, V, and C. What do you conclude?

23. You might hypothesize that a person's residence on a rural-urban continuum would be related to attitude about whether the government is spending the right amount on protection of the environment. Crosstabulate SRCBELT and NATENVIR, obtaining G, t_b and $\hat{d}_{yx}$. Set $\alpha = .05$. What do you conclude?

24. Liberals are often characterized as favoring government expenditures, compared to conservatives. Test this hypothesis by crosstabulating POLVIEWS and attitude about whether the government is spending the right amount on space exploration (NATSPAC). Set $\alpha = .01$. Compute G, t_b and $\hat{d}_{yx}$. What do you conclude?

25. Repeat the analysis in Problem 24, substituting NATENVIR for NATSPAC. Do the conclusions drawn from the results in Problem 24 change as a result of this analysis?

26. Belief in life after death might be expected to be related to attitude about abortion. To find out, crosstabulate POSTLIFE by ABSINGLE (attitude as to whether a single woman who becomes pregnant should be able to get an abortion). Set $\alpha = .01$. Compute χ^2 and ϕ. What do you conclude?

27. Repeat the analyses in Problem 26 by substituting attitude about abortion because the baby might have a serious defect (ABDEFECT) for ABSINGLE. Does the conclusion drawn in Problem 26 change?

Problems Requiring the 63-Cities Data

28. Determine, using the 63-cities data, whether per capita income (INCOMEPC) varies by region (REGION). (Recode income as follows: 242 thru 300 = 1; 301 thru 350 = 2; 351 thru HIGHEST = 3.) Set $\alpha = .05$. Compute χ^2, V, C, and λ, with per capita income as the dependent variable. Interpret the results.

29. Crosstabulate crime rate (CRIMRATE) by region (REGION) to see if the crime rate varies by region of the country. Recode CRIMRATE as follows: 8 thru 35 = 1; 36 thru 54 = 2; 55 thru HIGHEST = 3. Set $\alpha = .05$. Compute χ^2, V, C, and λ, with crime rate as the dependent variable.

30. Determine whether there is a relationship between per capita income (INCOMEPC) (recoded as in Problem 28) and percent white collar in the labor force (WCPCT) by crosstabulating the two variables. Recode WCPCT as follows: 0 thru 50 = 1; 51 thru 55 = 2; 56 thru HIGHEST = 3. Set $\alpha = .05$, and compute G, t_b and $\hat{d}_{yx}$, treating WCPCT as the dependent variable. What do you conclude?

31. Determine whether there is a relationship between the crime rate (CRIMRATE) (dependent variable) and per capita income (INCOMEPC) by crosstabulating the two variables. Use the recodes from Problems 28 and 29. Set $\alpha = .05$, and compute G, t_b and $\hat{d}_{yx}$. What do you conclude?

32. Determine whether Far West cities have a higher per capita income than the other regions of the country by crosstabulating REGION and INCOMEPC, after recoding the two variables. Recode REGION: 1, 2, 3 = 1 and 4 = 2; INCOMEPC:

0 thru 300 = 1 and 301 thru HIGHEST = 2. Set $\alpha = .05$, and compute χ^2, G, and ϕ. What do you conclude?

33. Determine whether Far West cities had grown more between 1960 and 1970 than cities in other parts of the country by crosstabulating REGION and POPCHANG. Recode REGION as in Problem 32 and POPCHANG as follows: -17 thru 0 = 1; 1 thru HIGHEST = 2. Set $\alpha = .05$, and compute χ^2, G, and ϕ. What do you conclude?

10

The Logic of Multivariate Contingency Analysis

In our examination of various aspects of the relationship between two variables in preceding chapters, we have discussed measures of association and tests of significance for both discrete and continuous variables. Tools have been presented to help you determine whether two variables systematically covary, and whether the covariation observed in sample data is likely to reflect that present in the population from which the sample was drawn.

For some research purposes, establishing that two variables significantly covary may be sufficient. In most instances the fact that, for example, on average men have higher earned incomes than women, even in the same occupation, hardly needs to be verified again with a new set of data. But the researcher may want to explore the income difference as a consequence of other social factors, such as amount of education, work experience, employment status (full or part time), and employer discrimination. In such cases, the research problem changes from the examination of a two-variable relationship to consideration of three or more variables, as their relationships bear upon some theoretical issue. In this chapter and the next, we present some of the basic procedures for conducting **multivariate contingency analysis**, or statistical analysis of data on three (or more) variables, using both discrete and continuous variables.

multivariate contingency analysis—statistical techniques for analyzing relationships among three or more discrete variables

10.1. Controlling Additional Variables

A basic reason for bringing additional variables into the analysis of the relationship between an independent and a dependent variable is to clarify the true relationship between them. Covariation between two variables can arise because of the confounding effects of other factors. To establish the true amount of covariation between two variables, we need to remove the part that is due to other factors.

In laboratory-type experiments the removal of the effects of other factors is accomplished by application of the experimental design. Some additional variables can literally be "held constant" by making sure they apply uniformly to all subjects under experimental and control conditions. For example, in studying the effects of different amounts of fertilizer on the productivity of seed crops, we hold constant the effects of soil, water, and sunlight by making sure that all experimental plots contain the same type of soil, receive the same amounts of water, and have the same exposure to sunlight. Then we can be fairly sure that any plant growth differences we find would not be due to differences in these rival factors that are known to affect crop growth. In a social experiment we might hold constant the manner in which stimuli are presented to subjects and their responses are recorded.

random assignment—in an experiment, the assignment of subjects to treatment levels on a chance basis

Variables which might disturb a bivariate relationship can be controlled in experiments by **random assignment** of subjects to the different experimental treatments. For example, to study the effects of marijuana on driving ability, we must try to eliminate such possible confounding effects as intelligence and motor skills. Clearly the results of the experiment would not be credible if all the better drivers had been exposed to low dosages and all the worst drivers to high dosage conditions. Since it would be very hard to obtain subjects who had identical IQ and manual dexterity scores, the preferred solution is to assign persons at random to smoke different amounts of marijuana before taking the driving test (including a control group which would smoke no marijuana). Random assignment to a treatment group might be made by flipping coins or by consulting a table of random numbers. The purpose of this procedure is to ensure that no experimental group differs, *on average,* from the others in driving abilities before subjects ingest different levels of the drug.

The technique of random assignment of subjects to experimental treatment groups helps to eliminate the confounding effect of rival factors. Thus it helps isolate the true impact of the independent variable on the dependent measure. Unfortunately, all social behavior cannot be studied experimentally. In naturally occurring data such as that collected through sample surveys, other techniques for eliminating rival factors must be used. These techniques consist of identifying the additional variables likely to affect a relationship, measuring these factors, and "holding constant" their effects through statistical manipulation of the data.

Statistical control of rival factors differs from experimental control in that the researcher has no direct physical ability to shape the attributes of the persons or objects studied. The methods for controlling additional variables statistically adjust the data in an attempt to render the respondents equivalent on the rival factors, thereby eliminating their impact on the bivariate relationship of interest. Statistical controls are less powerful than randomization in eliminating the confounding effects of other variables for two reasons: (1) measurement error, which is always present to some degree, reduces the precision with which statistical adjustments can be made; and (2) identification or inclusion of all potential rival factors may be impossible.

The second of these limitations suggests the extreme importance of theory in guiding social research. The decision to select some variables to be controlled statistically and not to consider others among the many possible candidates depends on our understanding of the role such variables play in the theory under investigation. For example, in studying the impact of religiosity on teenage premarital sexual behavior, we would be guided by theory and past research to control such factors as degree of parental supervision, academic ability, participation in school sex education programs, and peer group attitudes. But we would not try to adjust for the effects of physical appearance and food preferences.

The basic purpose of statistical controls is to eliminate, or at least reduce, the effect of confounding factors on a bivariate relationship. It is possible to distinguish three special applications, that provide various means to control for spuriousness, interpretation, and multiple causes.

10.1.1. Spuriousness

spuriousness—covariation between two variables due only to the effect of a third variable

An important point in controlling additional variables is that *establishing covariation between two variables is not equivalent to proving causation.* Even if the independent variable is shown to change in time prior to a change in the dependent variable, the conditions for inferring a causal relationship are not complete. Besides time order and covariation, a causal inference must eliminate the possibility of **spuriousness** in the observed relationship (see Section 12.1 in Chapter 12). Two variables are spuriously related when the only reason they correlate is that both are caused by one or more other variables.

A classic illustration of a spurious relationship is the observation that, in Holland, communities where many storks nest in chimneys have higher birth rates than communities where fewer storks nest. While covariation and (probably) temporal-order conditions can be reasonably met, we should not conclude that storks cause babies, even if we have no knowledge of human reproduction, because we have not ruled out the possibility that the observed pattern is a spurious consequence of one or more rival factors which are simultaneously affecting both the number of babies and the number of storks. Storks and babies are more prevalent in rural than in urban areas. Pollution and sanitation levels, community attitudes toward human and animal life, and historical patterns of selective migration may all combine to create a spurious correlation between these two variables.

Take another classic case: The number of fire trucks called to a fire and the subsequent amount of damage done to buildings covary together. Can you figure out what rival factor(s) makes this a spurious, noncausal relationship?

These examples present clear-cut cases of spurious relationships which disappear when the appropriate common cause of both variables is held constant (by techniques described later). Less obvious cases of spurious relationships may occur in social behavior.

> Where a covariation results because two variables have a common cause but are causally unrelated themselves, statistically holding constant the common factor (also called *partialling out* the effects due to other varia-

bles) eliminates the covariation observed in the data. If successive statistical controls for alleged common causes fail to alter the observed covariation, there are much firmer grounds than existed in the simple bivariate relationship for asserting the establishment of a causal relationship.

The variables chosen to control in these cases, of course, must be realistic candidates as causes of spuriousness. Social theory and past empirical research are indispensable sources of information about appropriate factors to control for when examining covariation for spuriousness.

10.1.2. Explanation

Another reason for bringing additional variables into the analysis of bivariate relationships is to attempt an **explanation** (or *interpretation*) of an observed association. Two variables may be causally related, but the process may be more complex than the simple correlation implies. Controlling the relationship for variables representing the *intervening process* connecting the independent and dependent variable can deepen our understanding of the bivariate relationship.

explanation (interpretation) of association—covariation between two variables due to an intervening third variable

For example, an inverse relationship between age and liberalism in personal morality is widely known to exist. That is, older persons tend to adhere to less permissive attitudes concerning child-rearing, sexuality, drug usage, and the like. One possible explanation for this relationship might be the degree of religiosity. Traditional religious values are more often found among older persons, both because they were raised in an era when such beliefs were strongly socialized (a generational or period process) and because as people age they may become more concerned about ultimate values (an aging process). Traditional religious doctrines contain many injunctions and prescriptions for belief and behavior in nonreligious matters. Hence, one reason older people may support less permissive morality is their greater religiosity, compared to younger people. (This explanation is commonly accepted by both religious leaders and social scientists.) By holding constant the level of religiosity, we might expect to reduce some or all of the observed covariation between age and personal morality beliefs.

FIGURE 10.1

Two Roles Played by a Third Variable in the Analysis of Bivariate Covariation

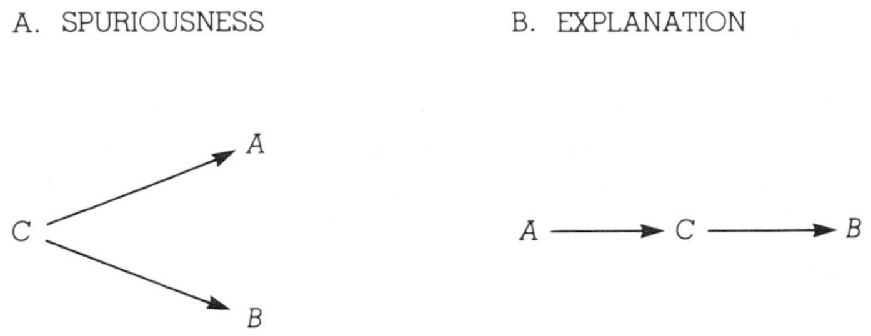

A. SPURIOUSNESS

$C \rightarrow A$

$C \rightarrow B$

B. EXPLANATION

$A \longrightarrow C \longrightarrow B$

Such an outcome would not imply that the two variables were spuriously related, but rather that the explanation for their covariation can be interpreted—at least in part—by a process in which greater age induces greater religiosity, which in turn restricts permissiveness in other areas. In a statistical sense, both spuriousness and explanation are similar in that holding constant other variables eliminates or reduces the original bivariate association. But in a substantive sense, the spuriousness and the interpretive analyses are quite different.

> A *spurious relationship* exists where the two original variables have no causal connection but are only dependent on a common cause. An *explanatory or interpretive relationship* exists where the two original variables have a causal connection and additional intervening variables elaborate the understanding of that connection.

Figure 10.1 illustrates the conceptual difference played by a third factor that completely accounts for the observed covariation between variables A and B according to the two types of effects, spuriousness and explanation. In the case of spuriousness, C is a common cause of both variables, while in the case of explanation, C is an intervening factor which shows in greater detail how the causal effect of A is transmitted to B. An understanding of the role a third variable plays in explicating the covariation between two others is not revealed by the statistical pattern. *The proper understanding of whether spuriousness or explanation is implied by an analysis depends on the researcher's ability to draw on past research and theory to conceptualize the roles that third variables play.*

10.1.3. Multiple Causes

Very few social theories, at least since the 19th century, have posited that human behavior is due to a single factor. Instead, many suggest **multiple causation;** that is, they identify several variables as causally important in the explanation of the variation of some dependent variable. Some factors may carry important policy implications because their values may be subject to manipulation, as through training programs or funding of services.

multiple causation—the view that social behavior is caused by more than one factor

In this type of research, the emphasis is not on the relationship between a single independent variable and the dependent measure, but on discovering the multiple, simultaneous relationship of several independent variables to the dependent measure(s) of interest. For example, in the study of social stratification, several researchers have developed and tested quite elaborate models of the status attainment process. An individual's earnings may be depicted as a consequence of the interplay of social background, education, occupation, industry, and other career variables. *By statistically controlling for several causal factors at the same time, the analyst can make some inferences about the relative importance of each factor for earnings, with spurious and explanatory relationships controlled.*

This chapter lays the groundwork by investigating the basic techniques for statistically controlling a single third variable in two-variable systems of categoric measures. The extension to more than three variable relationships is straightforward but is not covered in this text.

10.2 Controlling for a Third Variable in 2 × 2 Tables

As we noted in Chapter 4, the simplest bivariate relationship can be examined by the crosstabulation of two dichotomous variables. Similarly, the basic principles of examining a bivariate relationship for the effects of other variables utilize a *three-variable crosstabulation.* To present these principles, in this section we will describe the possible results of controlling for a third variable, illustrating different outcomes with purely hypothetical data.

10.2.1. A Hypothetical Example: Family Religiosity and Teenagers' Sex Activity

To make the hypothetical example more meaningful, we will assume a relationship between family religiosity (X) and premarital sexual intercourse among teenagers (Y). The hypothetical data in Table 10.1 classify 192 responses by whether or not teenagers

TABLE 10.1
Crosstabulation of Family Religiosity and Teenagers' Premarital Intercourse Variables, Hypothetical Data

		Family Highly Religious?		
		No	*Yes*	*Total*
Ever Premarital Sex?	*Yes*	42.9% (36)	16.0% (16)	27.1% (52)
	No	57.1% (56)	84.0% (84)	72.9% (140)
	Total	100.0% (92)	100.0% (100)	100.0% (192)

$r_{XY} = -0.26$

zero-order table—a crosstabulation of two variables in which no additional variables have been controlled

said they belonged to a "highly religious family" and whether or not they admitted to ever having had sexual relations. Table 10.1 can be called a **zero-order table**, where *order* refers to the number of other variables held constant.

phi coefficient—a measure of association for a 2 × 2 table, equivalent to the correlation coefficient for dichotomous variables

In this example, family religiosity is the independent variable and premarital sex the dependent variable. For this reason, we have percentaged the table on family religiosity, that is, the column percentages add to 100. As might be expected, nonreligious teenagers are considerably more likely to admit to premarital sex, at more than twice the rate of those who classify their families as religious. The correlation for this table is -0.26, which indicates a moderate inverse relationship. To compute this and other correlations in this chapter we have coded yes = 1 and no = 0. As suggested in Section 9.4.3 of Chapter 9, when the Pearson product moment correlation is applied to data in a 2 × 2 table, it could be referred to as a **phi coefficient** (ϕ).

We will consider whether this apparent causal relationship might be explained by the effects of other social processes which intervene between family religiosity and teenagers' premarital sexual experience. Suppose the provision of frequent opportunities to have sex is an important factor in determining whether a teenager actually has such an experience. Perhaps teenagers in more religiously oriented households have their activities restricted more by their parents. If this causal sequence exists, we would expect that when opportunity is statistically controlled, the relationship between religiosity and premarital sex might change.

Other possible outcomes could occur if the data in Table 10.1 were further crosstabulated by another dichotomy. In this hypothetical case, the control variable is whether or not the teenager has regular use of an automobile (*W*). Tables 10.2 to 10.5 illustrate four possible statistical patterns that might be found among these responses. Each result suggests a different interpretation about the original relationship. Each pair of tables is called a **first-order table,** since the number of other variables held constant is one.

first-order table—a subtable containing the crosstabulation or covariation between two variables, given a single outcome of a third, control variable

10.2.2 No Effect of Third Variable

In Table 10.2, the original 2 × 2 table has been split into two *subtables,* each resembling the original table in general form but with different cell frequencies. We now have a 2 × 2 × 2 table.

TABLE 10.2
Example of No Effect of Third Variable, Hypothetical Data

		Access to an Automobile?						
		No				***Yes***		
		Family Highly Religious?				***Family Highly Religious?***		
		No	***Yes***	***Total***		***No***	***Yes***	***Total***
Ever Premarital Sex?	***Yes***	39.7% (25)	15.9% (11)	27.3% (36)	***Yes***	37.9% (11)	16.1% (5)	26.7% (16)
	No	60.3% (38)	84.1% (58)	72.7% (96)	***No***	62.1% (18)	83.9% (26)	73.3% (44)
	Total	100.0% (63)	100.0% (69)	100.0% (132)	***Total***	100.0% (29)	100.0% (31)	100.0% (60)
		$r_{XY} = -0.27$				$r_{XY} = -0.25$		

Notice that the cell and marginal frequencies of the two subtables, when added pairwise, must exactly equal the frequencies in the original 2 × 2 table. As the total *N*s for both subtables show, about two-thirds of the teenage sample do not have regular use of an automobile. All respondents within the same subtable have the same level on the car access variable; that is, all 132 persons in the first subtable have no regular access and all 60 in the second subtable do have regular access. Thus, each subtable "holds constant" the variation in the third factor, permitting us to observe what happens to the covariation between the other two variables.

Note that in the example introduced with Table 10.1, the relationship between religiosity and sexual experience remains unchanged in Table 10.2. Both subtables of Table 10.2 show teens from highly religious families reported having sexual intercourse less than half as often as the other teenagers queried. The percentages are not noticeably different across the two subtables. Among persons without cars, $r_{XY} = -0.27$, and among persons with cars, $r_{XY} = -0.25$. These values are called **conditional correlation coefficients** because they refer to correlations under certain conditions of the third variable.

conditional correlation coefficients—correlation coefficients calculated between two crosstabulated continuous variables within each category of a third variable

If both first-order relationships in the two subtables are the same magnitude in the original zero-order relationship, we would conclude that opportunity—at least as operationalized by use of a car—has *no effect* on the original covariation. When such patterns are found, attention must be directed elsewhere for an explanation of why religiosity and premarital sex are related.

10.2.3. Partial Effect of Third Variable

Table 10.3 demonstrates another possible result when a third variable is held constant. In this case we can see that the magnitude of the association between religiosity and premarital sexual experience is the same in both subtables, but it is somewhat weaker than in the original 2 × 2 table. The conditional correlation coefficients are −0.16 and −0.17, respectively, for no regular access and regular access to a car. We can conclude that opportunity, as measured by regular use of an automobile, *partially accounts for the association* observed originally between the other two variables.

To understand what this partial explanation means, consider

TABLE 10.3
Example of Partial Effect of Third Variable, Hypothetical Data

		Access to an Automobile?						
		No				Yes		
		Family Highly Religious?				Family Highly Religious?		
		No	Yes	Total		No	Yes	Total
Ever Premarital Sex?	Yes	25.5% (12)	12.9% (11)	17.4% (23)	Yes	53.3% (24)	33.3% (5)	48.3% (29)
	No	74.5% (35)	87.1% (74)	82.6% (109)	No	46.7% (21)	66.7% (10)	51.7% (31)
	Total	100.0% (47)	100.0% (85)	100.0% (132)	Total	100.0% (45)	100.0% (15)	100.0% (60)
		$r_{XY} = -0.16$				$r_{XY} = -0.17$		

the percentages within each subtable. Among persons who are *not* from highly religious families, those able to use a car regularly are twice as likely to have had sex as those without access to a car (53.3% to 25.5%). Similarly, among teenagers from highly religious homes, those with access to a car are more than twice as likely to have had sex as those lacking such access (33.3% to 12.9%). However, the opportunity to use a car does not eliminate all the differences in sexual experience between the religious and nonreligious background teens. Teens from nonreligious families are still more likely to have had sex than teens from religious families, regardless of access to a car, but the differences have been diminished somewhat from that found in Table 10.1, where opportunity was not held constant. Hence, if these results were found, we would conclude that opportunity explains *part* of the association between religiosity and premarital sex, but considerable covariation still remains to be accounted for by other factors not yet held constant.

10.2.4 Complete Explanation by Third Variable

Table 10.4 shows what the data might look like if the third variable *completely accounts for the association* observed in the zero-order table. The correlations in both subtables (which you should check for yourself) are exactly zero, indicating that with each opportunity level held constant, no differences in sexual experience occur

TABLE 10.4.
Example of Complete Explanation by Third Variable, Hypothetical Data

		Access to an Automobile?						
		No				***Yes***		
		Family Highly Religious?				***Family Highly Religious?***		
		No	***Yes***	***Total***		***No***	***Yes***	***Total***
Ever Premarital Sex?	***Yes***	11.1% (4)	11.1% (8)	11.1% (12)	***Yes***	66.7% (32)	66.7% (8)	66.7% (40)
	No	88.9% (40)	88.9% (80)	88.9% (120)	***No***	33.3% (16)	33.3% (4)	33.3% (20)
	Total	100.0% (44)	100.0% (88)	100.0% (132)	***Total***	100.0% (48)	100.0% (12)	100.0% (60)
		$r_{XY} = 0.00$				$r_{XY} = 0.00$		

between the teens from highly religious and nonreligious families. Without regular access to a car, only 11.1% of the teenagers manage to have premarital intercourse, whereas among those with regular use of the car, two-thirds have had intercourse. The reason that the strong inverse relationship between family religiosity and sex appeared in the zero-order table is apparent from the marginal distributions in the two subtables: more than half the nonreligious background youths regularly use a car, but only one out of eight religious background teenagers have regular access to a car. If we found such clear-cut patterns in real data we would probably conclude that opportunity does explain the association. That is, a teenager's family religiosity determines the access he or she has to the family automobile (an opportunity process, in the sense that more religious parents are stricter about letting their children go on unchaperoned dates). In turn, the opportunity which use of an automobile permits is the major determinant of having premarital sexual intercourse.

What is the reason that holding access to automobiles constant explains the relationship of family religiosity to teenagers' premarital sex in Table 10.4 but not in Table 10.2? To find the answer, consider how access is related to the other two measures. As an exercise, form two tables, one showing the covariation between access and religiosity, the other showing the covariation between access and premarital sex. (Hint: use the marginal totals in the columns and the rows of the subtables.) You will find that in the "no effects" situation (Table 10.2), access is virtually uncorrelated with both other variables. Hence holding access constant can do nothing to alter the relationship between religiosity and premarital sex. But in the "explanation" situation (Table 10.4), access (W) is strongly correlated with both variables, positively with premarital sex ($r_{WY} = 0.60$) and inversely with family religiosity ($r_{WX} = -0.43$).

Thus, by holding constant the level of automobile use, the inverse covariation between the other two variables is reduced to zero. This situation is of greatest importance in trying to find statistical explanations for observed associations among pairs of variables. *A third variable can produce an explanation for the covariation between two others only if it has non-zero relationships with both the other variables.* If the original bivariate association is positive, the control variable must be positively associated with both the independent and dependent variables. If the original bivariate association is inverse, the control variable must be positively associated with one and inversely associated with the other, as in the example in Table 10.4.

Notice that the pattern in the first-order tables in Table 10.4 is the same that would be found if the covariation of the original two variables were spurious. That is, in both explanation and spuriousness, when the rival factor is held constant, the first-order associations between the two variables fall to zero. Whether you should regard such statistical results as indicating an explanation of the original association or as revealing a spurious original association depends on your understanding of the causal role of the factor which is held constant. In the hypothetical example, we regarded opportunity as an intervening process, that is, as a consequence of religiosity in the household and as a cause itself of premarital sexual experience. A different substantive variable which had a similar statistical effect of reducing the first-order correlations to zero might lead to an evaluation of the religiosity-sex association as spurious. For example, if we had achieved the same results as Table 10.4 where the variable held constant was family structure (such as, whether the teenager lived in an intact family or a so-called broken home), we probably would be inclined to describe the relationship as spurious. Intact families would tend to be more strict in all kinds of personal behavior—religious, sexual, educational—than would families lacking two parents. Hence, the asso-

TABLE 10.5
Example of Interaction Effect of Third Variable, Hypothetical Data

		Access to an Automobile?						
		No				*Yes*		
		Family Highly Religious?				*Family Highly Religious?*		
		No	*Yes*	*Total*		*No*	*Yes*	*Total*
Ever Premarital Sex?	*Yes*	50.0% (26)	11.8% (8)	28.3% (34)	*Yes*	25.0% (10)	25.0% (8)	25.0% (18)
	No	50.0% (26)	88.2% (60)	71.7% (86)	*No*	75.0% (30)	75.0% (24)	75.0% (54)
	Total	100.0% (52)	100.0% (68)	100.0% (120)	*Total*	100.0% (40)	100.0% (32)	100.0% (72)
		$r_{XY} = -0.42$				$r_{XY} = 0.00$		

ciation of religiosity with premarital sex would reflect not a causal relationship but a more general pattern of morality stemming from the family condition.

Our discussion of the differences between explanation and spuriousness in bivariate associations controlled for the effects of third variables should alert you to the fact that the finding is not determined by the statistical outcome. As with most situations, an understanding of social behavior requires thinking carefully about the meaning behind statistical relationships. A good grasp of basic social theory and previous empirical research, plus a little common sense, is helpful in this task.

10.2.5. Interaction Effect of Third Variable

Table 10.5 shows another result from controlling a 2 × 2 table which occurs with some frequency. The first-order associations differ substantially in the two subtables. In the example, the correlation between family religiosity and premarital sex for teens from highly religious families is zero, but the correlation is −0.42 among those from nonreligious families. Holding constant the third variable reduces the association in one subtable but increases it in another. At times we can even find conditional correlations with opposite signs.

FIGURE 10.2
Examples of Conditional Effects, Hypothetical Data (Tables 10.2 to 10.5)

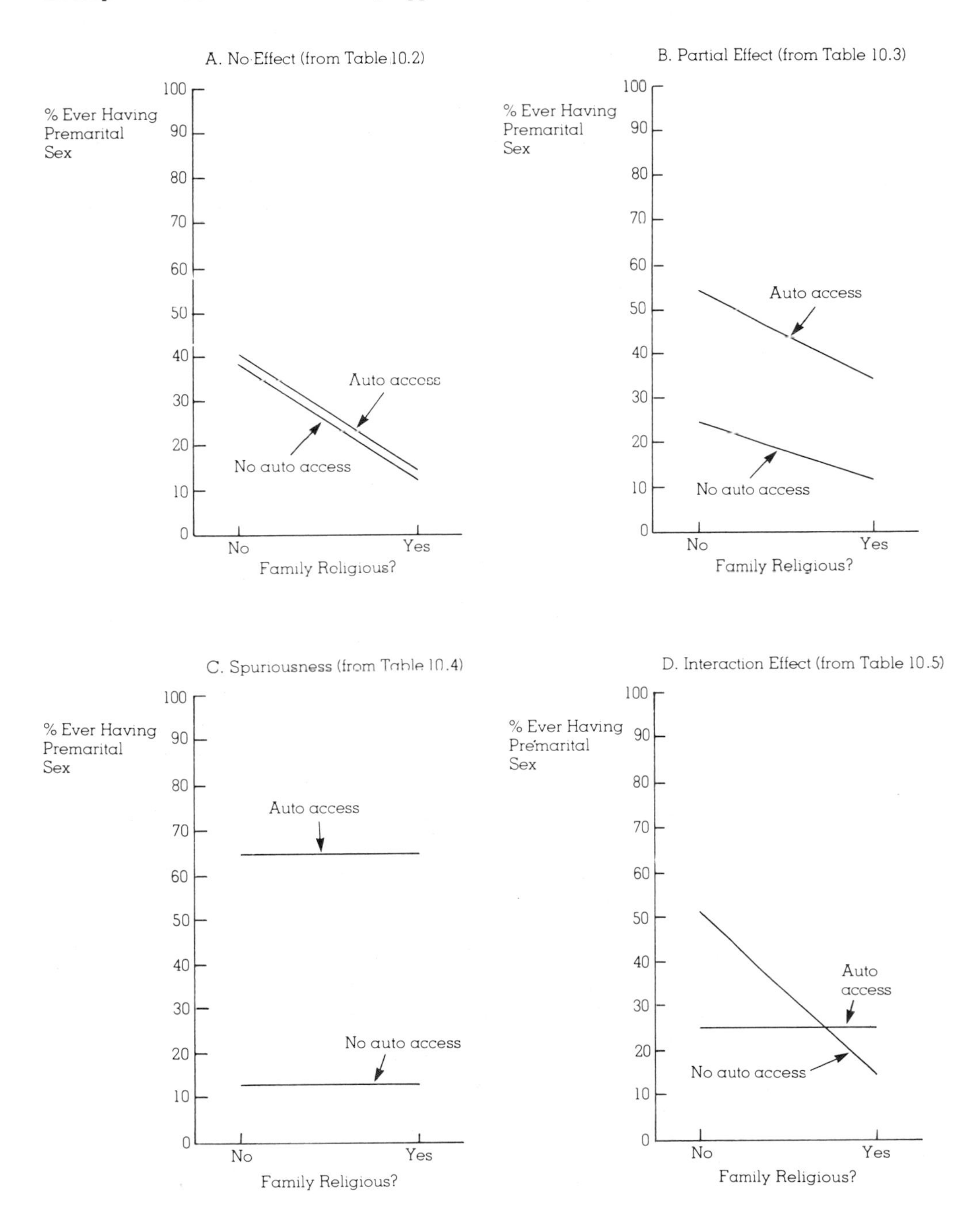

interaction effect—a difference in the relationship between two variables within categories of a control variable

Whenever the relationships in subtables are not the same, an **interaction effect** is present. That is, *the association between two variables in each partial table differs when controlling for a third variable.* We cannot describe the effect of the third variable, as we did in the other three examples, by reference to a single type of outcome. Instead, we must specify which subtable we are referring to in describing the effect of the controlled variable. Testing for interaction effects is described in Section 11.4.1 in Chapter 11.

The discovery of an interaction is often just the beginning for further analysis. It cries out for the researcher to provide an explanation of why the variables have different covariations for different levels of the third variable.

10.2.6. Summary of Conditional Effects

Because graphing results often makes tabled results clearer, the results found in Tables 10.2 to 10.5 are diagramed in Figure 10.2. Panel A, for example, shows clearly that having access to an automobile has no effect on the relationship between family religiosity and premarital sex—the lines are virtually identical for each category of access. Panel B, however, shows that *both* family religiosity and access to an automobile affect the rates of premarital sex. Both variables make independent contributions to explaining premarital sex. The large difference between the lines for access and no access indicates that availability of a car affects the percentage of teenagers ever having premarital sex. But the fact that each line still has a negative slope, across religious and nonreligious family membership, shows that religiosity has an independent effect on premarital sex. Therefore, automobile access provides only a partial explanation. In Panel C the relationship between family religiosity and premarital sex is seen to be spurious, because, when automobile access is controlled for, the slope of both lines is zero across family religiosity. An interaction effect is shown in Panel D; for those with automobile access, no relationship occurs between family religiosity and premarital sex. In contrast, for those without such access, there is a negative relationship between family religiosity and premarital sex.

One noteworthy point illustrated in Figure 10.2 is that *if there is no interaction effect, the relationship between the independent and dependent variables within categories of the control variables is the same.* Therefore when the relationship is graphed, it is shown by parallel lines. Thus the relationships in Panels A, B, and C are shown by two parallel lines because there is no interaction between the independent and control variables. In contrast, the lines in Panel D are *not* parallel, indicating that an interaction effect does exist.

The four examples above were deliberately chosen to give a clear-cut picture of different outcomes which might be found when a simple 2 × 2 relationship is controlled by a third variable. These illustrations were created to show ideal conditions. We cannot caution you too strongly that such exaggerated results are not likely to be found in most real data situations. In our personal experience, we have almost never discovered a situation such as that shown in Table 10.4, in which a strong bivariate relationship completely disappeared when a single third variable was held constant. Partial effects (Table 10.3) or interaction effects (Table 10.5) are more likely to be found. Social behavior is seldom so simple that controlling for a single additional variable fully accounts for the observed two-variable relationship.

We did not compute significance tests on the zero-order relationships reported in this section, but had the data been real we would have. This is done in the next section, where we show how statistical tests may be applied to three-variable crosstabulations, using examples from real GSS data.

10.3. The Partial Correlation Coefficient

Analyzing relationships of three variables together calls for computing each of the conditional correlation coefficients for each crosstabulation, as we did in Section 10.2. When no interaction effects are present, a single coefficient that is a weighted average of the conditional correlation coefficients can be computed. This is called the **partial correlation coefficient,** which is given by

partial correlation coefficient—a measure of association for continuous variables that shows the magnitude and direction of covariation between two variables that remains after the effects of a control variable have been held constant

$$r_{XY \cdot W} = \frac{r_{XY} - r_{XW} r_{YW}}{\sqrt{1 - r_{XW}^2}\sqrt{1 - r_{YW}^2}}$$

FIGURE 10.3
Venn Diagrams Showing Correlation between Two Variables

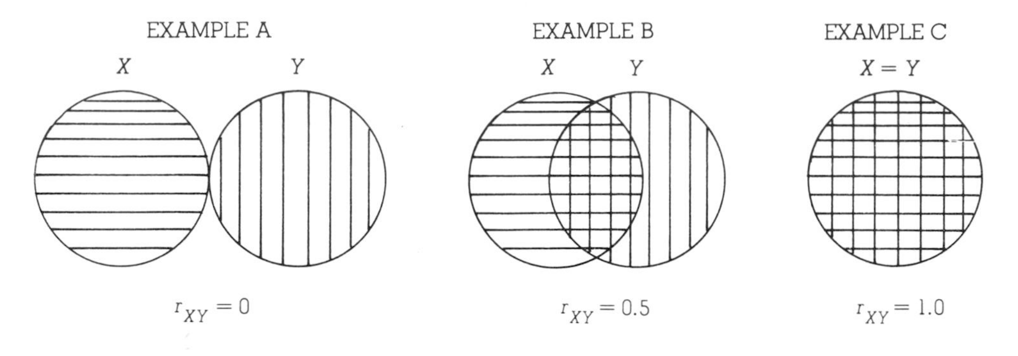

Read the term $r_{XY.W}$ as: "The partial correlation between X and Y, controlling for (or in the presence of) W." Notice that the computation of $r_{XY.W}$ requires first computing the three zero-order correlations, r_{XY}, r_{XW}, and r_{YW}.

Venn diagram—a type of graph that uses overlapping shaded circles to demonstrate relationships or covariation among a set of variables

A casual inspection of the formula for the partial correlation coefficient does not clearly reveal that it produces a correlation between two variables, controlling for a third. To clarify this measure of association we will use a visual presentation called the **Venn diagram**. In Figure 10.3 we have represented correlations of 0, 0.5, and 1.0 with three diagrams. In Panel A, where X and Y are uncorrelated, the two circles do not overlap at all. When the two are correlated $r_{XY} - 0.5$, as in Panel B, half the total area of Y overlaps with X. Perfect correlation is represented by the total overlap of X and Y in Panel C.

Now let's consider the interrelationship among three variables, W, X and Y, using the Venn diagrams in Figure 10.4. Note that the three circles in Panel A are drawn to indicate that X and Y and X and W are correlated, but W and Y are *un*correlated. This diagram suggests clearly that controlling for W should not reduce the correlation between X and Y, since W does not intersect the X-Y overlap. Now look at the numerator of the formula for the partial correlation coefficient, given above. If Y and W are uncorrelated, as in Panel A, the numerator is $r_{XY} - (r_{XW})(0) = r_{XY}$. That is, the intuition drawn from the Venn diagram is correct: To reduce

FIGURE 10.4
Venn Diagrams Showing Interpretation of $r_{XY.W}$

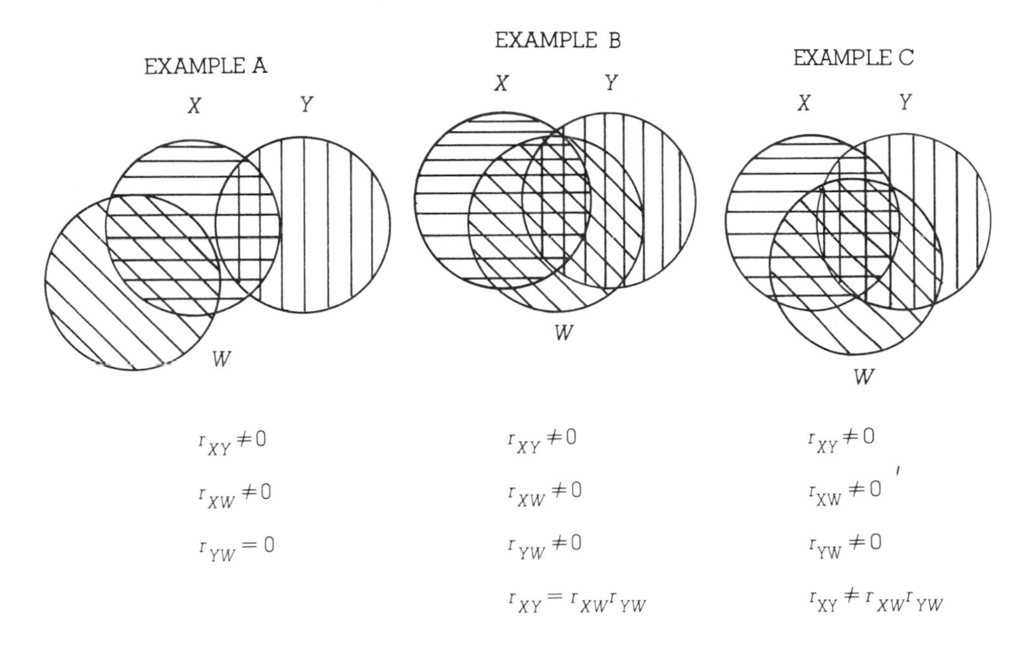

EXAMPLE A	EXAMPLE B	EXAMPLE C
$r_{XY} \neq 0$	$r_{XY} \neq 0$	$r_{XY} \neq 0$
$r_{XW} \neq 0$	$r_{XW} \neq 0$	$r_{XW} \neq 0$
$r_{YW} = 0$	$r_{YW} \neq 0$	$r_{YW} \neq 0$
	$r_{XY} = r_{XW} r_{YW}$	$r_{XY} \neq r_{XW} r_{YW}$

or "explain" the *X*-*Y* correlation, it is necessary for *W* to correlate with *both* *X* and *Y*.

Now consider Panel B in Figure 10.4. In this case the entire overlap between *X* and *Y* is also intersected by *W*. In correlational terms, this means that the entire correlation between *X* and *Y* can be explained by taking *W* into account. This occurs in partial correlation whenever $r_{XY} = r_{XW}r_{YW}$, since in this situation the numerator of the partial correlation coefficient equation, which is $r_{XY} - r_{XW}r_{YW}$, equals zero.

The typical situation is shown in Panel C in Figure 10.4. In this case, one part of *W* intersects the *X*-*Y* crosshatch. Therefore, controlling for *W* does not entirely account for the correlation between *X* and *Y*. While the partial correlation, $r_{XY.W}$, will be smaller than the zero-order correlation, r_{XY}, it will not be zero.

The squared partial correlation coefficient indicates the amount of variance in Y "explained" by X, controlling for W. If W is uncorrelated with either X or Y, r^2_{XY} and $r^2_{XY.W}$ will be virtually the same. If, however, W is correlated with both X and Y, $r^2_{XY.W}$ will usually be smaller than r^2_{XY}. That is, the amount of variance X can explain in Y is reduced, with W being taken into account.

We will discuss the conceptual meaning of partial correlation again in Chapter 12. The tools introduced here should aid your understanding of this meaning.

10.3.1. An Example: Relationships among Three Variables

To show how to use the partial correlation coefficient, we will consider the relationships between public drinking, premarital sexual attitudes, and gender from the 1977 GSS data set. Example A in Table 10.6 presents the zero-order relationship between visiting a public drinking place (X), such as bars and taverns, and attitude toward premarital sexual intercourse (Y). The latter, originally a four-category response, has been collapsed into a dichotomy between those who think it is "always" or "almost always" wrong and those who are more tolerant.[1]

As Table 10.6 shows, these two variables are strongly related. Suppose we set $\alpha = .05$. The zero-order correlation coefficient is $r_{XY} = -0.337$, with $t = -13.75$. Therefore we reject the null hypothesis that in the population $\rho_{XY} = 0$ and conclude instead that those who claim never to drink in public places are considerably more likely to view premarital sex as wrong. But perhaps women are intolerant of both public drinking and premarital sex. To test for this possibility we introduce gender (W) as a control variable. These zero-order tables are also shown in Table 10.6. In Example B, the zero-order correlation of $r_{YW} = 0.156$ indicates that females are more likely than men to see premarital sex as wrong. And in Example C, the zero-order correlation of $r_{XW} = -0.205$ suggests that females are less likely to visit a bar than men are. (Both of these zero-order correlations are significant at the $\alpha = .05$ level, with t values of 6.07 and -8.05.)

1. We have collapsed this variable only to simplify our illustration. In an actual analysis of these data we would not have done so, since we thereby lose some valuable information.

Now we can compute the partial correlation coefficient:

$$r_{XY \cdot W} = \frac{-0.337 - (-0.205)(0.156)}{\sqrt{1 - (-0.205)^2}\sqrt{1 - 0.156^2}} = -0.316$$

That is, controlling for gender reduces the zero-order relationship between public drinking and attitudes toward premarital sex from −0.337 to −0.316—a very small amount. Hence we would conclude that the relationship between public drinking and attitude toward premarital sex cannot be explained by gender.

10.3.2. Testing the Partial Correlation for Significance

We have computed a partial correlation coefficient, but we have not tested whether it is statistically significant. To do this we establish the null hypothesis, $H_0: \rho_{XY \cdot W} \geq 0$, and the alternative hypothesis, $H_1: \rho_{XY \cdot W} < 0$. Thus we expect the relationship to be negative.

The test for the partial correlation is:

$$F_{1,N-3} = \frac{r^2_{XY \cdot W}(N - 3)}{1 - r^2_{XY \cdot W}}$$

Or, by taking the square root of the right side of the equation, the expression is for a t test:

$$t_{N-3} = \frac{r_{XY \cdot W}\sqrt{N - 3}}{\sqrt{1 - r^2_{XY \cdot W}}}$$

Using the results from the example in Section 10.3.1 and setting $\alpha = .05$, we find:

$$F_{1,1475} = \frac{(-0.337)^2(1{,}478 - 3)}{1 - (-0.337)^2} = 188.98$$

or

$$t_{1475} = -13.75$$

Since the critical value for the t test is -1.65 for a one-tailed test, the null hypothesis can be confidently rejected. That is, we infer that in the population, visiting bars and premarital sexual attitude are correlated, even after holding constant gender.

TABLE 10.6
Zero-Order Relationships between Public Drinking, Premarital Sex, and Gender

EXAMPLE A

		Visit Bar? (X)		
		Never	***Yes***	***Total***
Premarital Sex (Y)	***Wrong***	57.2% (419)	24.1% (180)	40.5% (599)
	Not Wrong	42.8% (313)	75.9% (566)	59.5% (879)
	Total	100.0% (732)	100.0% (746)	100.0% (1,478)

$r_{XY} = -0.337$

EXAMPLE B

		Gender (W)		
		Male	***Female***	***Total***
Premarital Sex (Y)	***Wrong***	32.1% (214)	47.5% (385)	40.5% (599)
	Not Wrong	67.9% (453)	52.5% (426)	59.5% (879)
	Total	100.0% (667)	100.0% (811)	100.0% (1,478)

$r_{YW} = 0.156$

EXAMPLE C

		Gender (W)		
		Male	***Female***	***Total***
Visit Bar (X)	***Yes***	61.8% (412)	41.2% (334)	50.5% (746)
	Never	38.2% (255)	58.8% (477)	49.5% (732)
	Total	100.0% (667)	100.0% (811)	100.0% (1,478)

$r_{XW_p} = -0.205$

Source: 1977 General Social Survey.

10.4. Multivariate Contingency Analysis in Larger Tables

While we have been concerned only with the analysis of 2×2 tables in this chapter, the same principles apply to larger $R \times C$ tables. When variables have only two outcomes, we can always use the correlation coefficient to estimate the relationship between them. When two variables have a larger number of categories, however, the correlation coefficient should be used only when there is reason to believe that the underlying variables being analyzed are continuous. In these circumstances, calculating zero-order and partial correlation coefficients is perfectly reasonable. Indeed, this is the thrust of the material covered in the next chapter. If the variables are measured at the discrete level and are larger than 2×2 in size, however, the proper tools for analysis are those presented in Chapter 9.

Review of Key Concepts

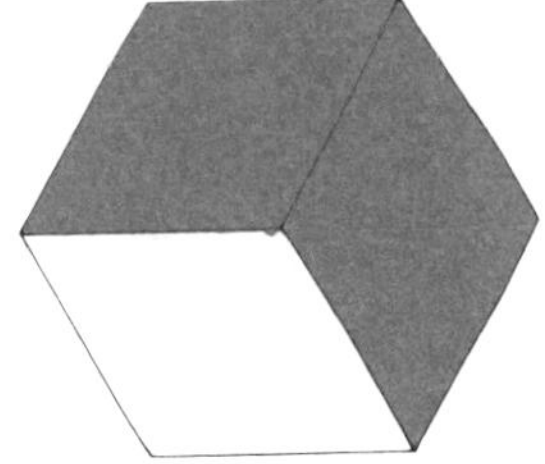

This list of the key concepts introduced in Chapter 10 is in the order of appearance in the text. Combined with the definitions in the margins, it will help you review the material and can serve as a self-test for mastery of the concepts.

multivariate contingency analysis
random assignment
spuriousness
explanation (interpretation) of association
multiple causation
zero-order table
phi coefficient
first-order table
conditional correlation coefficients
interaction effect
partial correlation coefficient
Venn diagram

PROBLEMS

General Problems

1. These hypotheses might be found in comparative sociology:

    Among developing nations, the more an economy relies on a single raw export product, the more vulnerable the labor force is to periodic high unemployment.

    Repressive political regimes, such as military dictatorships, are more likely to take over when developing nations experience economic instability.

    What kind of causal relationship is expressed by these hypotheses? Diagram the relationship.

2. Suggest how an inverse covariation between education and political conservatism could be spurious due to age. Show this hypothesis as a causal diagram. What would you expect to happen to the correlation coefficient between education and conservatism when age is held constant, if the spurious hypothesis is correct?

3. Suggest three interpretive relationships for a causal connection in which an increase in labor force participation by women leads to an increase in the divorce rate.

4. Show the relationships you developed in Problem 3 as causal diagrams.

5. When various groups were asked about the most important characteristics of a job, the following percentages said high income was the primary consideration: young manual workers, 77%; older manual workers, 84%; young nonmanual workers, 34%; older nonmanual workers, 60%. What kind of relationship is present? Diagram these percentages as in Figure 10.2.

6. A three-way crosstabulation of gender, wages, and work status yields the following partial frequency tables:

		Full-Time Work			*Part-Time Work*	
		Men	*Women*		*Men*	*Women*
Wages	*High*	50	20	*High*	10	15
	Low	20	20	*Low*	10	25

Calculate the correlation coefficients for the zero-order relation between gender and wages. What kind of relationship is revealed by holding work status constant?

7. A sex researcher hypothesizes that men patronize pornographic films more often than women because they are more likely to believe that such films do not lead to a breakdown in moral behavior. She observes the following partial tables:

Do X-rated movies lead to moral breakdown?

Did you attend an X-rated movie last year?	*No*: *Men*	*No*: *Women*		*Yes*: *Men*	*Yes*: *Women*
Yes	20	8	*Yes*	5	10
No	11	4	*No*	25	60

Reconstruct the zero-order film attendance by gender table, calculate the partial table correlations, and determine whether the sexologist's hypothesis is correct. What kind of relationship do you find?

8. Panel A is the zero-order table for the relationship of age to attitude towards legalization of marijuana, and Panel B is the partial table of responses among highly religious persons.

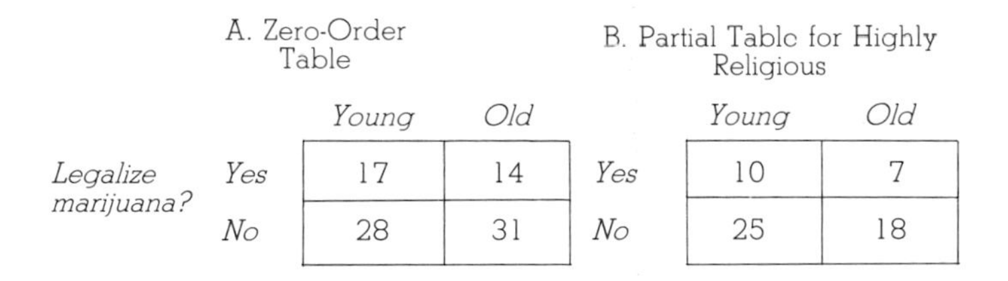

A. Zero-Order Table

Legalize marijuana?	*Young*	*Old*
Yes	17	14
No	28	31

B. Partial Table for Highly Religious

	Young	*Old*
Yes	10	7
No	25	18

Calculate the zero-order correlation coefficient and the conditional correlation coefficients. What kind of effect is observed when religiosity is held constant?

9. For the data in Problem 8, calculate the correlation of religiosity with the other two variables.

10. To study whether interracial contacts might be a function of opportunities as mediated by attitudes, in the 1980 GSS white respondents were asked about the residential proximity of blacks and whether they had invited a black to dinner recently. The zero-order relationship was:

		Brought a black home for dinner recently?	
		No	Yes
Blacks live close by?	Yes	242	144
	No	668	177

Upon holding constant the respondents' opinions about whether or not blacks try to push themselves where they aren't wanted, the following partial tables were observed:

		Blacks are pushy		Blacks aren't pushy	
		Brought a black home?		Brought a black home?	
		No	Yes	No	Yes
Blacks live close by?	Yes	174	75	68	69
	No	478	118	190	59

Do these results support the hypothesis that the relationship between proximity and behavior can be completely explained by an intervening attitude?

11. Another analysis of the hypothesis in Problem 10 produced the following partial tables, in which the control variable is now the respondents' preferences on national spending to improve the conditions of blacks:

		Too little is spent		Too much or the right amount is spent	
		Brought a black home?		Brought a black home?	
		No	Yes	No	Yes
Blacks live close by?	Yes	35	46	194	96
	No	115	35	501	130

Do these results provide any different support for the hypothesis?

12. In a structural theory of social association, Peter Blau asserted that "The larger the difference in size between two groups, the greater is the discrepancy in the rates of intergroup associations between them." (Peter M. Blau, *Inequality and Heterogeneity* [New York: Free Press, 1977], p. 23.) This hy-

pothesis predicts an interaction effect in which the relationship of group membership to intergroup association differs according to the relative size of the two groups. Suppose the following data were observed:

		Larger Minority			*Smaller Minority*	
		Group Membership			*Group Membership*	
		Majority	*Minority*		*Majority*	*Minority*
Married to person of opposite group?	*Yes*	20	20	*Yes*	10	10
	No	40	20	*No*	70	10

Notice that the relative frequency of minority group intermarriage is identical in both settings ($p = 0.50$). Calculate the conditional correlation coefficients. Do they support Blau's hypothesis? Calculate the zero-order relationship between group membership and marital behavior. What kind of effect does holding constant the relative group size have?

13. In a set of data, $r_{XZ} = -0.39$ and $r_{YZ} = 0.57$. What must be the sign of the correlation between X and Y if variable Z is to explain at least some of the covariation between X and Y? If Z completely explains the correlation between X and Y, what is the value of that correlation?

14. Calculate the partial correlation coefficients between X and Y after controlling for Z, i.e., $r_{XY \cdot Z}$

	r_{XY}	r_{XZ}	r_{YZ}
a.	0.30	0.40	0.70
b.	−0.45	0.24	−0.30
c.	0.15	0.20	−0.75
d.	−0.60	−0.40	−0.40

15. Find the partial correlation for:
    a. Gender and wages in Problem 6, controlling work status.
    b. Gender and movie attendance in Problem 7, controlling belief.
    c. Proximity and behavior in Problem 10, controlling attitude.
    d. Proximity and behavior in Problem 11, controlling attitude.

16. Test the significance of these partial correlation coefficients where H_0: $\rho_{XY \cdot W} = 0$ with $\alpha = .01$, two-tailed alternative:

	$r_{XY \cdot W}$	N
a.	−0.40	23
b.	0.12	200
c.	0.33	38
d.	0.53	14

Give t ratios and your decision.

17. Give the proportion of variance in Y explained by X that remains after controlling for W. What percentage of the original covariation is this?

	r_{XY}	$r_{YX \cdot W}$
a.	0.60	0.30
b.	−0.40	−0.30
c.	0.30	0.20
d.	−0.80	−0.80

18. Assortative mating means that marriages usually take place between persons of similar social backgrounds. The degree of assortative mating may differ by social context, however. Suppose you found the following:

		Urban Origin		*Rural Origin*	
		Husband's Religion		*Husband's Religion*	
		Catholic	*Protestant*	*Catholic*	*Protestant*
Wife's Religion	*Protestant*	30	60	10	80
	Catholic	30	20	20	5

a. Calculate: (1) the zero-order correlation between spouses' religions; (2) the correlation between religions in the two subtables; (3) the partial correlation between religions, holding constant origin; and (4) the t ratio for the significance of the partial r at $\alpha = .001$.

b. What kind of effect does urban-rural origin have on assortative mating?

19. The concept of statistical control of tabular data can be extended beyond dichotomous variables. For example, consider a control variable, income, that has been divided into four categories. Within each category, the crosstabulations of dichotomous measures of social and economic attitudes are formed.

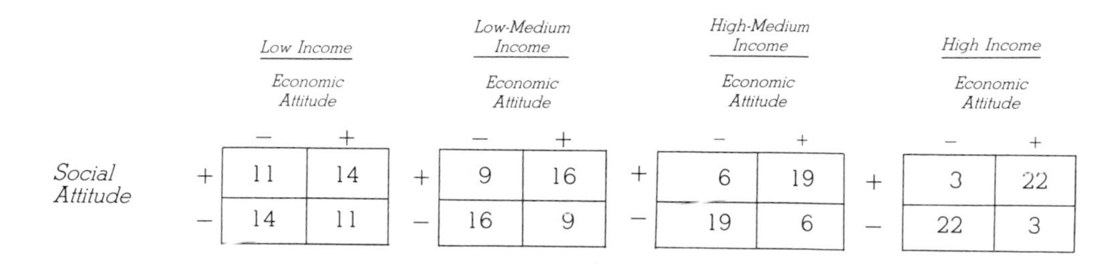

		Low Income		Low-Medium Income		High-Medium Income		High Income	
		Economic Attitude		Economic Attitude		Economic Attitude		Economic Attitude	
		−	+	−	+	−	+	−	+
Social Attitude	+	11	14	9	16	6	19	3	22
	−	14	11	16	9	19	6	22	3

a. Calculate: (1) the zero-order correlation of social and economic attitudes; (2) the first-order correlations for each subtable; (3) the correlation for income and economic attitude; (4) the correlation for income and social attitude; (5) the partial correlation between social and economic attitudes controlling for income; and (6) the t ratio for the significance of the partial correlation for $\alpha = .001$.

b. Give an interpretation of these results.

Note: In your calculations, assume income is coded from 1–4 for low to high.

20. The relationship between family size (number of children) and satisfaction with family life may depend on the age at which women began having their children. Here are two partial tables, one for women who started their families while still teenagers, and the other for women who began childbearing at a later age:

		Began in Teens			Began Later		
		Size of Family			Size of Family		
		Low	Medium	High	Low	Medium	High
Satisfaction with Family	High	15	10	5	60	30	15
	Medium	5	15	20	20	15	15
	Low	5	5	25	20	15	10

Analyze these data, giving: *(a)* zero-order correlation; *(b)* both first-order correlations; *(c)* the partial correlation between satisfaction and size of family, controlling for age at which childbearing began; and *(d)* t ratio for the significance of the partial correlation for $\alpha = .05$.

Note: In your calculations, assume the categories are coded Low = 0, Medium = 1, High = 2.

Problems Requiring the 1980 General Social Survey Data

21. Political analysts have noted that self-described liberals tend to be concentrated among the college educated, but this phenomenon may be a recent development. Test this hypothesis by dichotomizing POLVIEWS into liberal and nonliberal categories, EDUC into college graduates and nongraduates, and AGE into those 45 or older and those younger. What happens to the POLVIEWS-EDUC correlation when AGE is held constant?

22. Attitudes toward abortion for poor women (ABPOOR) covary with religiosity (RELITEN), but is the strength of association the same for Catholics as for Protestants? *Note:* Control for RELIG after excluding other categories.

23. The "status attainment" model argues that a person's education determines his or her occupation, which in turn affects the amount of income earned. Thus, if occupational prestige is held constant, the observed correlation between education and income should become zero. See if this hypothesis is supported by using EDUC, PRESTIGE, and RINCOME variables, dichotomized at or near their median values.

24. What are the partial correlations between subjective social class (CLASS) and attitude toward working if the financial reward were unnecessary (RICHWORK), controlling separately for education, occupational prestige, and income as dichotomized in Problem 23?

25. Test Blau's hypothesis in Problem 12 using RACE (excluding those not white or black) as group, RACHOME as the intergroup association measure, and four categories of community size, XNORCSIZ (1, 2 = 1; 3, 4 = 2; 7, 8 = 3; 5, 6, 9, 10 = 4) to control for relative size of the minority group. Does the pattern of conditional correlations support the expectation that differences in the rate of intergroup contact increase as minority size is smaller?

Problems Requiring the 63-Cities Data

26. The urban fiscal crises of the 1970s came mostly in older cities that had accumulated large debts they were increas-

ingly unable to handle. First, show this relationship by cross-tabulating a dichotomy of CITYAGE (under 80 years versus 81 years and older) with a dichotomy of STRESS74 (under $350 per capita versus $351 and over). Then, determine whether the covariation between these two variables is the same across all four REGIONS.

27. The number of municipal functions a city performs depends on the formal structure of its government, with reform government cities generally performing fewer functions. But this covariation may be spurious, because of a common dependence on city size (smaller cities are both less reformed and less likely to undertake many tasks, compared to bigger cities). Control the FUNCTION by REFORMGV crosstabulation by a dichotomy of POPULAT (under 300 thousand residents versus 300 thousand or more). What happens to the zero-order correlation when the partial correlation for size is calculated?

28. Repeat the analysis in Problem 27, but controlling for a dichotomy of CITYAGE (under 80 years versus 81 years and older). Is there any difference in the ability of age or size to account for the covariation of number of functions performed and type of government?

29. What is the zero-order relationship between municipal decision making (DECINDEX) and expenditures on police (POLICEXP), and what kind of effect do you find when controlling for crime rate (CRIMRATE)? Use dichotomies (divide DECINDEX at 6.66; POLICEXP at 25; and CRIMRATE at 40).

30. Population change may be related to the demographic composition of the cities. Specifically, were cities that were losing population between 1960 and 1970 (POPCHANG) more likely to have large black populations (BLACKPCT of 14% or higher) compared to cities that did not lose population? What happens to the zero-order relationship when controlling for level of white-collar labor force (WCPCT of 50% or higher)?

Multiple Regression Analysis

11

This chapter introduces regression methods for analyzing problems with more than one independent variable. This is probably the most important chapter in this book for you to master, since most research in the social sciences examines hypotheses in which a dependent variable is assumed to have multiple causes. For example, we could assume that people's occupational aspirations are determined by the aspirations of their peers, parents' educations, gender, and so on. Or the likelihood that an adolescent will become involved in juvenile delinquency might be assumed to be a function of the individual's school performance, the involvement of peers in delinquency, the average income of households in the home neighborhood, and so on. Indeed, we would find it difficult to suggest any social phenomena that can be assumed to be a function of a single cause.

The technique introduced in this chapter, which is called **multiple regression analysis**, is a logical extension of the materials presented in Chapters 8 to 10. Before presenting the regression model and the technical details involved in using it, we will describe a substantive problem for which regression techniques are appropriate.

multiple regression analysis—a statistical technique for estimating the relationship between a continuous dependent variable and two or more continuous or discrete independent variables

TABLE 11.1
Frequency Distributions for Three Items in Sexual Permissiveness Index

A. *Premarital Sexual Relations*

. . . If a man and woman have sexual relations before marriage, do you think it is always wrong, . . . or not wrong at all?

B. *Extramarital Sexual Relations*

What is your opinion about a *married* person having sexual relations with someone *other* than the marriage partner—is it always wrong, . . . or not wrong at all?

C. *Homosexual Relations*

What about sexual relations between two *adults* of the *same* sex—do you think it is always wrong, . . . or not wrong at all?

		A. Premarital Sex		*B. Extramarital Sex*		*C. Homosexual Sex*	
Code	*Response*	*N*	*%*	*N*	*%*	*N*	*%*
1	Always wrong	459	30.0	1103	72.1	1044	68.2
2	Almost always wrong	141	9.2	206	13.5	84	5.5
3	Wrong only sometimes	340	22.2	153	10.0	109	7.1
4	Not wrong at all	541	35.4	48	3.1	216	14.1
8	Don't know*	39	2.5	13	0.8	69	4.5
9	No answer*	10	0.7	7	0.5	8	0.5
	Total	1,530	100.0%	1,530	100.0%	1,530	99.9%†

*Treated as missing data in the analysis.

†Does not total to 100.0% due to rounding.

Source: 1977 General Social Survey.

11.1. An Example: Explaining Sexual Permissiveness

Ira Reiss, a prominent family sociologist, has developed a theory of sexual permissiveness that has been rather widely accepted among sociologists. The basic proposition of his theory is: To the extent that a particular group has an autonomous, free courtship system, the likelihood of accepting sexual permissiveness is increased.[1] Anything that promotes autonomy from the social control institutions of society promotes higher levels of sexual permissiveness.

1. I. L. Reiss, and B. Miller, "Heterosexual Permissiveness: A Theoretical Analysis," in W. Burr, R. Hill, I. Nye, and I. L. Reiss (Eds.), *Contemporary Theories about the Family: Research-Based Theories,* Vol. 1 (New York: Free Press, 1979); I. L. Reiss, *Family Systems in America* (3rd ed.) (New York: Holt, Rinehart & Winston, 1980), Chap. 7.

Two factors that Reiss examines in his work with a national adult sample are religion and education. Religions embody many of the major sexual norms against premarital and extramarital sexuality in society, so the first derived proposition we will examine is:

> P1: The greater the religiosity, the less the sexual permissiveness.

Since education can lead students to challenge and explore commonly accepted values, it should encourage autonomy of thought and action. The second derived proposition, therefore, is:

> P2: The higher the individual's educational attainment, the greater the sexual permissiveness.

We will test these two hypotheses using the 1977 GSS data, in which sexual permissiveness is measured by a three-item index. An **index** is a variable which is a composite of other variables that are assumed to reflect some underlying construct. In this case the index is the sum of each individual's responses to three items divided by three. The three items deal with attitudes toward premarital sex, extramarital sex, and homosexual relations. The actual items, along with their frequency distributions, are shown in Table 11.1.

index—a variable that is a summed composite of other variables that are assumed to reflect some underlying construct.

FIGURE 11.1
Effects of Latent, Unobserved Variable on Covariation among a Set of Observed Indicators

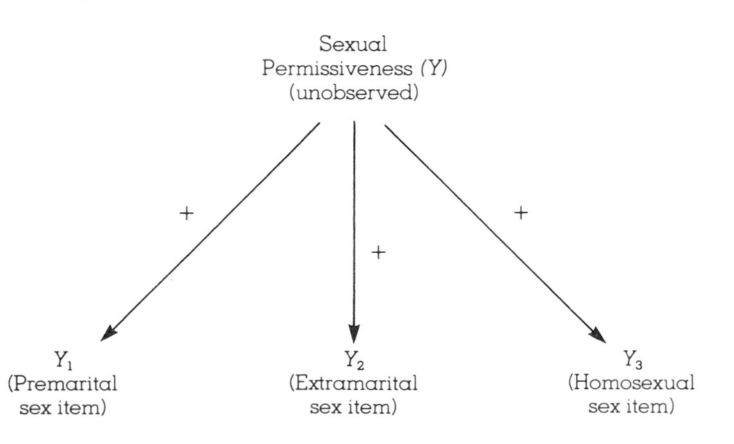

BOX 11.1 An Introduction to Index Construction

Social scientists often use indexes constructed from other variables to reflect some underlying latent unobservable variable. Perhaps the most common example is the IQ test, with which you are familiar. These test constructors assume that the responses to many individual items can be summed into a total score called IQ. Indices are often proxies for constructs, which social scientists find useful to explain observations. We don't really know if there is something called IQ, or religiosity, for example, but by positing such constructs we can account for observations in a succinct, logical, summary way.

When a set of indicators is posited to reflect an underlying construct, the items should be substantially correlated with one another. The higher the correlation among the items, the more confident we can be that the items are measuring the same construct. Furthermore, for a given level of correlation among the measures, the greater the number of indicators, the more confidence we can have in the index constructed from them. If five different persons measure the length of an open field and average the five observations, we will be more confident of what the true size of the field is than if a single person takes the measure. For the same reason, our confidence in an index increases with the number of independent measures taken. The assumption, however, is that *all* of the items do in fact reflect the same underlying construct. To summarize, *the quality of an index can be judged by the average intercorrelation among the indicators and the number of indicators that comprise it.*

A statistic that summarizes the reliability of an index is called Cronbach's alpha. Alpha is a measure of the internal consistency of a set of items, and it ranges from zero (no internal consistency) to unity (perfect internal consistency). The usual computing formula for alpha is too complex to present here. But alpha reduces to a rather simple formula if the items to be summed into an index (and perhaps divided by the numbers of items, to create an average response across items) have equal or near-equal variances, as

Box 11.1 (continued)

they will if all items use the same response category format. This formula also can be used if all items are converted to Z scores before summing. It is:

$$\alpha = \frac{k\bar{r}}{1 + (k - 1)\bar{r}}$$

Where:

k = The number of indicators in the index.
$\bar{r}$ = The average intercorrelation among the k items comprising the index.

We can show how to use alpha with the three-item sexual permissiveness index example used in the text. The $k = 3$ items intercorrelate 0.40, 0.40, and 0.39. Therefore the average intercorrelation is $\bar{r} = (0.40 + 0.40 + 0.39)/3 = 0.397$. And:

$$\hat{\alpha} = \frac{3(0.397)}{1 + (3 - 1)(0.397)} = 0.664$$

where the caret (^) is used to indicate an estimate of the underlying population parameter.

Since this index is composed of only three items, an $\hat{\alpha}$ of 0.664 is barely acceptable. We strive for indices with alphas of 0.70 or higher. If we had five or six items with an average correlation of 0.40, we could use the formula above to verify that $\hat{\alpha}$ would be 0.77. This demonstrates that the number of indicators, as well as their average intercorrelation, contributes to our confidence in an index.

To be able to construct an index, we must assume that the items comprising it reflect some underlying latent, unobservable characteristic. In this case, we are assuming that people who hold a sexually restrictive attitude (the latent, unobserved variable) are more likely to say that premarital and extramarital sex and homosexual relations are wrong. In contrast, those who hold a permissive attitude would be expected to argue that these behaviors are not wrong. Figure 11.1 is a model which shows that we conceive of the

construct—unobserved concept used by social scientists to explain observations

indicator—observable measure of underlying unobservable theoretical construct

Cronbach's alpha—a measure of internal reliability for multi-item summed indexes

three observed variables as a function of a latent, unobserved theoretical **construct**—sexual permissiveness. Since the three observed variables, or **indicators**, as they are often called, are presumed to reflect a single underlying variable, we expect them to be positively correlated with each other. Indeed they are. In the 1977 GSS data, correlation between attitudes about premarital and extramarital sex is 0.40, between premarital sex and homosexual relations it is 0.40, and between extramarital sex and homosexual relations it is 0.39. These data provide some evidence for the presumption that the three items can be used as a single index of sexual permissiveness. More details on index construction of this sort are given in Box 11.1, which also introduces the concept of **Cronbach's alpha**, a measure of internal reliability.

Religiosity is measured by church attendance, which in the GSS data ranges from "never" (0) to "several times a week" (9). Education is coded from 0, for no formal schooling, to 20, for eight years of college. The two hypotheses drawn from P1 and P2 to be tested are:

H1: The greater the church attendance, the less the sexual permissiveness.

H2: The more the years of education, the higher the sexual permissiveness.

11.2 The Three-Variable Regression Model

As in the bivariate regression model introduced in Section 8.2, Chapter 8, in multiple regression the dependent variable is assumed to be linearly related to the independent variables. In the case of two independent variables:

The three-variable *population regression equation* (see Section 8.6.2) is:

$$Y_i = \alpha + \beta_1 X_{1i} + \beta_2 X_{2i} + e_i$$

Where:

X_1 and X_2 = The two independent variables.
e = The error or residual term.

The *population prediction equation* is:

$$\hat{Y}_i = \alpha + \beta_1 X_{1i} + \beta_2 X_{2i}$$

With two independent variables, also, the *sample regression* and *prediction equations* are given by:

$$Y_i = a + b_1 X_{1i} + b_2 X_{2i} + e_i$$

$$\hat{Y}_i = a + b_1 X_{1i} + b_2 X_{2i}$$

As in bivariate regression, we use ordinary least squares (OLS) methods to estimate values for α, β_1, and β_2, such that Σe_i^2 is a minimum. We state without proof that the following sample values are unbiased estimates of the population parameters:

$$a = \bar{Y} - (b_1 \bar{X}_1 + b_2 \bar{X}_2)$$

$$b_1 = \left(\frac{s_Y}{s_{X_1}} \right) \frac{r_{YX_1} - r_{YX_2} r_{X_1X_2}}{1 - r_{X_1X_2}^2}$$

$$b_2 = \left(\frac{s_Y}{s_{X_2}} \right) \frac{r_{YX_2} - r_{YX_1} r_{X_1X_2}}{1 - r_{X_1X_2}^2}$$

Notice that the regression estimates can all be obtained if we simply know the sample means, standard deviations, and zero-order correlation coefficients for the three variables under investigation. Therefore, when doing regression analysis, we must always compute and report these statistics in a summary table.

We have done this for the GSS sexual permissiveness example in Table 11.2. Notice that we have tabled only half the correlations. Since $r_{XY} = r_{YX}$ for all X and Y, to show the bottom half of the matrix would be redundant. We also have tabled the sample data for age and gender of respondent. These data will be used in examples later in the chapter.

TABLE 11.2

Correlations, Means, and Standard Deviations among Sexual Permissiveness, Religiosity, Education, Age, and Gender

Variable	(1)	(2)	(3)	(4)	(5)	Mean	Standard Deviation
1 Sexual Permissiveness	1.000	−0.318	0.190	−0.233	−0.107	2.069	1.048
2 Religiosity		1.000	0.033	0.127	0.164	3.967	2.681
3 Education			1.000	−0.326	−0.061	11.683	3.222
4 Age				1.000	0.043	44.578	16.944
5 Gender					1.000	0.550	0.498

Note: $N = 1{,}499$.

Source: 1977 General Social Survey.

listwise deletion—in multiple regression analysis, the removal of all cases that have missing values on *any* of the variables

pairwise deletion—in multiple regression analysis, the removal of a case from the calculation of a correlation coefficient only if it has missing values for one of the variables

In making these and all other computations in the subsequent examples, we have used a **listwise deletion** procedure; that is, we have used an observation only if data on *all* variables being analyzed are present. The 1977 GSS has $N = 1{,}530$ cases. By using a listwise procedure we have lost 31 cases, resulting in $N = 1{,}499$ in Table 11.2. The alternative to the listwise procedure is **pairwise deletion**. Since regression analysis depends on correlations for estimating the regression coefficients, and because correlations are based on pairs of observations, we can use the available cases for each individual correlation coefficient.

There are two possible problems with the pairwise procedure. First, since the number of cases available varies from correlation to correlation, the N for any given problem is unclear. The N is crucial in computing tests of significance. Second, the matrix (table) of correlations generated by pairwise deletion is sometimes poorly conditioned, so that the configuration of correlations observed may not logically occur in real data. The technical details of this condition are beyond the scope of this text. While an ill-conditioned correlation matrix is rarely generated when pairwise missing data procedures are used, it can happen. When it does, the computer will not be able to obtain a solution for the regression estimates sought.

We recommend that pairwise deletion procedures only be used when there will be a large loss in cases (roughly 5% or more) due to use of a listwise procedure. The danger of using a listwise procedure in such instances is the nonrandom loss of cases, which can provide biased estimates of population parameters. When there is a substantial missing data problem for a variable, the researcher needs to show whether and how the missing cases might affect the analysis being done. The usual procedure is to show, using t tests, that the missing observations do not differ significantly from the observations that are present on major demographic characteristics such as sex, education, race, and so on. Showing no significant differences on these variables is only partially reassuring, however, since we can never know whether the missing data have been generated for random or systematic reasons. Can you think of systematic ways that data loss can occur?

We can now use the formulas presented above to estimate the intercept and two regression coefficients in the sexual permissiveness example. Since we need to know b_1 and b_2 to estimate the intercept, we estimate the regression coefficients, b_1 and b_2, prior to estimating a:

$$b_1 = \left(\frac{1.048}{2.681} \right) \frac{-0.318 - (0.190)(0.033)}{1 - 0.033^2} = -0.127$$

$$b_2 = \left(\frac{1.048}{3.222} \right) \frac{0.190 - (-0.318)(0.033)}{1 - 0.033^2} = 0.065$$

and we estimate a:

$$a = 2.069 - ((-0.127)(3.967) + (0.065)(11.683)) = 1.813$$

Therefore the prediction equation for this example is:

$$\hat{Y} = 1.813 - 0.127X_1 + 0.065X_2$$

11.2.1. Interpretation of b_1 and b_2

The regression coefficients b_1 and b_2 have the following interpretation:

> A **multiple regression coefficient** measures the amount of increase or decrease in the dependent variable for a one-unit difference in the independent variable, controlling for the other independent variable or variables in the equation.

multiple regression coefficient—a measure of association showing the amount of increase or decrease in a continuous dependent variable for a one-unit difference in the independent variable, controlling for the other independent variable(s)

In this example, the dependent variable ranges from 1 = "always wrong" to 4 = "not wrong at all." Thus $b_1 = -0.127$

(established above) means that for a one-unit change on the religious attendance measure, a respondent's score on the sexual permissiveness scale is expected to drop by -0.127, on average. And for each additional year of education, the respondent's permissiveness score is expected to change an average of 0.065, since $b_2 = 0.065$.

As we noted in describing the concept of control in Chapter 10, if two independent variables are uncorrelated, controlling for one of them will not affect the relationship between the other independent variable and the dependent variable. This can be seen by noting that if $r_{X_1X_2} = 0$:

$$b_1 = \left(\frac{s_Y}{s_{X_1}}\right) \frac{r_{YX_1} - (r_{YX_2})(0)}{1 - 0^2}$$

$$= \frac{s_Y}{s_{X_1}} r_{YX_1}$$

This equals the bivariate regression coefficient, *b*, introduced in Chapter 8. That is, when the two independent variables are uncorrelated, the relationship between one of them and the dependent variable is unchanged when controlling for the other independent variable. In the example above, $r_{X_1X_2} = 0.033$, which is very close to zero. This suggests that the bivariate coefficients relating the independent variables to the dependent variable should not differ much from the multiple regression coefficients. That they do not is indicated in the table below:

Independent Variable	*Bivariate Coefficients*	*Multiple Regression Coefficients*
Religiosity	−0.124	−0.127
Education	0.062	0.065

11.2.2 Standardized Regression Coefficients (Beta Weights)

Since the measurement units of the dependent variable and religious attendance are both arbitrary, the interpretation of the regression coefficients is less clear than might be wanted. For this reason we suggest that the standardized regression (or beta) coefficients introduced in Section 8.5 in Chapter 8 for the bivariate case should be computed as well. These are the regression coefficients, or *beta weights*, we would obtain if the three variables were converted to *Z* scores prior to estimating the regression equation.

Regardless of the number of independent variables, the following relationship holds between the metric coefficients and standardized regression coefficients (beta weights):

$$\beta_j^* = \left(\frac{s_{X_j}}{s_Y} \right) b_j$$

We simply multiply the metric regression coefficient by the ratio of the standard deviation of the independent variable X_j to the dependent variable Y. Hence in the case of two independent variables, X_1 and X_2:

$$\beta_1^* = \left(\frac{s_{X_1}}{s_Y} \right) b_1$$

$$\beta_2^* = \left(\frac{s_{X_2}}{s_Y} \right) b_2$$

Using the data from the sexual permissiveness example:

$$\beta_1^* = \left(\frac{2.681}{1.048} \right) (-0.127) = -0.325$$

$$\beta_2^* = \left(\frac{3.222}{1.048} \right) \left(0.065 \right) = 0.200$$

Since the means of Z-transformed variables are zero, the intercept for the standardized regression equation is zero. Hence:

$$\hat{Z}_Y = -0.325Z_1 + 0.200Z_2$$

The two independent variables are now in the same metric, and so we can determine their relative ability to predict sexual permissiveness by examining which of the variables is larger in absolute value. Religiosity is a more potent predictor ($\beta_1^* = -0.325$) than education ($\beta_2^* = 0.200$). For each standard deviation difference in religiosity, controlling for education, we can expect, on average, a -0.325 standard deviation change in the sexual permissiveness variable. In contrast, a standard deviation change in education, controlling for religiosity, results in a 0.200 standard deviation change, on average, in sexual permissiveness.

We still do not know how much of the variance in sexual permissiveness these two variables can account for. We also do not know whether the two regression coefficients are statistically significant. The next three sections consider these issues.

11.2.3 The Coefficient of Determination in the Three-Variable Case

The coefficient of determination, $R^2_{Y \cdot X}$, was introduced in Section 8.4.1 of Chapter 8 as the sum of squares due to regression ($SS_{\text{REGRESSION}}$), divided by the total sum of squares (SS_{TOTAL}). The equation is:

$$R^2_{Y \cdot X} = \frac{SS_{\text{REGRESSION}}}{SS_{\text{TOTAL}}}$$

We can use this same formulation to determine how much variance X_1 and X_2 can "explain" in the dependent variable Y.

An alternative formulation for the coefficient of determination which was presented in Section 8.4.1 in Chapter 8 is:

$$R^2_{Y \cdot X} = \frac{\Sigma(Y_i - \overline{Y})^2 - \Sigma(Y_i - \hat{Y}_i)^2}{\Sigma(Y_i - \overline{Y})^2}$$

$$= \frac{SS_{\text{TOTAL}} - SS_{\text{ERROR}}}{SS_{\text{TOTAL}}}$$

since $SS_{\text{TOTAL}} = SS_{\text{REGRESSION}} + SS_{\text{ERROR}}$, and $SS_{\text{REGRESSION}} = SS_{\text{TOTAL}} - SS_{\text{ERROR}}$. In the bivariate case $\hat{Y}_i = a + bX_i$, where in the case of two independent variables, $\hat{Y}_i = a + b_1 X_{1i} + b_2 X_{2i}$. This latter equation, therefore, must be substituted in the formula above for $\hat{Y}_i$ in order to determine $SS_{\text{REGRESSION}}$.

Ordinarily the sum of squares due to regression will be larger in the multivariate than in the bivariate case. It can *never* be smaller, since one reason for including additional variables is to explain additional variance in the dependent variable. There are several computational formulas for $R^2_{Y \cdot X_1 X_2}$, four of which are presented in this section. You should convince yourself they all give the same answer.

The first two formulas for the coefficient of determination in the three-variable case are:

$$R^2_{Y \cdot X_1X_2} = \frac{r^2_{YX_1} + r^2_{YX_2} - 2r_{YX_1}r_{YX_2}r_{X_1X_2}}{1 - r^2_{X_1X_2}}$$

$$R^2_{Y \cdot X_1X_2} = \beta^*_1 r_{YX_1} + \beta^*_2 r_{YX_2}$$

We have added subscripts to R^2 to clarify which independent variables are being used to predict the dependent variable. In the subscript, the variable being predicted (Y) is to the left of the dot and the predictor variables, X_1 and X_2, are to the right of the dot.

We can use either of these equations to compute the amount of variance in sexual permissiveness that can be accounted for by religiosity and education. We will use the second one because it involves fewer computations. Using the GSS data in Table 11.2 above and the beta weights calculated in the preceding section:

$$R^2_{Y \cdot X_1X_2} = (-0.325)(-0.318) + (0.200)(0.190)$$
$$= 0.1414$$

Thus these two variables together account for 14.1% of the variance in sexual permissiveness. Since 86% of the variance remains unexplained, a more complex theory than these two simple propositions is obviously needed.

Two other formulations of the coefficient of determination are informative, not as computational formulas, but because they shed light on the meaning of the coefficient. They are:

$$R^2_{Y \cdot X_1X_2} = r^2_{YX_1} + (r^2_{YX_2 \cdot X_1})(1 - r^2_{YX_1})$$

$$R^2_{Y \cdot X_1X_2} = r^2_{YX_2} + (r^2_{YX_1 \cdot X_2})(1 - r^2_{YX_2})$$

These two equations indicate that with two independent variables, the coefficient of determination can be divided into two components. In the first equation the first component is the amount of variance in Y that X_1 alone can account for—$r^2_{YX_1}$. The second component is the additional amount of variance in Y that X_2 can

FIGURE 11.2

Venn Diagrams Showing Two Different But Equivalent Decompositions of $R^2_{Y \cdot X_1 X_2}$

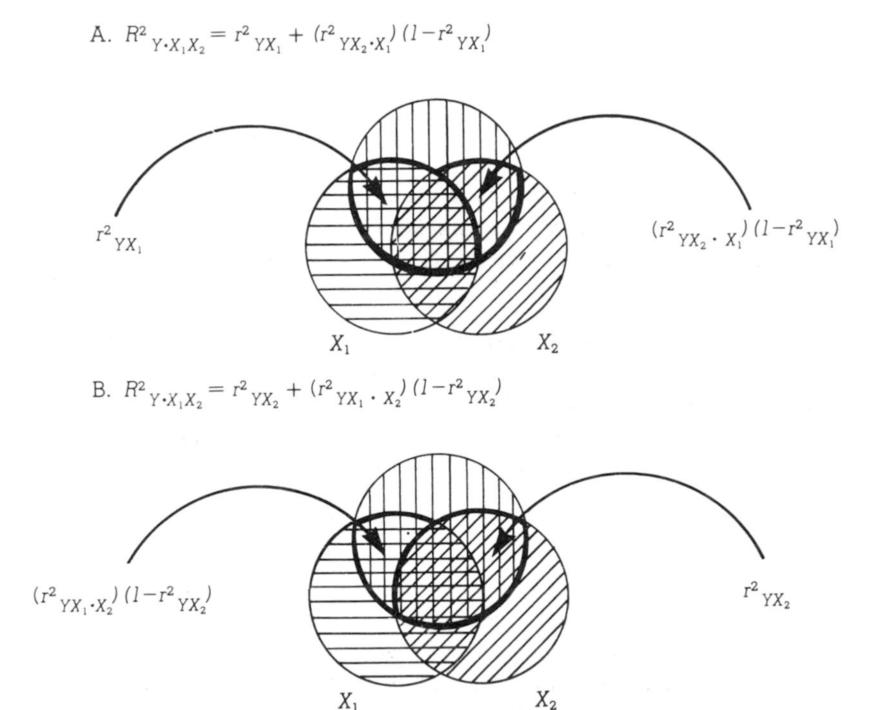

explain, after controlling for X_1. This term, $(r^2_{YX_2 \cdot X_1})(1 - r^2_{YX_1})$, is the **part correlation** squared between Y and X_2, controlling for X_1. (Notice that the part correlation is the square of the partial correlation $(r_{YX_2 \cdot X_1})$, multiplied by the term $(1 - r^2_{YX_1})$. Hence the part and partial correlations are intimately related.)

part correlation—a measure of the proportion of variance in a dependent variable that an independent variable can explain, when squared, after controlling for the other independent variable in a multiple regression equation

Figure 11.2 examines how these two components relate to the coefficient of determination, using Venn diagrams. Panel A, which diagrams the first equation, shows $r^2_{YX_1}$ as the area of overlap between the Y and X_1 circles, *including* that part of the Y and X_1 overlap that also overlaps X_2. And $(r^2_{YX_2 \cdot X_1})(1 - r^2_{YX_1})$ is shown as the overlap between Y and X_2, *excluding* that part of the Y and X_2 overlap that also overlaps X_1. This diagram indicates that the coefficient of determination can be thought of as the amount of variance explained in Y by X_1, *plus* the amount of Y accounted for by X_2, controlling for X_1. In other words, we have first allocated to X all the variance in Y that is *jointly contributed* by X_1 and X_2; then we have added on the variance in Y that is *uniquely* explained by X_2.

An alternative but equivalent decomposition is shown in the second equation above and in Panel B of Figure 11.2. This decomposition shows $r^2_{YX_2}$ as the overlap between the Y and X_2 circles, including that part of the Y and X_2 overlap that also includes X_1. As in Panel A, the coefficient of determination also includes the amount of variance in Y that can be accounted for by X_1, controlling for X_2; that is, $(r^2_{YX_1 \cdot X_2})(1 - r^2_{YX_2})$.

These two alternative but equivalent formulas for $R^2_{Y \cdot X_1 X_2}$ make an important point:

> When both X_1 and X_2 correlate with the dependent variable and are themselves intercorrelated, there is no unique way to partition the amount of variance in Y due to the two independent variables. However, when the two independent variables are uncorrelated, as they are in an experiment (or should be if the assignment of subjects to conditions has indeed been random), then we can uniquely partition the amount of variance in Y to X_1 and X_2.

This is one reason experiments are preferred to nonexperimental research in the social sciences. When an experiment is possible and can be ethically justified, it offers a better opportunity to determine the effects of independent variables on a dependent variable.

When X_1 and X_2 are uncorrelated, the total variance in Y can be uniquely partitioned into two segments, one due to X_1 and the second due to X_2. If, in the first formula for $R^2_{Y \cdot X_1 X_2}$ presented above, $r_{X_1 X_2} = 0$, then:

$$R^2_{Y \cdot X_1 X_2} = \frac{r^2_{YX_1} + r^2_{YX_2} - 2r_{YX_1}r_{YX_2}(0)}{1 - 0^2}$$

$$= r^2_{YX_1} + r^2_{YX_2}$$

FIGURE 11.3
Venn Diagrams Showing the Decomposition of $R^2_{Y \cdot X_1X_2}$ When X_1 and X_2 Are Uncorrelated

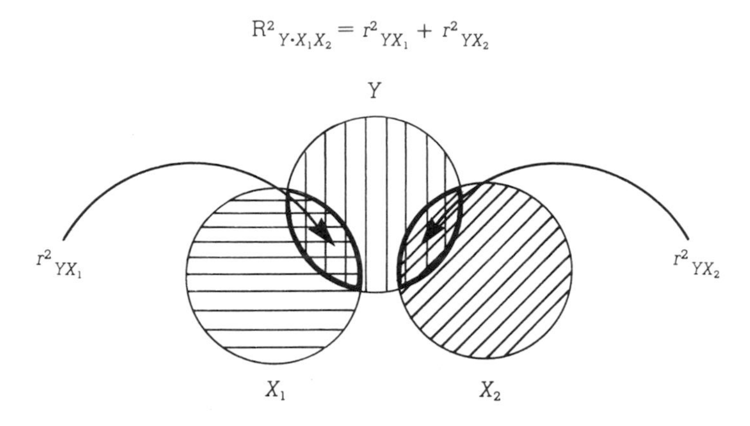

That is, if X_1 and X_2 are uncorrelated, the amount of variance each explains in Y is simply $r^2_{YX_1}$ and $r^2_{YX_2}$, respectively, and these two components sum to the coefficient of determination—$R^2_{Y \cdot X_1X_2}$. This is shown with Venn diagrams in Figure 11.3.

multiple correlation coefficient—the coefficient for a multiple regression equation, which, when squared, equals the ratio of the sum of squares due to regression to the total sum of squares

The square root of the coefficient of determination is called the **multiple correlation coefficient**. While $R_{Y \cdot X_1X_2}$ is often reported in research reports, it is less useful than $R^2_{Y \cdot X_1X_2}$ as an interpretive tool, since it has no clear meaning.

11.2.4. Testing the Significance of the Coefficient of Determination with Two Independent Variables

In Section 8.6.1 of Chapter 8, for the bivariate case, we tested $R^2_{Y \cdot X}$ for significance with an F test with 1 and $N - 2$ degrees of freedom. Where $\rho^2_{Y \cdot X_1X_2}$ is the coefficient of determination in the population with two independent variables, the null hypothesis is H_0: $\rho^2_{Y \cdot X_1X_2} = 0$. We will also test this hypothesis with an F test, though the degrees of freedom will differ from those in the bivariate case. The df associated with SS_{TOTAL} are $N - 1$, regardless of the number of independent variables. In the case of two independent variables, $SS_{REGRESSION}$ is estimated from the two regression coefficients and hence has 2 df associated with it. Since we know that in general $df_{\text{TOTAL}} = df_{\text{REGRESSION}} + df_{\text{ERROR}}$, by subtraction $df_{\text{ERROR}} = N - 3$.

We can compute the mean squares needed for the F test by dividing the sums of squares by the appropriate degrees of freedom. That is, with two independent variables:

$$MS_{\text{REGRESSION}} = \frac{SS_{\text{REGRESSION}}}{2}$$

$$MS_{\text{ERROR}} = \frac{SS_{\text{ERROR}}}{N - 3}$$

If the null hypothesis that $\rho^2_{Y \cdot X_1 X_2} = 0$ in the population is true, both $MS_{\text{REGRESSION}}$ and MS_{ERROR} are unbiased estimates of the variance of the errors of prediction, σ^2_e. If, however, $\rho^2_{Y \cdot X_1 X_2}$ is greater than zero in the population, then $MS_{\text{REGRESSION}}$ is greater than MS_{ERROR} as well. If in a given sample the ratio of $MS_{\text{REGRESSION}}$ to MS_{ERROR} is larger than some predetermined critical value, we reject the hypothesis that $\rho^2_{Y \cdot X_1 X_2} = 0$ in the population.

Specifically, we choose an α level and calculate the test statistic:

$$F_{2,N-3} = \frac{MS_{\text{REGRESSION}}}{MS_{\text{ERROR}}}$$

We next look in Appendix E to determine the critical value for an F with 2 and $N - 3$ degrees of freedom. If the test statistic is as large or larger than the critical value, we reject the null hypothesis; otherwise we do not.

To calculate the mean squares, we need to know the terms in the numerators $SS_{\text{REGRESSION}}$ and SS_{ERROR}. To calculate these sums of squares we use the same logic as we did in the bivariate case (see Section 8.6.1 in Chapter 8). Therefore:

$$SS_{\text{REGRESSION}} = R^2_{Y \cdot X_1 X_2} SS_{\text{TOTAL}}$$

and

$$SS_{\text{ERROR}} = SS_{\text{TOTAL}} - SS_{\text{REGRESSION}}$$

We can now ask whether the coefficient of determination observed in Section 11.2.3, 0.1414, is significantly different from zero in the population. We set $\alpha = .01$. To calculate the sums of squares, we note from Table 11.2 that $s_Y = 1.048$; therefore the variance of variable Y in the sample, $s_Y^2 = (1.048)^2 = 1.098$, with $N = 1{,}499$. Since in general $SS_{\text{TOTAL}} = s_Y^2(N - 1)$, it follows that in these data:

$$SS_{\text{TOTAL}} = (1.098)(1{,}499 - 1) = 1{,}645.259$$

$$SS_{\text{REGRESSION}} = (0.1414)(1{,}645.26) = 232.640$$

$$SS_{\text{ERROR}} = 1{,}645.259 - 232.640 = 1{,}412.619$$

We now can divide these figures by their degrees of freedom to determine the mean squares:

$$MS_{\text{REGRESSION}} = \frac{232.640}{2} = 116.320$$

$$MS_{\text{ERROR}} = \frac{1{,}412.619}{1{,}499 - 3} = 0.944$$

Hence the test statistic is:

$$F_{2,1496} = \frac{116.320}{0.944} = 123.22$$

An examination of Appendix E indicates that the critical value for 2 and 1,496 *df* is 4.61. Hence we can reject with great confidence the null hypothesis that $\rho^2_{Y \cdot X_1X_2} = 0$ in the population.

11.2.5 Testing b_1 and b_2 for Significance

Two null hypotheses can be tested to determine whether or not the two regression coefficients are zero in the population. These hypotheses assert that the population parameters are zero: H_0: $\beta_1 = 0$ and H_0: $\beta_2 = 0$. We will show how to test whether the sample regression coefficient, b_1, is statistically significant first and return to the test of b_2 later. The two tests are obviously very similar, however.

To test whether the observed b_1 differs statistically from zero we construct a t variable, just as we did in the bivariate case. In the three-variable case, this is:

$$t = \frac{b_1 - 0}{s_{b_1}}$$

To obtain the t value, we need to estimate the standard error of the regression coefficient, s_{b_1}. If we assume that in the population being sampled the dependent variable, Y, is normally distributed for any *joint outcome* of X_1 and X_2, the sampling distribution of b_1 (and of b_2 as well) is normally distributed as N gets large. Furthermore, the mean of the sampling distribution of b_1 will equal β_1; that is, $E(b_1) = \beta_1$. The variance of the sampling distribution of b_1 is:

$$\sigma^2_{b_1} = \frac{\sigma^2_e}{\Sigma(X_{1i} - \bar{X}_1)^2(1 - \rho^2_{X_1 \cdot X_2})}$$

As was true for the bivariate case, we can estimate the numerator, σ^2_e, with MS_{ERROR}. The term $\rho^2_{X_1 \cdot X_2}$ is the correlation squared for predicting X_1 from X_2. It is estimated by $R^2_{X_1 \cdot X_2}$. Now that we have estimates of σ^2_e and $\rho^2_{X_1 \cdot X_2}$, we can obtain a sample estimate of $\sigma^2_{b_1}$. For the sample statistic, this is:

$$s^2_{b_1} = \frac{MS_{\text{ERROR}}}{\Sigma(X_{1i} - \bar{X}_1)^2(1 - R^2_{X_1 \cdot X_2})}$$

The square root is used in the denominator of the t ratio. The t ratio will have $N - 3$ degrees of freedom, since there are $N - 3$ *df* associated with MS_{ERROR}. Hence for b_1 we have:

$$t_{N-3} = \frac{b_1}{s_{b_1}} = \frac{b_1}{\sqrt{\dfrac{MS_{\text{ERROR}}}{\Sigma(X_{1_i} - \overline{X}_1)^2\,(1 - R^2_{X_1 \cdot X_2})}}}$$

In the preceding section we determined that MS_{ERROR} is 0.944. Since $s^2_{X_1} = \Sigma(X_{1_i} - \overline{X}_1)^2/(N - 1)$, we also know that $\Sigma(X_{1_i} - \overline{X}_1)^2 = s^2_{X_1}\,(N - 1)$. And since Table 11.2 shows that $s_{X_1} = 2.681$, we can see that $s^2_{X_1} = (2.681)^2 = 7.1878$. With $N = 1{,}499$, it follows that $\Sigma(X_{1_i} - \overline{X}_1)^2 = (7.1878)(1{,}499 - 1) = 10{,}767.266$. And we see from Table 11.2 that $r_{X_1X_2} = .033$. Hence $R^2_{X_1 \cdot X_2} = .0011$.

Now we set $\alpha = .01$ and calculate:

$$t_{1496} = \frac{-0.127}{\sqrt{\dfrac{0.944}{(10{,}767.266)(1 - 0.0011)}}}$$

$$= \frac{-0.127}{0.0094} = -13.56$$

For $\alpha = .01$ and 1,496 *df*, we see from Appendix D that the critical value is -2.33 for a one-tailed test. Since the observed test statistic is much larger than the c.v., we reject the null hypothesis that β_1 equals zero.

The test for a b_2 regression coefficient is very similar to that for b_1. The t test for calculating the test statistic is:

$$t_{N-3} = \frac{b_2}{s_{b_2}} = \frac{b_2}{\sqrt{\dfrac{MS_{\text{ERROR}}}{\Sigma(X_{2i}-\overline{X}_2)^2\,(1-R^2_{X_2 \cdot X_1})}}}$$

Where:

$$R^2_{X_2 \cdot X_1} = r^2_{X_2 X_1}$$

We know that $MS_{\text{ERROR}} = 0.944$, and from Table 11.2 we can see that $s^2_{X_2} = 3.222^2 = 10.381$, $r_{X_1 X_2} = .033$, with $N = 1{,}499$. Therefore $\Sigma(X_{2i} - \overline{X}_2)^2 = s^2_{X_2}(N - 1) = (10.381)(1{,}498) = 15{,}550.74$. Furthermore, $R^2_{X_2 \cdot X_1} = .033^2 = .0011$. If we set $\alpha = .01$ we can compute the test statistic:

$$t_{1496} = \frac{0.0650}{\sqrt{\dfrac{.944}{(15{,}550.74)(1-.0011)}}}$$

$$= \frac{0.0650}{0.0078} = 8.34$$

Appendix D indicates that the critical value for a one-tailed hypothesis test with 1,496 degrees of freedom is 2.33. Since this test statistic is 8.34, we reject the null hypothesis.

Table 11.3, which summarizes the results for testing the two propositions about sexual permissiveness, shows support for both propositions. That is, as religiosity increases, sexual permissive-

TABLE 11.3

Results of Regression of Sexual Permissiveness Index on Religious Attendance and Educational Attainment Variables

Independent Variable	β_j^*	b_j	s_{b_j}	t
Religious attendance	−0.325	−0.127	0.0094	−13.56*
Educational attainment	0.200	0.065	0.0078	8.34*
Intercept		1.813		
		$R^2_{Y \cdot X_1X_2} = 0.141$*		

*Significant for $\alpha = .01$.

ness decreases, controlling for educational attainment. And as educational attainment increases, so does sexual permissiveness.[2] Before accepting these propositions, however, we need to be certain that no other variables might account for the observed relationships. One such variable is the gender of the respondent. In Section 11.3.3 we will examine whether adding gender to the regression equation changes the results.

11.2.6. Confidence Intervals for b_1 and b_2

As in the bivariate case, we can use the standard errors of b_1 and b_2 to construct confidence intervals around the point estimates. If we pick $\alpha = .01$, the upper and lower limits for the 99% confidence interval for the population parameter, β_1, are $b_1 + s_{b_1} 2.58$ and $b_1 - s_{b_1} 2.58$. Since we saw in Table 11.3 that $s_{b_1} = 0.0094$ and $b_1 = -0.127$, therefore we have $-0.127 + (0.0094)(2.58) = -0.103$ and $-0.127 - (0.0094)(2.58) = -0.151$. Hence the 99% confidence interval for β_1 is bounded by -0.151 and -0.103.

The upper and lower limits to a 99% confidence interval for β_2 are given by $b_2 + s_{b_2}2.58$. and $b_2 - s_{b_2}2.58$. Convince yourself that in the sexual permissiveness data these limits are .045 and 0.085.

2. It is also possible to test a, the intercept for statistical significance, but the formula for doing so is omitted here because it is too complex. Most statistics computer programs routinely calculate the t test for you.

11.2.7 Partial Correlation in the Three-Variable Case

Partial correlation methods were introduced in Chapter 10. For three variables, Y, X_i, and X_j:

$$r_{YX_j \cdot X_i} = \frac{r_{YX_j} - r_{YX_i} r_{X_i X_j}}{\sqrt{1 - r^2_{YX_i}}\sqrt{1 - r^2_{X_i X_j}}}$$

The partial correlation coefficient is an estimate of the relationship between a variable, Y, and a second variable, X_j, controlling for a third variable, X_i. And the square of the partial correlation coefficient is the amount of variance in Y that can be accounted for by X_j, controlling for (or taking into account) X_i. As we showed in Sections 11.2.2 and 11.2.3, the standardized regression coefficient, β^*_j, is given by:

$$\beta^*_j = \left(\frac{s_{X_j}}{s_Y} \right) b_j = \frac{r_{YX_j} - r_{YX_i} r_{X_i X_j}}{1 - r^2_{X_i X_j}}$$

Notice that the beta weight in this equation and the partial correlation coefficient given above have the same numerator. This means both will always have the same sign, and they usually are very similar in size. Their similarity has led some researchers and statisticians to refer to the regression coefficients from multiple regression analysis as *partial regression coefficients.* The similarity between β^*_j and $r_{YX_j \cdot X_i}$ can be illustrated for the example where sexual permissiveness was regressed on religious attendance and education. As we found in Section 11.2.2, $\beta^*_1 = -0.325$, and:

$$r_{YX_1 \cdot X_2} = \frac{-0.318 - (0.190)(0.033)}{\sqrt{1 - 0.190^2}\sqrt{1 - 0.033^2}} = -0.330$$

Since the results from the regression and partial correlation analyses are very similar, one of them is superfluous. Ordinarily the regression coefficients are of greater interest, so partial correlation coefficients are usually not presented or discussed in research reports. One interesting and important application of the partial correlation coefficient will be shown in Chapter 12, which describes a technique called *path analysis.*

> The tests of significance for the partial correlation coefficients are identical to those used in testing the regression coefficients. Therefore if you test the regression coefficients for significance you have also indirectly tested the partial correlation coefficients for significance.

11.3 Multiple Regression with Several Independent Variables

In the sexual permissiveness example above, we examined only two of several possible propositions. Reiss, for example, indicates that females tend to be less sexually permissive than males.[3] He explains these differences as a function of the fact that females traditionally have been more closely attached to the family than men, although, as he surely is aware, this orientation is changing rapidly. For most females, the primary goal for years was to be good wives and mothers. Women also have been perceived by society as the principal socialization agent for children. Together, these traditional attitudes and values help account for women's less permissive attitudes towards sex. A third derived proposition, therefore, is:

P3: Females are less sexually permissive than males.

In addition to gender, Reiss also examines the roles of race, marital status, love, and so on. In other words, a comprehensive theory of sexual permissiveness will contain many independent variables. We will examine a model with only three independent

3. Reiss and Miller, "Heterosexual Permissiveness"; Reiss, *Family Systems in America.*

variables in this section, but this model will show you how to estimate more complex and more complete regression models.

For up to k independent variables, the general *population regression model* is:

$$Y_i = \alpha + \beta_1 X_{1i} + \beta_2 X_{2i} + \ldots + \beta_k X_{ki} + e_i$$

The *sample regression equation* and the *prediction equation,* for estimating the population regression equation with k independent variables, are:

$$Y_i = a + b_1 X_{1i} + b_2 X_{2i} + \ldots + b_k X_{ki} + e_i$$

$$\hat{Y}_i = a + b_1 X_{1i} + b_2 X_{2i} + \ldots + b_k X_{ki}$$

When there are more than two independent variables, mathematics beyond the scope of this book is required to show computational formulas for the regression coefficients, the intercept, and their standard errors. However, the interpretation follows straightforwardly from the case of two independent variables. That is, a multiple regression coefficient measures the amount of increase or decrease in the dependent variable for a one-unit difference in the independent variable, controlling for the other $k - 1$ independent variables.

The relationship between the beta weights and the raw regression coefficients is also identical to what it is in the case of two independent variables, namely, $\beta_j^* = (s_{X_j}/s_Y)b_j$.

11.3.1. Testing the Coefficient of Determination for Several Independent Variables

The coefficient of determination for k independent variables is:

$$R^2_{Y \cdot X_1 X_2 \ldots X_k} = \beta_1^* r_{YX_1} + \beta_2^* r_{YX_2} + \ldots + \beta_K^* r_{YX_k}$$

BOX 11.2 Deriving the *F* Test for the Coefficient of Determination with Several Independent Variables

When the coefficient of determination is tested with two or more independent variables,

$$MS_{\text{REGRESSION}} = SS_{\text{REGRESSION}}/df_{\text{REGRESSION}}$$

and $MS_{\text{ERROR}} = SS_{\text{ERROR}}/df_{\text{ERROR}}$. In the case of k independent variables, there are k degrees of freedom associated with $SS_{\text{REGRESSION}}$ and $N - k - 1$ with SS_{ERROR}. Therefore:

$$F_{k,N-k-1} = \frac{SS_{\text{REGRESSION}}/k}{SS_{\text{ERROR}}/(N - k - 1)}$$

As we showed in Section 8.6 (Chapter 8):

$$SS_{\text{REGRESSION}} = R^2_{Y.X} SS_{\text{TOTAL}}$$

and:

$$1 = R^2_{Y.X} + \frac{SS_{\text{ERROR}}}{SS_{\text{TOTAL}}}$$

From this it follows that:

$$SS_{\text{ERROR}} = (1 - R^2_{Y.X})SS_{\text{TOTAL}}$$

When we substitute these two results for the case of k independent variables in the equation for F, we obtain:

$$F_{k,N-k-1} = \frac{(R^2_{Y.X_1X_2\ldots X_k})SS_{\text{TOTAL}}/k}{(1 - R^2_{Y.X_1X_2\ldots X_k})SS_{\text{TOTAL}}/(N-k-1)}$$

$$= \frac{R^2_{Y.X_1X_2\ldots X_k}/k}{(1 - R^2_{Y.X_1X_2\ldots X_k})/(N-k-1)}$$

since SS_{TOTAL} cancels in both the numerator and the denominator.

The test of significance associated with the coefficient of determination is:

$$F_{k,N-k-1} = \frac{MS_{\text{REGRESSION}}}{MS_{\text{ERROR}}}$$

$$= \frac{R^2_{Y.X_1X_2\ldots X_k}/k}{(1 - R^2_{Y.X_1X_2\ldots X_k})/(N - k - 1)}$$

This new formulation for testing the significance of the coefficient of determination is a very handy one which is commonly used. The derivation of this formula is presented in Box 11.2.

11.3.2 Testing Regression Coefficients for Several Independent Variables

The formulas for estimating the standard errors of the regression coefficients, the σ_{b_j}, are too complex to present here. Modern computers, however, can easily estimate them with statistical programs such as SPSS (see Preface). We do not present the exact formulas for estimating the standard errors, but the procedure to be followed in determining statistical significance is identical to that shown thus far in this chapter. We calculate the k t ratios, each with $N - k - 1$ degrees of freedom; that is:

$$t_{N-k-1} = \frac{b_j}{s_{b_j}}$$

Then we compare the resulting t against the critical value for a given α level.

Two notes of caution are in order here. First, unless the k independent variables are uncorrelated with each other, the k t ratios used to test for significance are not independent of each other. This means that reported levels of statistical significance may be slightly biased. However, in most cases the degree of bias should be small enough so as not to cause any practical concern.

Second, and more important, the standard errors of the b_j are affected by the degree of intercorrelation among the k independent variables. In particular, to the degree that an independent variable is predictable from the other $k - 1$ independent variables, the standard error increases in size. We call this situation **multicollinearity**. If one of the independent variables is perfectly predictable from the others (that is, the multiple correlation is 1.00), the result is perfect multicollinearity. Much more common is the case where 50% to 75% of the variance in one or more of the variables can be accounted for by the other $k - 1$ variables. In these cases one or more of the independent variables should be eliminated from the regression equation.

multicollinearity—a condition of high or near perfect correlation among the independent variables in a multiple regression equation

There are several ways to test for multicollinearity, but except for doing a series of multiple regressions among the independent variables, they are beyond the scope of an introductory text. Something that can be easily done, however, is to inspect the matrix of intercorrelations among the variables. If high correlations are observed among the independent variables (e.g., 0.80 or higher), then any regressions based on these data run the risk of multicollinearity and resultant large standard errors for the regression coefficients.

After a computer has estimated the standard errors, they can also be used to construct confidence intervals around the β_j, as we did above in the case of two independent variables.

11.3.3. An Example: Examining the Effects of Gender on Sexual Permissiveness

Using the 1977 GSS data, we can expand the original problem from $k = 2$ to $k = 3$ independent variables. The research hypothesis is:

H3: Females score lower on the sexual permissiveness index than males do.

As we will show in more detail in the next section, a common and useful way to enter a discrete variable into a regression equation as an independent variable is to create one or more dichotomous variables coded 0 and 1, to represent the categories of the discrete variable. To enter gender into the regression equation, therefore, we create a dichotomous variable with female = 1 and male = 0.

In this example the sample equation to be estimated by multiple regression is:

$$\hat{Y} = a + b_1X_1 + b_2X_2 + b_3X_3$$

Where:

Y = The sexual permissiveness index.
X_1 = Church attendance.
X_2 = Educational attainment.
X_3 = Gender (1 = female, 0 = male).

Using SPSS, we obtain the following estimates, with the standard errors of the regression coefficients in parentheses below each one:

$$\hat{Y} = \underset{(0.104)}{1.859} - \underset{(0.009)}{0.124X_1} + \underset{(0.008)}{0.064X_2} - \underset{(0.051)}{0.089X_3}$$

with $R^2_{Y \cdot X_1X_2X_3} = 0.1430$. For $\alpha = .01$ the critical value for all four t tests is ± 2.33, depending on the direction of the hypothesis. The t ratios for b_1, b_2, and b_3 and a are -13.78, 8.00, -1.75 and 17.88, respectively. In other words, controlling for gender leaves the coefficients for religion and education highly significant. Indeed the coefficients themselves are virtually unchanged, since the coefficients without gender controlled were -0.127 and 0.065. The coefficient for gender is not statistically significant, meaning that P3 is not supported.

We can also test the coefficient of determination for statistical significance. The degrees of freedom for the F test are 3 and 1,495. For $\alpha = .01$, the critical value is 3.78, as can be verified from Appendix E. The test statistic is:

$$F_{3,1495} = \frac{0.1430/3}{(1 - 0.1430)/1{,}495} = 83.15$$

This indicates that the coefficient of determination is clearly significant.

The standardized regression equation for this example is:

$$\hat{Z}_Y = -0.317Z_1 + 0.198Z_2 - 0.042Z_3$$

As was true for the raw regression coefficients, the standardized coefficients for religion and education are not very different from

FIGURE 11.4

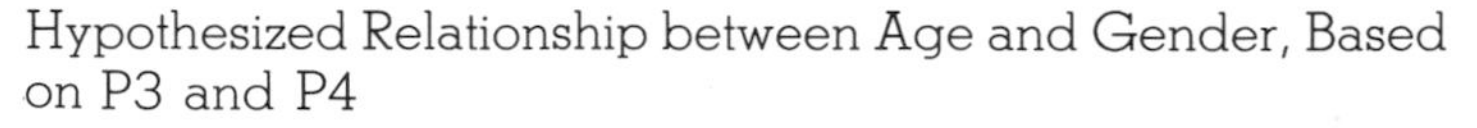

Hypothesized Relationship between Age and Gender, Based on P3 and P4

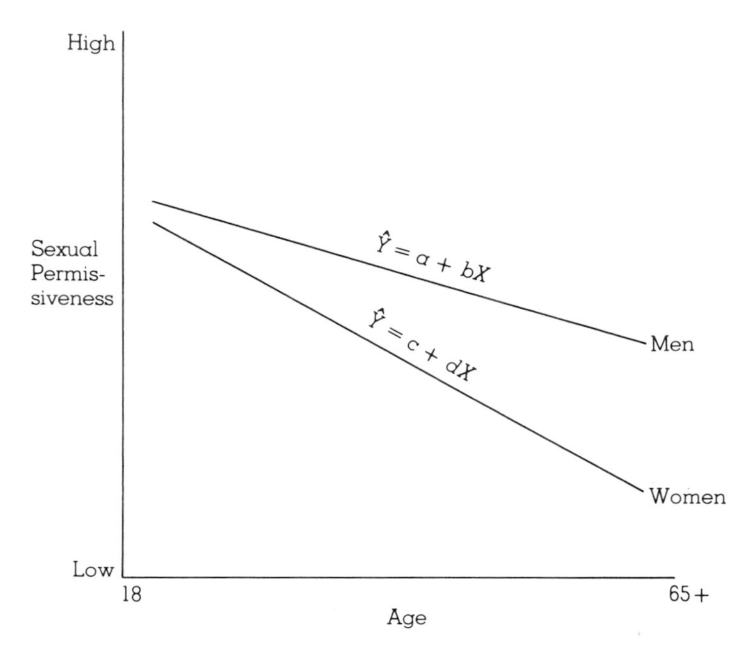

those obtained before gender was included in the equation, and the beta weight for gender is very small.

11.4 Dummy Variable Regression Analysis

In the preceding section we found that gender has a small, insignificant effect on sexual permissiveness in the 1977 GSS data. One question that can be explored is whether women's attitudes toward sexual freedom have changed in recent years. Most casual observers of American society would probably say that younger women's attitudes about sex are more permissive than those of older women, and younger men's attitudes are more permissive than those of older men. There is also a popular perception that as a result of recent sex role changes, younger men and women hold more similar views about sex. The logic of these arguments leads to two additional propositions:

P4: The younger a person is, the more sexually permissive he or she is.

P5: Young women are closer in sexual permissiveness to young men than older women are to older men.

The last proposition suggests an *interaction effect* between age and gender in the explanation of sexual permissiveness. If P5 is true, the relationship between age and sexual permissiveness should vary for men and women. The general form of these hypothetical relationships is shown in Figure 11.4. First, the higher the age, the lower the permissiveness, as shown by the negatively sloping lines for both sexes. Second, the gap between the sexual permissiveness of men and women increases with age, as revealed by the increasing distance between the lines at greater ages.

To test these hypotheses we will again use the 1977 GSS data. Hypotheses H4 and H5 are identical to P4 and P5, and we will not restate them. Age, X_4, is coded 1 = 10–19, 2 = 20–29, and so on, to 8 = 80 or over. The estimated equation for H4 is:

$$\hat{Y} = \underset{(0.0741)}{2.712} - \underset{(0.00155)}{0.0144}X_4$$

and $R^2_{Y \cdot X_4} = 0.0544$. For $\alpha = .01$, the null hypothesis that $\beta_4 = 0$ is rejected, since $t = -9.29$, which is far beyond the critical value of -2.33 required for 1,497 *df*. Furthermore, we can also reject the hypothesis that $\rho^2_{Y \cdot X_4} = 0$ in the population, since:

$$F_{1,1497} = \frac{0.0544}{(1 - 0.0544)/(1{,}499 - 1 - 1)} = 86.12$$

This is far beyond the c.v. of 3.78 for $\alpha = .01$. The standardized regression equation is:

$$\hat{Z}_Y = -0.2332 Z_{X_4}$$

Clearly, H4 receives support in the 1977 GSS data. The younger a person is, the more sexually permissive he or she is.

dummy variable—a variable coded 1 to indicate the presence of an attribute and 0 in its absence

To test H5 we need to consider the notion of a dummy variable more carefully. A **dummy variable**, D_j, is a variable which is coded 1 to indicate the presence of an attribute and 0 in its absence.

Dummy variables are especially useful for representing the effects of a discrete *independent* variable in a regression equation. Consider the variable religion. Suppose it is coded:

1 = Protestant.
2 = Catholic.
3 = Jew.
4 = None.

Clearly this variable cannot be used in a meaningful way in regression analysis without recoding. One way to do this is to create a set of four dummy variables, *one for each category* of religion. That is,

D_1 = 1 if a respondent is Protestant, 0 if not.
D_2 = 1 if a respondent is Catholic, 0 if not.
D_3 = 1 if a respondent if Jewish, 0 if not.
D_4 = 1 if a respondent is none, 0 otherwise.

We could then enter *any three of the four* dummy variables as independent variables in a regression equation to predict a dependent variable. We can only enter three because all information is contained in any subset of three of the variables. That is, if we know a person is 1 on D_1, 0 on D_2 and 0 on D_3, that person must be 0 on D_4. A person cannot be both a Protestant and none, in other words. Similarly, if we know that D_1, D_2, and D_3 are all scored 0—that is, if the respondent is not a Protestant, Catholic, or Jew—then $D_4 = 1$, since the person must have no religious identification. In general, *if a variable has J categories, then any $J - 1$ dummy variables created from it can be used in a regression equation.* Note that in testing H_4 above we implicitly used a dummy variable for gender.

Dummy variables are also very useful in testing for interaction effects. We will show how this can be done to test H5. Since gender has two outcomes (i.e., $J = 2$), we need only one dummy variable, D, to capture the effects of gender in a regression equation. In this case $D = 1$ if the respondent is a female and 0 if not. (This is equivalent to the coding we used for X_3 in the preceding section, and therefore $D = X_3$ in this example.) We now need to consider an important principle:

> Interaction effects between two independent variables are measured by multiplying the two independent variables together and including the product term in the regression equation, along with the components that comprise the product.

In other words, if we want to estimate sexual permissiveness as a function of age (X_1) and gender (D), but we also expect an interaction effect due to age and sex, the appropriate regression equation to estimate this effect is:

$$\hat{Y} = a + b_1X_1 + b_2D + b_3DX_1$$

In this equation, the term DX_1 is the product of gender and age used to estimate the interaction between them. (We have used X_1 to represent age, even though it was used for religious attendance in the earlier example, because otherwise there would have been a discrepancy between the subscripts of the variables and their associated regression coefficients.)

When $D = 0$, that is, when a respondent is a male, the equation becomes:

$$\hat{Y}_{\text{male}} = a + bX_1 + b_2(0) + b_3(0)X_1$$
$$= a + b_1X_1$$

In contrast, when $D = 1$, that is, when a respondent is a female, the equation is:

$$\hat{Y}_{\text{female}} = a + b_1X_1 + b_2(1) + b_3(1)X_1$$
$$= (a + b_2) + (b_1 + b_3)X_1$$

If $c = a + b_2$ and $d = b_1 + b_3$, this is:

$$\hat{Y}_{\text{female}} = c + dX_1$$

Thus, by using a dummy variable, we have in effect *two separate regression equations*—one for predicting the sexual permissiveness scores of the males and a second for predicting the scores of the females. That is what we mean by an interaction effect, as Figure 11.4 makes clear. The slopes and intercepts are different for the two sexes. The intercepts are a for the males and $(a + b_2) = c$ for the females. And the regression coefficient is b_1 for the males and $(b_1 + b_3) = d$ for the females.

Researchers often use many dummy variables in a single regression equation. For this reason we will begin to drop the detailed subscripts associated with the coefficient of determination (R^2). But we will need to be clear about which coefficient of determination we are referring to, as will become clear shortly. We will use subscripts such as 1, 2, and 3 to do so; that is, R_1^2, R_2^2, and R_3^2, to refer to the coefficients of determination associated with different equations having different combinations of independent variables.

Using the 1977 GSS data (not given here) to test H5—that there is an interaction between age and gender in predicting sexual permissiveness—we find:

$$\hat{Y} = \underset{(0.1101)}{2.7552} - \underset{(0.0024)}{0.0129X_1} - \underset{(0.1485)}{0.1011D} - \underset{(0.0031)}{0.0023DX_1}$$

with $R_2^2 = 0.0641$. (We will make clear why we have used the subscript 2 on R^2 below.)

The two equations for males and females are:

$$\hat{Y}_{\text{male}} = 2.7555 - 0.0129X_1$$

$$\hat{Y}_{\text{female}} = (2.7555 - 0.1011) + (-0.0129 - 0.0023)X_1$$

$$= 2.6544 - 0.0152X_1$$

11.4.1 Testing for Interaction Effects

In final form, the two equations above for males and females look very similar. So much so, in fact, that a statistical test is needed to see whether or not they do in fact differ. To test for such interaction effects (see Section 10.2.5 in Chapter 10), *the standard errors of the regression coefficients are not examined.* Instead, the coefficient of determination of the equation with the interaction term(s) included (R_2^2) is compared with the coefficient of determination for the equation with the interaction terms omitted (R_1^2).

To test whether R_2^2 is significantly larger than R_1^2, we use the following F test:

$$F_{(k_2-k_1),\,(N-k_2-1)} = \frac{(R_2^2 - R_1^2)/(k_2 - k_1)}{(1 - R_2^2)/(N - k_2 - 1)}$$

Where:

k_2 = The number of independent variables in the equation used to estimate R_2^2.

k_1 = The number of independent variables in the equation used to estimate R_1^2.

In this equation, $k_2 > k_1$, and since R_2^2 is based on more variables than R_1^2, it is always true that R_2^2 is greater than R_1^2.

To test for interaction effects, the null hypothesis is H_0: $\rho_2^2 - \rho_1^2 = 0$ in the population. If this hypothesis is rejected, we conclude that there are interaction effects. If the hypothesis cannot be rejected, we conclude that there are no interaction effects.

For the GSS data given in Section 11.4, we reestimate the equation without the interaction term:

FIGURE 11.5
Estimated Relationships among Sexual Permissiveness, Age, and Gender

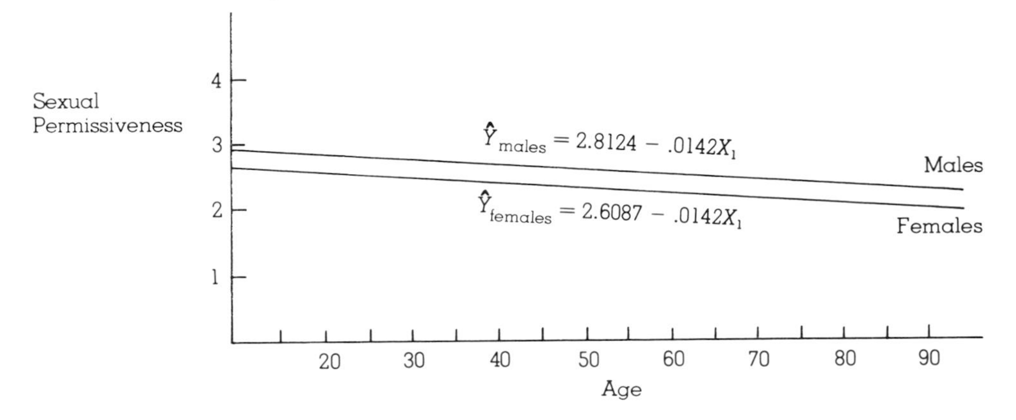

$$\hat{Y} = \underset{(0.0782)}{2.8124} - \underset{(0.0015)}{0.0142X_1} - \underset{(0.0527)}{0.2037D}$$

with $R_1^2 = 0.0637$. Now $k_2 - k_1 = 3 - 2 = 1$, and since $N =$ 1,499, $N - k_2 - 1 = (1{,}499 - 3 - 1) = 1{,}495$. If we set $\alpha = .01$, the test of whether $\rho_2^2 - \rho_1^2 = 0$ in the population is given by:

$$F_{1,1495} = \frac{(0.0641 - 0.0637)/1}{(1 - 0.0641)/1{,}495} = 0.639$$

Since for $\alpha = .01$ with 1 and 1,495 degrees of freedom the critical value is 6.63, we *cannot* reject the null hypothesis. Hence we conclude that age and gender do *not* interact in predicting sexual permissiveness.

Since the regression equation above does not have any interaction terms in it, we can use the standard error to determine whether the coefficients for age (X_1) and gender (D) are statistically significant. The two t ratios, each with 1,496 *df*, are -9.47 and -3.87, respectively, far greater in absolute value than the c.v. of 1.65 needed to reject the null hypothesis. We therefore conclude that age and gender combine in an *additive* (rather than interactive) way in the explanation of sexual permissiveness.

We still have separate equations for males and females, but since there is no interaction between age and gender, they share a *common slope.* The two equations are:

$$\hat{Y}_{\text{male}} = 2.8124 - 0.0142X_1$$

$$\hat{Y}_{\text{female}} = (2.8124 - 0.2037) - 0.0142X_1$$

$$= 2.6087 - 0.0142X_1$$

These two equations are graphed in Figure 11.5. The figure makes clear that sexual permissiveness decreases with age, and that females are less permissive than males, regardless of age.

When we estimate an equation with one or more dummy variables along with a continuous variable (or several continuous variables) and there are *no interaction terms,* the analysis is called an **analysis of covariance.** The continuous variables are called the **covariates**; the dummy variables are called **treatment levels**.

analysis of covariance—a multiple regression equation including one or more dummy variables, with a continuous independent variable and no interaction terms

covariates—the continuous variables in an analysis of covariance

treatment level—a term in experimental research to indicate the experimental group to which a subject has been assigned

11.4.2. An Example with Three Dummy Variables and a Continuous Variable

Let's examine a slightly more complex situation. Suppose we ask whether the relationship between religious attendance and sexual permissiveness can be explained by a person's religious identification. That is, does being Catholic or Jewish, for example, explain degree of permissiveness, or is permissiveness due to religiosity per se (as indexed by attendance at religious services)?

To answer this question we construct three dummy variables from the 1977 GSS data on religious identification, or the variable RELIG. They are:

D_1 = 1 if of no religious identification (NRI), otherwise 0.
D_2 = 1 if Jewish, otherwise 0.
D_3 = 1 if Protestant, otherwise 0.

We do not need a dummy variable for Catholics, since we need only $J - 1$ dummy variables to represent J groups or treatment levels.

Now to test whether there is an interaction between religious attendance and religious identification, we must construct *three interaction terms*—one for each dummy variable. If we let religious attendance be X_1, the three interaction terms are D_1X_1, D_2X_1 and D_3X_1. The equation to be estimated is:

TABLE 11.4
Correlations, Means, and Standard Deviations for Sexual Permissiveness, Religious Identification, and Religious Attendance Variables

Variable	*(1)*	*(2)*	*(3)*	*(4)*	*(5)*	*(6)*	*(7)*	*(8)*	*Mean*	*S.D.*
(1) Sexual Permissiveness	1.000	−0.124	0.085	0.171	−0.314	0.057	0.042	−0.263	2.069	1.048
(2) Protestant		1.000	−0.212	−0.385	0.018	−0.219	−0.165	0.665	0.661	0.474
(3) Jewish			1.000	−0.042	−0.075	−0.024	0.778	−0.141	0.023	0.148
(4) No religious identification				1.000	−0.258	0.569	−0.033	−0.256	0.071	0.257
(5) Attendance					1.000	0.021	0.016	0.607	3.979	2.680
(6) (2) × (5)						1.000	−0.019	−0.146	0.105	0.666
(7) (3) × (5)							1.000	−0.110	0.060	0.505
(8) (4) × (5)								1.000	2.651	2.859

Note: N = 1,511.
Source: 1977 General Social Survey.

$$\hat{Y} = a + b_1X_1 + b_2D_1 + b_3D_2 + b_4D_3 + b_5D_1X_1 + b_6D_2X_1 + b_7D_3X_1$$

The 1977 GSS data, with $N = 1{,}511$, was used to estimate this equation. The correlation matrix, means, and standard deviation are shown in Table 11.4. Using these data, we find that $R_3^2 = 0.1176$, with $k_3 = 7$ independent variables. We then estimate a second equation without the three interaction terms. That is, we estimate:

$$\hat{Y} = a + b_1X_1 + b_2D_1 + b_3D_2 + b_4D_3$$

and find $R_2^2 = 0.1173$, with $k_2 = 4$ independent variables.

Now the test of whether $\rho_3^2 - \rho_2^2 = 0$ in the population is given by:

$$F_{(7-4),\,(1511-7-1)} = \frac{(0.1176 - 0.1173)/(7 - 4)}{(1 - 0.1176)/(1{,}511-7-1)}$$

$$= 0.170$$

For $\alpha = .01$ with 3 and 1,503 degrees of freedom, the critical value is 3.78. Therefore we cannot reject the null hypothesis. Hence *we conclude that no interaction occurs between religiosity and religious identification.*

We can now ask whether adding attendance, X_1, to the three dummy variables significantly increases the amount of explained variance. The answer can be found by computing the coefficient of determination (R_1^2) for an equation containing only the three dummy variables for religion and comparing it to R_2^2, based on the equation that contains the attendance variable, X_1, as well.

The equation containing only the three religion dummy variables is:

$$\hat{Y} = a + b_1D_1 + b_2D_2 + b_3D_3$$

Using the GSS data, the estimated equation yields $R_1^2 = 0.0396$. Now we can test whether the null hypothesis that $\rho_2^2 - \rho_1^2 = 0$ in the population can be rejected. Since there are $k_1 = 3$ independent variables in the equation immediately above, the test statistic is given by:

$$F_{(4-3),(1511-4-1)} = \frac{(0.1173 - 0.0396)/(4-3)}{(1-0.1173)/(1{,}511-4-1)}$$

$$= 132.57$$

In Appendix E we find the critical value for $\alpha = .01$, with 1 and 1,506 *df*, is 5.43. Hence, we conclude that religiosity explains additional variance in sexual permissiveness, above the variance accounted for by religious identification.

The estimated regression equation that includes religious attendance and the three dummy variables for religious identification is:

$$\hat{Y} = \underset{(0.0694)}{2.6172} - \underset{(0.0099)}{0.1141X_1} + \underset{(0.1129)}{0.2709D_1} + \underset{(0.1779)}{0.3417D_2} - \underset{(0.0633)}{0.1832D_3}$$

When one of the D_j is 1, the other two are 0. As a result, separate regression equations for all four religious groups can be derived from the single equation above, as follows. When $D_1 = 1$, the resulting equation is:

$$\hat{Y}_{\text{NRI}} = (2.6172 + 0.2709) - 0.1141X_1$$
$$= 2.8881 - 0.1141X_1$$

When $D_2 = 1$, the resulting equation is:

$$\hat{Y}_{\text{JEW}} = (2.6172 + 0.3417) - 0.1141X_1$$
$$= 2.9589 - 0.1141X_1$$

When $D_3 = 1$, the equation is:

$$\hat{Y}_{\text{PROT}} = (2.6172 - 0.1832) - 0.1141X_1$$
$$= 2.4340 - 0.1141X_1$$

Finally, when neither D_1 nor D_2 nor D_3 is 1, we get the equation for the omitted category—the Catholics, in this case:

$$\hat{Y}_{\text{CATH}} = 2.6172 - 0.1141X_1$$

Note that the intercept, *a*, is also the intercept for the omitted group.

As an exercise, convince yourself that regardless of which category is omitted, *the coefficient of determination is the same, and the regression equations for the four groups are also the same.*

The t test for equations with dummy variables has a special interpretation:

> The t test associated with a given dummy variable included in a regression analysis is a test of the difference between two conditional means on Y. The first conditional mean is that associated with category j, and the second is that associated with the omitted category. The test also controls for the continuous variable(s) in the equation.

The implication of this result is that you should be careful to select for the omitted category a substantially or theoretically important group, so that meaningful statistical tests can be performed. Of course, in many instances no such criteria may exist.

Thus the t test associated with D_1, where $t = 0.2709/0.1129 = 2.40$, tells us that, if we choose $\alpha = .05$, persons of no religious identification are more sexually permissive than Catholics (the omitted category), controlling for religious attendance, since the critical value is 1.96 for a two-tailed test. The t values for comparing Jews and Protestants to Catholics are 1.92 and 2.89, respectively. Hence Protestants are significantly different from Catholics in this respect, but Jews are not. Somewhat surprisingly, Protestants are found to be *less* sexually permissive than Catholics in this data set.

TABLE 11.5

Sexual Permissiveness as a Function of Religious Identification, Using Analysis of Variance

A. Summary Statistics

Treatment Category	*Mean*	*Standard Deviation*	*N*
No religious identification (NRI)	2.703	1.202	110
Jewish	2.657	0.993	34
Protestant	1.978	1.002	1,000
Catholic	2.078	1.047	372
			$N =$ 1,516

B. ANOVA Summary Table

Source	*SS*	*df*	*MS*	*F*
Between groups	64.339	3	21.446	20.274*
Within groups	1,599.425	1,512	1.058	
Total	1,663.764	1,515		

*Significant at $\alpha = .01$ level.

Had we chosen those of no religious identification (NRI) to be the omitted dummy variable and included a dummy for Catholics in the analysis, the resulting tests of significance would have compared those of no religious identification with the other three groups on sexual permissiveness. *The t tests associated with the dummy variables, therefore, are not to be interpreted in the same way as t tests associated with continuous variables.*

11.4.3. Analysis of Variance with Dummy Variable Regression

You may have guessed that if we regress a dependent variable, Y, on $J - 1$ dummy variables, chosen to represent J treatment conditions, the results ought to be identical with those obtained from an analysis of variance (ANOVA). *And indeed the results are identical.* Rather than showing you how this identity is derived mathematically, we will demonstrate it by example.

Earlier in this chapter we used the 1977 GSS data to examine the mean level of sexual permissiveness by religious identification. Table 11.5 gives the results of this analysis obtained with the SPSS BREAKDOWN subprogram (see Preface). This subprogram, which performs the analysis of variance (ANOVA) introduced in Chapter 7, generates the means, standard deviations, and an overall F test indicating that for $\alpha = .01$, the null hypothesis that all means are equal in the population sampled must be rejected. Furthermore, we can estimate the strength of the relationship between sexual permissiveness and religious identification by computing an estimate of eta squared ($\hat{\eta}^2$). As we showed in Section 7.8 of Chapter 7, $\hat{\eta}^2 = SS_{\text{BETWEEN}}/SS_{\text{TOTAL}}$. In these data $\hat{\eta}^2 = 64.399/1{,}663.764 = 0.0387$. That is, only 3.9% of the variance in sexual permissiveness can be explained by knowing respondents' religious identifications.

Table 11.6 shows the results of regressing the sexual permissiveness index on three dummy variables constructed for those of no religious identification (NRI), Jews, and Protestants. The omitted category of the religious identification variable again is Catholic. The estimated regression equation is:

$$\begin{array}{lllll} \hat{Y} = & 2.078 + & 0.625D_1 + & 0.579D_2 - & 0.100D_3 \\ & (0.053) & (0.112) & (0.184) & (0.062) \end{array}$$

with $R^2_{Y \cdot D_1 D_2 D_3} = 0.0387$. Note that the coefficient of determination (R^2) from the dummy variable regression analysis is identical to

TABLE 11.6

Sexual Permissiveness as a Function of Religious Identification, Using Regression Analysis

Independent Variable	*Regression Coefficient*	*Standard Error*	*t*
No religious identification	0.625*	0.112	5.58
Jewish	0.579*	0.184	3.15
Protestant	−0.100	0.062	−1.61
Intercept	2.078	0.053	38.97
		$R^2_Y = .0387$*	

*Significant for $\alpha = .01$.

Note: $N = 1{,}516$.

$\hat{\eta}^2$ from the ANOVA. Both analyses indicate that 3.9% of the variance in permissiveness can be explained by religious identification.

If we substitute 0 for 1 in the regression equation, to indicate the presence or absence of a given religious identification, we obtain four equations, one for each of the religious identification variables:

$$\hat{Y}_{\text{NRI}} = 2.078 + 0.625(1) + 0.579(0) - 0.100(0)$$
$$= 2.703 = \overline{Y}_{\text{NRI}}$$
$$\hat{Y}_{\text{JEWISH}} = 2.078 + 0.625(0) + 0.579(1) - 0.100(0)$$
$$= 2.657 = \overline{Y}_{\text{JEWISH}}$$
$$\hat{Y}_{\text{PROT}} = 2.078 + 0.625(0) + 0.579(0) - 0.100(1)$$
$$= 1.978 = \overline{Y}_{\text{PROT}}$$
$$\hat{Y}_{\text{CATH}} = 2.078 + 625(0) + 0.579(0) - 0.100(0)$$
$$= 2.078 = \overline{Y}_{\text{CATH}}$$

Thus the conditional means for each religious group can be obtained from the regression analysis. Note that these means are identical to those obtained from the ANOVA shown in Table 11.5. We could use either analysis to get estimates of the "treatment" effects, since $\hat{\alpha}_j = \hat{Y}_j - \overline{Y}$. We get a windfall benefit from the regression analysis, however: t tests contrasting the means of the $k - 1$ dummy variables with the mean of the omitted category. In this example, since the critical value for $\alpha = .01$ is ± 2.33, we find that the mean sexual permissiveness of the NRIs and Jews is significantly higher than that of the Catholics. The difference between the Catholics and Protestants is not significant, however.

These final examples using dummy variables help demonstrate the power of multiple regression analysis for testing a variety of hypotheses. Given a continuous dependent variable, we can estimate the effects of continuous or discrete independent variables. Furthermore, any combination of independent variables measured at discrete and continuous levels can be included in the regression analysis, along with product terms that allow for the testing of interaction effects. And both the analysis of variance and the analysis of covariance can be performed within a regression framework.

We will describe another application of regression analysis when we introduce *path analysis* in Chapter 12. This is a powerful tool which deserves your serious attention to master its applications.

Review of Key Concepts

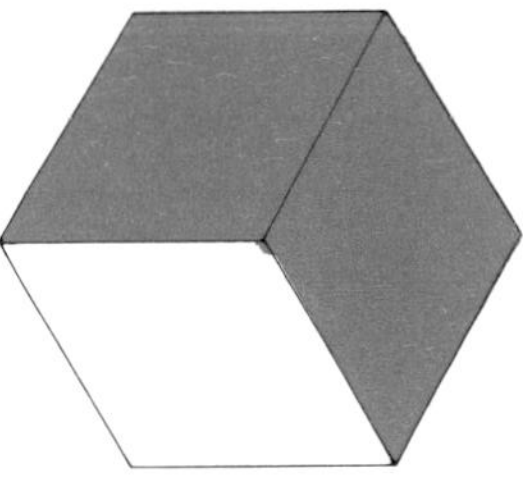

This list of the key concepts introduced in Chapter 11 is in the order of appearance in the text. Combined with the definitions in the margins, it will help you review the material and can serve as a self-test for mastery of the concepts.

multiple regression analysis
index
construct
indicator
Cronbach's alpha
listwise deletion
pairwise deletion
multiple regression coefficient
part correlation
multiple correlation coefficient
multicollinearity
dummy variable
analysis of covariance
covariates
treatment level

PROBLEMS

General Problems

1. With the following information, compute the a, b_1, and b_2 coefficients for the regression of Y on X_1 and X_2:

Variable	*Mean*	*Standard Deviation*	*Correlations*	
			X_2	Y
X_1	7	4	0.30	0.50
X_2	30	10	—	−0.40
Y	50	25	—	—

2. Given the following data, compute the a, b_1, and b_2 coefficients for the regression of Y on X_1 and X_2:

Variable	*Mean*	*Standard Deviation*	*Correlations*	
			X_2	Y
X_1	24	3	0.02	0.30
X_2	16	4		0.40
Y	4	1		

3. Using the data in Problem 2:
   a. Compute the zero-order regression coefficients b_{YX_1} and b_{YX_2} and compare them to b_1 and b_2 from Problem 2.
   b. What do the results suggest?
   c. What fact accounts for this result?

4. Using the data from Problem 2:
   a. substitute $r_{X_1X_2} = 0.70$ for $r_{X_1X_2} = 0.02$ and recompute a, b_1, and b_2.
   b. Compare b_1 and b_2 to the zero-order regression coefficients b_{YX_1} and b_{YX_2}. What do these results, combined with those from Problem 3, illustrate?

5. a. Compute the beta weights for the data shown in Problem 1.
   b. Write out the regression equation for standardized variables.

6. a. Compute the beta weights for the data shown in Problem 2.
   b. Write out the regression equations for the standardized variables.

7. Compute $R^2_{Y \cdot X_1 X_2}$ for the data in Problem 1.
8. Compute $R^2_{Y \cdot X_1 X_2}$ for the data in Problem 2.
9. Standardize the coefficients in the following equation, given that $s_Y = 4.50$; $s_{X_1} = 1.50$; $s_{X_2} = 0.75$; and $s_{X_3} = 4.10$:

$$\hat{Y} = 37.43 - 1.53\,X_1 + 4.29\,X_2 + 0.74\,X_3$$

Which independent variable has the largest effect and which the smallest?

10. Find $R^2_{Y \cdot X_1 X_2}$ for the following sets of correlations:

	$r_{X_1 X_2}$	$r_{X_1 Y}$	$r_{X_2 Y}$
a.	0.60	0.70	0.15
b.	0.25	−0.50	−0.30
c.	0.00	0.63	0.45
d.	−0.70	0.50	−0.20

11. Using survey data from a sample of 227 respondents, a regression equation predicting racial prejudice scores with five social and psychological independent variables has a coefficient of determination of 0.113. If the variance of the racial prejudice scale is 35.4, is the coefficient of determination significantly different from zero for $\alpha = .01$?
12. If $N = 100$ and $\alpha = .05$, is $R^2_{Y \cdot X_1 X_2}$ in Problem 7 statistically significant?
13. If $N = 10$ and $\alpha = .05$, is $R^2_{Y \cdot X_1 X_2}$ in Problem 8 statistically significant?
14. When approval of the death penalty *(Y)* is regressed on the opinion of 86 respondents that the courts are too lenient toward criminals (X_1) and the frequency of their church attendance (X_2), the following statistics are found: $b_1 = 1.33$, $s_{b_1} = .75$; $b_2 = -0.06$, $s_{b_2} = .04$. Test both b_js for statistical significance using $\alpha = .05$, one-tailed.
15. When belief in God is regressed on education (X_1) and age (X_2) for $N = 200$, the following statistics are found:

$$s_1 = 3.1 \qquad \beta_1^* = -0.21 \qquad s_{b_1} = 0.31$$
$$s_2 = 10.0 \qquad \beta_2^* = 0.17 \qquad s_{b_2} = 0.85$$
$$s_y = 1.3$$

Test both b_js for statistical significance, using $\alpha = .05$, one-tailed tests.

16. When $N = 100$, $k = 2$, $b_1 = 1.61$, $MS_{\text{ERROR}} = 2.07$, $\Sigma (X_{1_i} - \bar{X}_1)^2 = 4.36$, and $R^2_{X_1 \cdot X_2} = 0.130$, is b_1 significant for $\alpha = .01$ for a one-tailed test?

17. When $N = 50$, $k = 2$, $b_2 = -2.70$, $MS_{\text{ERROR}} = 3.16$, $\Sigma (X_{2_i} - \bar{X}_2)^2 = 7.12$, and $R^2_{X_2 \cdot X_1} = 0.274$, is b_2 significant for $\alpha = .05$ for a one-tailed test?

18. Build a 95% confidence interval around the b_j in Problem 16.

19. Build a 99% confidence interval around the b_j in Problem 17.

20. Given the zero-order correlations shown in Problem 1, compute $r_{YX_1 \cdot X_2}$ and $r_{YX_2 \cdot X_1}$.

21. Given the zero-order correlations shown in Problem 2, compute $r_{YX_1 \cdot X_2}$ and $r_{YX_2 \cdot X_1}$.

22. Give a verbal description of the following unstandardized regression equation, where Y is a 10-point civil liberties index, X_1 is the age in years, X_2 is education in years, and X_3 is a dummy variable for place of residence (1 = rural, 0 = urban):

$$\hat{Y} = 6.25 - 0.031\ X_1 + 0.123\ X_2 - 0.546\ X_3$$

23. Construct dummy variable codes for marital status measured with these categories: married; widowed; divorced; separated; never married; no answer.

24. Assume that D_1 = Republican, D_2 = Democrat, D_3 = Independent, D_4 = Other, X is a measure of alienation, and Y is a measure of attitude about voting in national elections. Set up two equations necessary to test whether political identification and alienation interact in the prediction of attitude about voting.

25. In a sample of 310 persons, an index for liking of country and western music (Y) is predicted by education (X_1) and three dummy variables for place of residence: D_1 = city; D_2 = town; and D_3 = country—plus the interactions. The results of three regression analyses are:

$$\hat{Y} = 4.37 + 1.92\ D_2 + 3.68\ D_3 \qquad R^2 = 0.296$$

$$\hat{Y} = 4.10 + 1.74\ D_2 + 3.84\ D_3 - 0.08X_1 \qquad R^2 = 0.312$$

$$\hat{Y} = 2.77 + 1.54\ D_2 + 3.70\ D_3 - 0.02X_1 - 0.23X_1D_2 + 0.14\ X_1D_3 \qquad R^2 = 0.357$$

a. Is there a significant interaction between education and liking of country music? Set $\alpha = .01$.
b. What are the predicted liking scores for persons with 12 years of education in each of the three types of communities?
c. Give a verbal interpretation of the relationship.

Problems Requiring the 1980 General Social Survey Data

26. Regress self-reported health (HEALTH) on INCOME and years of education (EDUC).
    a. Present the regression coefficients above with their tests of significance and the beta weights.
    b. Present the coefficient of determination with its test of significance.
    c. Interpret the results.
    Note: Set $\alpha = .01$ for all tests (two-tailed).

27. Regress respondent's occupational prestige (PRESTIGE) on respondent's education (EDUC) and father's occupational prestige (PAPRES16). What do the results suggest about the relative effects of the two independent variables on the respondent's occupational prestige?

28. Add respondent's sex (SEX) to the variables used to explain the respondents' health in Problem 26.
    a. Does sex have an impact on health, independent of the other two independent variables in explaining health?
    b. Is there a significant sex-by-income interaction in explaining health?
    Note: Set $\alpha = .01$ for both tests.

29. Add respondent's race (RACE) to the list of independent variables in Problem 27.
    a. Is there an effect of race on prestige net of the other two variables, at $\alpha = .01$?
    b. Is there a significant race-by-education interaction in explaining prestige? Set $\alpha = .01$ for all tests.
    Note: Recode blacks and others into a single nonwhite category.

30. Build a satisfaction index from SATCITY, SATFAM and SATFRND. Treat an index score as missing data if more than

one item has missing data. (See p. 116 of the SPSS manual for an example of how to construct an index.)

a. Compute alpha for the index.

b. Regress the index on a set of four dummy variables constructed from MARITAL status. (Omit the "never married" dummy variable.) Show the regression equation and R^2.

c. Construct an ANOVA summary table from the results (set $\alpha = .01$). Also show mean satisfaction level for each marital status.

d. What does $\hat{\eta}^2$ equal?

31. Using the results from Problem 30, which marital statuses are significantly different than the "never marrieds" on the satisfaction index? Specify how each status differs in satisfaction as well. Set $\alpha = .05$.

32. Construct an index for attitude toward abortion from ABNOMORE, ABPOOR and ABSINGLE. Treat an index score as missing if more than one item has missing data (see p. 116 of the SPSS manual for an example of how to construct an index).

    a. Compute alpha for the index.

    b. Regress the index on a set of three dummy variables constructed from respondent's religion (RELIG). (Omit the "none other" dummy variable from the analysis.) Show the regression equation and R^2.

    c. Construct an ANOVA summary table from the results in step *b* (set $\alpha = .01$). Also show mean attitude toward abortion for each religion.

    d. What does $\hat{\eta}^2$ equal?

33. Using the results from Problem 32, which religions differ significantly from the "none" category on the abortion index? Set $\alpha = .05$.

34. Show that SPSS ONEWAY can be used to obtain the same results as the regression analysis for Problems 30 and 31.

35. Show that SPSS ONEWAY can be used to obtain the same results as the regression analysis for Problems 32 and 33.

Problems Requiring the 63-Cities Data

36. *a.* Estimate the multiple regression equation for predicting the percent of housing built before 1950 (HOUSEPCT)

from a city's per capita expenditures (EXPENDIT) and the number of functions it performs (FUNCTION). Compute R^2 and test it for statistical significance for $\alpha = .05$, one-tailed test.

b. Does a city's population size (POPULAT) and percentage of black residents (BLACKPCT) add significantly ($\alpha = .05$) to the variance explained by EXPENDIT and FUNCTION?

37. Estimate the standardized regression coefficients (beta weights) for the prediction of decentralized decision making in cities (DECINDEX) from the number of functions performed (FUNCTION), city age (CITYAGE), and the percentage of residents in white-collar occupations (WCPCT). Discuss the relative effects of the three independent variables, based on their relative sizes and their statistical significance. (Set $\alpha = .05$.)

38. *a.* Estimate cities' fiscal stress in 1970 (STRESS70) from the number of functions performed (FUNCTION) and the age of the city (CITYAGE), reporting significance tests for the regression coefficients and R^2 (set $\alpha = .05$). In estimating this equation, dichotomize FUNCTION by the recode: 1 = 1 thru 4 functions and 2 = 5 thru 8 functions. Interpret the results.

    b. Test whether there is an interaction between number of functions performed and city age in explaining fiscal stress in 1970. (Set $\alpha = .05$.) Interpret the results.

39. *a.* Regress the percent of work force in white-collar occupations (WCPCT) on a set of three dummy variables constructed from REGION. (Omit the "Far West" dummy variable from the analysis.)

    b. Set $\alpha = .05$ and construct an ANOVA summary table from the results. Also show the mean percent of the labor force in white-collar occupations by region using the regression results.

    c. What does $\hat{\eta}^2$ equal?

    d. Does the Far West have significantly more white-collar workers than any of the other three regions?

40. *a.* Does CITYAGE add significantly ($\alpha = .05$) to the variance explained in Problem 39?

b. Is there a significant interaction effect between CITYAGE and REGION in explaining the percent of work force in white-collar occupations?

41. Show that SPSS ONEWAY can be used to obtain the same results as the regression analysis for Problem 39.

Statistics for
Social Data
Analysis

Causal Models and Path Analysis

12

The idea that social theory consists of propositions that implicitly connect two or more variables in a causal relationship has been stressed throughout this text. This chapter elaborates on the *causal* mode of thinking about social behavior and describes a basic technique—path analysis—for representing causal relations among quantitative variables. Causal reasoning is increasingly coming to dominate many areas of social research, but the foundations of these conceptualizations are often not carefully articulated. Considerable pseudo-philosophical mumbo-jumbo may accompany efforts to explain social phenomena in cause-and-effect terms. The first section of this chapter states the basic assumptions that must be met before causal explanations can be seriously entertained.

12.1. Causal Assumptions

Propositions or hypotheses of the form "If A, then B," or "The higher the A, the higher (lower) the B," merely state an expectation that variables A and B are related. Such hypotheses are in fact *covariation statements,* in which differences or changes in one variable are systematically related to differences or changes in a second variable. Often the author of a research proposition obviously intends a causal effect to be present. For example, in the hypothesis, "The higher the level of test anxiety, the lower the performance on a course final exam," it is quite evident that the test anxiety precedes and is a cause of poor scores. A statement like "The greater the practice of irrigation, the more centralized the state authority in early civilizations," however, is unclear about which variable—technology or political structure—is the cause and which is the effect. *A genuine causal hypothesis should explicitly state the researcher's expectations.* Statements of the format, "An increase in variable *A* causes an increase (decrease) in variable *B*," leave no doubt as to the author's intentions.

> A proposition in causal form is more informative than one which is only covariational. Causation is typically (but not necessarily) asymmetrical, in the sense that a change in the cause *creates* a change in the effect, whereas the reverse is not true—changing the dependent variable should leave the independent variable unaltered.

Consider a causal hypothesis from agricultural economics: "Greater rainfall causes higher crop yields" (everything else being equal, which it seldom is). Because extensive experimental and observational data have been collected, the covariation of moisture and productivity is well established. Knowledge of plant physiology provides a basis of inference as to which variable is cause and which is effect. A student who asserted that sowing more seeds per acre would increase a locality's rainfall would soon be laughed out of agriculture school. This trivial but true hypothesis, therefore, is unambiguously causal in intent and in consequence.

Many theoretical statements in social science are not so clear. Sometimes theorists and researchers are simply confused about the "chicken-and-egg" sequence of the phenomena they study. More often, social reality is so incredibly complex that disentangling a causal process is almost impossible. Rarely does any interesting social behavior have a single cause which can be easily isolated. Rather, human activity is governed by a variety of influences, not the least of which is *intentionality* (purpose). For example, race riots in the United States may erupt from a complex interplay of police insensitivity, summer heat, poor housing, ghetto size, unemployment, black power ideology, white political incompetence, "outside" agitators (including mass media reports)—the list of potential causes is endless. To make an assertion that a single variable has a discernible impact is a bold step that should not be lightly undertaken.

As it has matured, social science has come a long way from the simplistic monocausal thinking of the 19th century to the contemporary emphasis on multicausal theories or models. An important recent development—path analysis—is providing a way for social scientists to cast their theories into explicit multicausal frameworks.

Three basic conditions are necessary to establish causal priority among variables, and none is sufficient by itself. The conditions of causality are covariation, time order, and nonspuriousness, as Section 10.1.1 in Chapter 10 suggested.

12.1.1. Covariation

For causal relationships to be present among a pair of variables, there must be **covariation** between the independent and dependent variables. Systematic changes or differences in one variable must accompany systematic changes or differences in the other. Covariation can take several forms: positive or negative linear association, or several kinds of nonlinear association.

covariation—joint variation, or association, between a pair of variables

In a linear association, as the values in one variable rise, the values of the other variable either rise or fall in linear fashion. When both variables vary together in the same direction (rising or falling together), the relation is *positive* or *direct linear covariation* or association. When a rise in one variable is linearly related to a decline in the other, *negative* or *inverse linear covariation* or association exists.

Nonlinear associations might take any of several forms. The simplest is a curvilinear pattern which, when graphed, looks like a parabola. That is, as variable X increases in numerical value, variable Y first increases, then decreases (or vice versa). More complex curvilinear covariation might resemble cubic equations, sine waves, or other patterns. Few social theories are so sophisticated that they attempt to specify anything more than simple linear covariation, however. An exception is psychophysics, which is based on the discovery that nonlinear "power curves" seem to describe perception and learning patterns best.

12.1.2. Time Order

time order—the necessary condition that changes in a purported independent variable must precede in time the change in the dependent measure, when a causal relationship between the two is assumed.

The **time order** condition of causality is a "metatheoretical" assumption which is shared almost unconsciously by most Western peoples. *For causality to occur, the change in the purported independent variable must precede in time the change in the alleged dependent measure.* This time order helps to establish the essential asymmetry of a causal relationship. The example of rainfall and crop growth has an implicit time order: In predicting a bumper crop we point to the previous winter and spring rains, not to rains that come after the harvest. Knowing something about the historical sequence of events is a great help in time-ordering variables for causal explanations. You might well assert that the race riots in summer 1968 were triggered by Martin Luther King's assassination, for example, but you'd be foolish to attribute the 1964 Watts outbreak to the same cause.

Causal research on social mobility and socioeconomic attainment, for example, has been developed rapidly in recent years by applying knowledge about the time ordering among parental status and events of the offspring's occupational career. Causal explanations of attitude structures, however, have been frustrated by inability to determine temporal sequences among expressed attitudes recorded during a single interview. Researchers frequently assume that certain background characteristics of survey respondents—race, education, religion, or occupation, for example—were formed sufficiently in advance of later behaviors—such as voting, drinking, marrying, or divorcing—that the characteristics can safely be assumed to cause the variable behaviors.

12.1.3. Nonspuriousness

Even if two variables covary and a temporal order can be ascertained, a third condition must be satisfied before a causal relationship can seriously be considered to exist: *The pattern of association between variables Y and X must not arise from other, common causal factors.* The classic example, introduced in Section 10.1 in Chapter 10, is the observation that in Holland, communities where many storks nest in chimneys have higher birth rates than communities where fewer storks nest. While covariation and temporal-order conditions can reasonably be met in this example, various "rival factors" might affect both the number of nesting storks and the number of human babies. The predominance of rural areas, pollution and sanitation levels, community attitudes, and patterns of selective migration may all combine to produce a spurious correlation between the two variables, as we noted in Chapter 10.

The establishment of **nonspuriousness** in a causal relationship is one of the most difficult problems to solve in social research. We cannot literally examine every possible alternative explanation for why two variables are related. Various research methods and statistical techniques that reduce much of the opportunity for spurious covariation to remain undetected have been developed. *Controlled experiments represent the most effective way to control rival factors.* When subjects are randomly assigned to various experimental and control conditions, all factors except the manipulated independent variable can be expected to be held constant. Nonexperimental research such as field studies or systematic interviews provide fewer means to control potential common causes, and hence the conclusions about causal relationships among variables that can be derived are far more tentative.

nonspuriousness— covariation between two variables that is causal and not due to the effects of a third variable

12.2. Causal Diagrams

The three basic assumptions described in the preceding section, which must be satisfied before causality can be imputed among variables, can be graphically displayed in a **causal diagram** that reveals the hypothesized cause-and-effect relationships. The conventions for causal diagraming are indispensable aids in thinking through problems of causal reasoning, as well as communicating your ideas to other researchers.

causal diagram— a visual representation of the cause-and-effect relationships among variables, using keyword names and directed arrows

In a causal diagram, variables are represented by "keyword" names or letters. Annual household unit income might be labeled "Income" in a diagram. Scores on the Stanford-Binet intelligence test could be shortened to "IQ." Temporal order is conventionally organized from left to right, as in Western-culture reading and writing. Variables placed to the left in a diagram, therefore, are considered antecedents to those located toward the right. If the researcher posits a causal connection between two variables, this relationship is represented by a single-headed arrow, with the tail emerging from the cause and the head pointed at the effect. Two antecedent variables that are not causally analyzed but are *correlated* are shown by linking the keyword names with a curved,

BOX 12.1 Rules for Constructing Causal Diagrams

1. Variable names are represented either by short keywords or letters.
2. Variables placed to the left in a diagram are assumed to be causally prior to those on the right.
3. Causal relationships between variables are represented by single-headed arrows.
4. Variables assumed to be correlated but not causally related are linked by a curved, double-headed arrow.
5. Variables assumed to be correlated but not causally related should be at the same point on the *horizontal* axis of the causal diagram.
6. The direction of causality presumed between two variables is indicated by placing + or − signs along the causal arrows to show how increases or decreases in one variable offset the other.

double-headed arrow. If two dependent variables are not hypothesized to have a direct causal connection, no arrow is drawn to directly link the two. The direction of causality between linked variables is indicated by placing signs—plus or minus for positive or negative covariation—along the causal arrows. These rules are summarized in Box 12.1.

Figure 12.1 displays some elementary types of causal diagrams using these diagrammatic conventions. Letters rather than substantive variable names are used. Panel A, the bivariate pattern, shows a simple **direct effect**, or causal relationship, between *A* (the independent variable) and *B* (the dependent variable). By adding a third variable *(C)* in temporal sequence, the simple causal chain in Panel B shows that increases in levels of *A* raise the value of *B,* but higher levels of *B* in turn reduce the level of *C.* By inference, then, the higher the amount of *A,* the less the amount of *C.* The sign of this **indirect effect** of *A* on *C* (via *B*) can also be calculated by multiplying the signs of the intervening paths. A positive (+) times a negative (−) is a negative, so the indirect effect of *A* on *C* is negative.

direct effect—
a connecting path in a causal model between two variables without an intervening third variable

indirect effect—
a compound path connecting two variables in a causal model through an intervening third variable

Panel C, the direct-and-indirect-effects model, depicts *A* as

FIGURE 12.1
Some Elementary Causal Diagrams

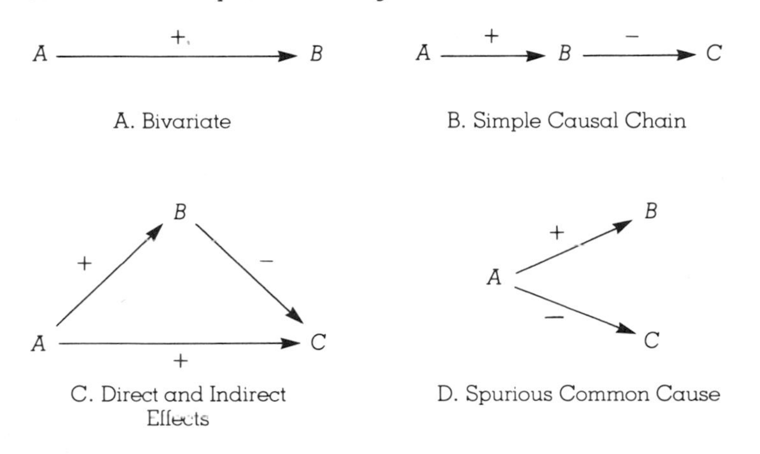

having both types of impact on C, but opposite in sign. The indirect effect through B is still negative, while the direct effect—holding constant the contributions of B—is for A to raise the level of C. This diagram does not give enough information to determine whether the negative or the positive effect will be stronger. Later, in the discussion of path analysis, we will show how to estimate actual values of the causal effects, in order to answer this question.

The spurious common cause model, Panel D, shows how an observed covariation between B and C might come about without any direct causation between these two variables. Variable A is a common cause of both, raising the level of B and decreasing the magnitude of C, thereby generating an inverse covariation between both dependent variables. If B is the number of trucks at a fire and C is the amount of undamaged property, what variable might A be?

You now should have enough information about causal thinking and diagraming to begin to put it to use in your own work with social theory and hypothesis generation. We will work through one simple example to show you how all the pieces fit together. This is a mini model of the academic achievement process which is presumed to operate in most American schools. Two causal propositions can be stated at an abstract level:

P1: The higher a child's family status, the greater his or her academic achievement.

P2: The higher a child's intelligence, the greater his or her academic achievement.

Next, selecting empirical indicators for each of the three concepts, we can recast the model in hypothesis form:

H1: The higher the annual parental income, the higher a student's grade point average.

H2: The higher the student's IQ score, the higher his or her GPA.

Figure 12.2 displays the causal relations at both the analytic, abstract level and the empirical, observable level. The double-headed curved arrows between status and intelligence at the abstract, unobservable level and between income and IQ at the concrete, observable level indicate that no causal assumptions are

FIGURE 12.2
Causal Diagram for Mini Model of Scholastic Achievement (also shows the link between the unobservable constructs and the observable indicators)

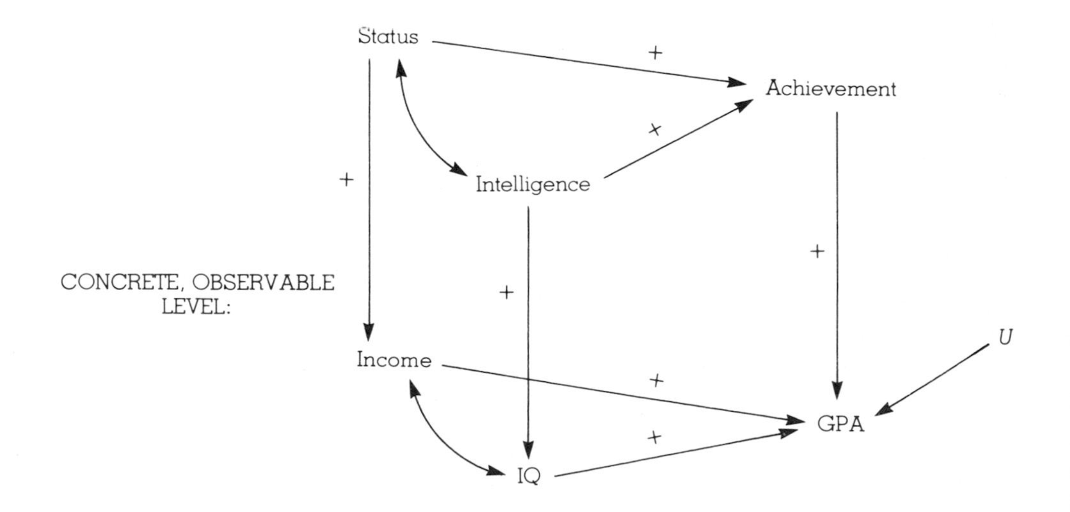

made about these pairs. These pairs of variables are called **predetermined variables** in this particular model, because their causes remain unspecified and outside the interest of the model.

predetermined variable —a variable in a path model whose causes are not specified

Figure 12.2 introduces, at the concrete level, another convention of causal diagrams. An arrow is drawn from an unmeasured variable *(U)* to the dependent variable. This factor, called a **residual variable,** represents the belief that the variation in GPA is not completely explained by the two causal factors hypothesized by the model. This model, therefore, is **probabilistic,** or *stochastic,* rather than **deterministic.**

residual variable—an unmeasured variable in a path model that is posited as causing the unexplained part of an observed variable

probabilistic —a causal relationship in which change in one variable produces change in another variable, with a certain probability of occurrence

deterministic—a causal relationship in which change in one variable always produces a constant change in another variable

The model in Figure 12.2 is also **recursive.** That is, *all of the causal influences are assumed to be in one direction and one direction only.* If X causes Y, then we do not allow for Y to also cause X. Models which allow for bidirectional causality are said to be **nonrecursive.** The estimation of nonrecursive models is complex, and the discussion in this chapter will be restricted to recursive models. Note that a residual cause is shown as uncorrelated and not causally related to any independent or residual variables in the diagram. In the following sections on path analysis, we will show why we need to make these assumptions. We will also show how the magnitude of the residual effect can be estimated.

recursive model—a model in which all the causal influences are assumed to be asymmetric (one-way).

nonrecursive model—a model in which causal influences between dependent variables may occur in both directions

Figure 12.2 also shows vertical arrows linking the abstract concepts to concrete, empirical variables. These arrows depict **epistemic relationships** (from the Greek word for *knowledge.*). They demonstrate how the abstract concepts in the theoretical propositions were operationalized with measurable variables whose patterns of covariation can be studied with data. Most social researchers tend to construct causal diagrams only at the concrete level of empirical indicators, omitting the connection to the theoretical concepts. In a later example, we follow this practice. But displaying the full set of relationships involved in a causal system, as in this figure, is important in order to make clear that estimating a set of parameters in path analysis can also be generalized to a set of unobservable constructs.

epistemic relationships —the relationship between abstract, theoretical (unobserved) concepts and their corresponding operational (observed) measurements

12.3. Path Analysis

path analysis—a method for analyzing correlations that yields empirical estimates of effects in a hypothesized causal system

Path analysis is a method for analyzing quantitative data which yields empirical estimates of the effects of variables in an hypothe-

FIGURE 12.3
Causal Diagram for Drinking Behavior Model

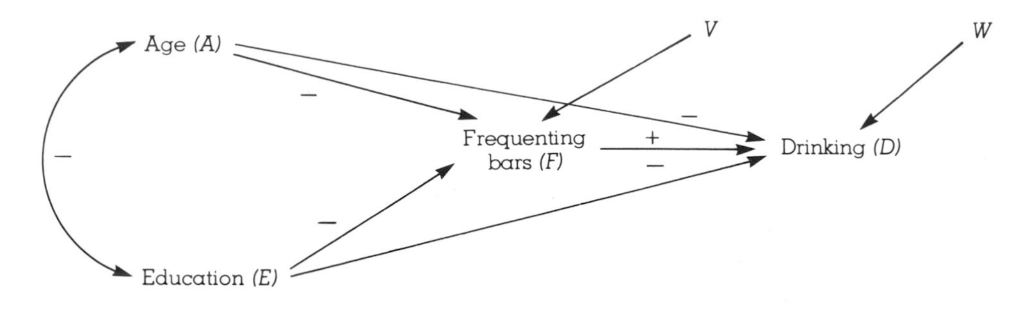

sized causal system. Originally developed by the geneticist Sewell Wright, path analysis and its variants have gained wide currency in the social sciences in recent decades. The technique requires all the causal assumptions discussed earlier in this chapter, and it makes extensive use of diagrams to represent the cause-and-effect relationships among empirical indicators. Moreover, before you can follow the discussion in this section, you must be familiar with standardized multiple regression, as described in Chapter 11.

12.3.1. An Example: The Drinking Behavior Model

The example we will use to illustrate path analytic principles in this chapter is a simple four-variable causal system representing some of the hypothesized causes of drinking behavior, as shown in Figure 12.3. The pluses (+) and minuses (−) indicate the direction of the hypothesized relationships. A respondent's age and education are posited as predetermined variables and are thus placed to the left in the diagram. The curved double-headed arrow indicates we are not interested in the causal relationship between these two measures. Both greater age *(A)* and more education *(E)* are hypothesized to decrease the frequency with which respondents visit bars and taverns *(F)*. Age, education, and frequenting bars all are hypothesized to have direct causal impacts on the amount of drinking *(D)* reported by respondents. The direct effects of age and education are posited to be negative, the direct effect of frequenting bars, positive. Both dependent variables in the diagram have arrows drawn from uncorrelated residual factors, (*V* and *W*), to indicate that there are other causes of drinking and of frequenting bars than those shown.

The operationalization of these variables is described in Section 12.3.6, which discusses how to estimate the model.

12.3.2. Structural Equations

Path analysis begins with a set of **structural equations** which represent the structure of interrelated hypotheses in a theory. These equations bear a one-to-one relationship with a causal diagram such as Figure 12.3. Typically the variables in a path analysis have been put in standardized form, or *Z* scores. Although the raw scores can also be analyzed, we will follow the tradition of analyzing *Z* scores in this chapter.

structural equation—a mathematical equation representing the structure of hypothesized relationships among variables in a social theory

The four variables—age (*A*), education (*E*), frequenting bars (*F*), and drinking (*D*)—can be represented by two equations. The first captures the hypothesized relationships between age, education, and the frequenting of bars, making the assumption that the two independent variables are linearly related to the dependent variable. If *p* represents the **path coefficients**, the equation is:

path coefficient—a numerical estimate of the causal relationships between two variables in a path analysis

$$F = p_{FA}A + p_{FE}E + p_{FV}V$$

FIGURE 12.4
Path Diagram with Coefficient Symbols for Drinking Behavior Model

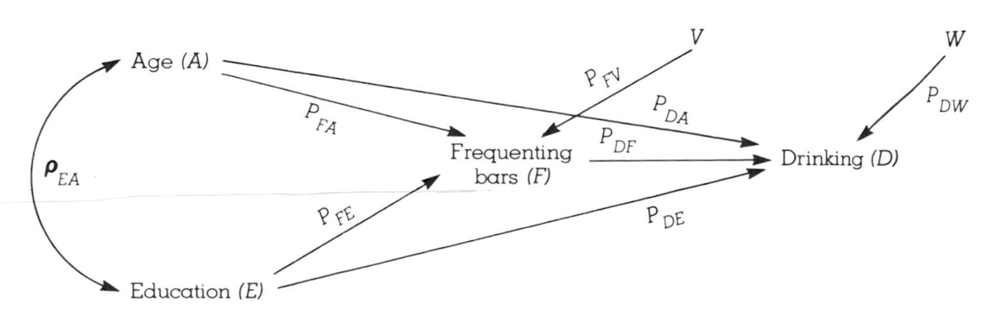

Notice that a path coefficient links age to the frequenting of bars (p_{FA}) and another such coefficient links education to the frequenting of bars (p_{FE}). We have also included a path coefficient (p_{FV}) to capture the relationship between the unobservable residual term, V, and the dependent variable, F.

The second equation relates age, education, and the frequenting of bars to drinking (*D*):

$$D = p_{DF}F + p_{DA}A + p_{DE}E + p_{DW}W$$

Again we include a path coefficient for each of the variables assumed to cause drinking behavior, plus one from the residual variable W. Figure 12.4 shows the relationships between the equations and the causal arrows.

> There are two major problems in path analysis. The first is to estimate the path coefficients. The second is to show how the correlations between the independent and dependent variables are implied by the presumed causal population parameters. These problems are considered in the next two sections.

12.3.3. Estimating Path Coefficients

There is a rather straightforward solution to the problem of estimating path coefficients for the causal model shown in Figure 12.4. All we need do is presume that a dependent variable, frequenting of bars (F), is linearly related to two independent variables in the first equation. To get estimates of the coefficients, we regress F on A and E. Since the variables are in standardized form, our estimates of *the path coefficients for A and E are simply the two beta weights.* That is:

$$\hat{p}_{FA} = \beta^*_{FA}$$

$$\hat{p}_{FE} = \beta^*_{FE}$$

We use the double subscripts *FA* and *FE* to make clear which dependent variable is being predicted, because age and education are also assumed to be causally related to drinking behavior. We also state without proof that:

$$\hat{p}_{FV} = \sqrt{1 - R^2_{F \cdot AE}}$$

That is, *the path coefficient from the residual variable to the dependent variable is simply the square root of the coefficient of nondetermination.* We have put carets (^) on the path coefficients to make clear that they are sample estimates of the population path coefficient parameters.

The path coefficients for the second equation can be estimated similarly by regressing drinking *(D)* on age *(A)*, education *(E)*, and the frequenting of bars *(F)*. In this case:

$$\hat{p}_{DA} = \beta^*_{DA}$$

$$\hat{p}_{DE} = \beta^*_{DE}$$

$$\hat{p}_{DF} = \beta^*_{DF}$$

$$\hat{p}_{DW} = \sqrt{1 - R^2_{D \cdot FAE}}$$

12.3.4. Decomposing Implied Correlations into Causal Parameters

The solution to the second problem, showing how the causal population parameters imply correlation between the variables, is less straightforward. It involves **decomposition,** or division of a correlation coefficient into its components.

decomposition—the division of a correlation coefficient into its component parts, involving direct effects, indirect effects, and dependence on common causes

As a first step in this process, we must express the correlations in terms of path coefficients. Some algebra with which you may not be familiar is required. The theorem you need to know is: $\Sigma aX = a\Sigma X$, where *a* is a constant and *X* is a variable. You also need to know that $\Sigma(X + Y) = \Sigma X + \Sigma Y$. That is, the sum of *X* plus *Y* equals the sum of *X* plus the sum of *Y*. The detail of these theorems is given in Appendix A as Rules 2 and 5.

Now consider the first equation given in Section 12.2.1:

$$F = p_{FA}A + p_{FE}E + p_{FV}V$$

We first multiply both sides by A, to find the correlation between age and frequenting of bars:

$$AF = A(p_{FA}A + p_{FE}E + p_{FV}V)$$

$$= p_{FA}A^2 + p_{FE}AE + p_{FV}AV$$

Then we take the sum of both sides:

$$\Sigma AF = \Sigma(p_{FA}A^2 + p_{FE}AE + p_{FV}AV)$$

$$= \Sigma p_{FA}A^2 + \Sigma p_{FE}AE + \Sigma p_{FV}AV$$

(This last step follows from Rule 5 in Appendix A.)

Now we divide the left-hand term by N, as well as the terms on the right-hand side, and bring the ps outside the Σ signs (i.e., we apply Rule 2):

$$\frac{\Sigma AF}{N} = p_{FA}\frac{\Sigma A^2}{N} + p_{FE}\frac{\Sigma AE}{N} + p_{FV}\frac{\Sigma AV}{N}$$

As we showed in Section 8.4.3 of Chapter 8, however, the sum of the product of two Z scores divided by $N - 1$ in the sample or N in the population (as here) is just their correlation. That is:

$$\frac{\Sigma AF}{N} = \rho_{AF}$$

$$\frac{\Sigma AE}{N} = \rho_{AE}$$

$$\frac{\Sigma AV}{N} = \rho_{AV}$$

These terms appear in the equation below. Further, $\Sigma A^2/N = \Sigma AA/N = \rho_{AA}$. But the correlation of a variable with itself equals 1.0, and hence $\rho_{AA} = 1.0$.

Substituting this information into the equation above yields:

$$\rho_{AF} = p_{FA} + p_{FE}\rho_{AE} + p_{FV}\rho_{AV}$$

In the correlation between age and the residual term V, ρ_{AV}, the V is unobservable, and so it is not measured. This means we have no way of estimating ρ_{AV}. Therefore *we assume that the correlations between the residual variable and the independent variables in a given equation are zero.* That is, in the first equation we assume that $\rho_{AV} = 0$.

Substituting this assumption into the equation above, and adding the prime (′) to make clear that this is the correlation *implied* by the model and not necessarily the observed correlation, we have:

$$\rho'_{AF} = p_{FA} + p_{FE}\rho_{AE}$$

That is, the implied correlation between A and F, ρ'_{AF}, is due to a direct path from age to frequenting bars, p_{FA}, and the product of the path from education to frequenting bars times the correlation between age and education (i.e., $p_{FE}\rho_{AE}$).

We can also show how the correlation between education and the frequenting of bars is implied by the causal mechanism suggested. We again begin with the equation:

$$F = p_{FA}A + p_{FE}E + p_{FV}V$$

This time we multiply both sides by E, since we want to analyze ρ_{EF}:

$$\begin{aligned} EF &= E(p_{FA}A + p_{FE}E + p_{FV}V) \\ &= p_{FA}EA + p_{FE}E^2 + p_{FV}EV \end{aligned}$$

We sum both sides:

$$\begin{aligned} \Sigma EF &= \Sigma(p_{FA}EA + p_{FE}E^2 + p_{FV}EV) \\ &= \Sigma p_{FA}EA + \Sigma p_{FE}E^2 + \Sigma p_{FV}EV \end{aligned}$$

and divide both sides by N. Then we bring the path coefficients outside the summation signs:

$$\frac{\Sigma EF}{N} = p_{FA}\frac{\Sigma EA}{N} + p_{FE}\frac{\Sigma E^2}{N} + p_{FV}\frac{\Sigma EV}{N}$$

Now $\Sigma EF/N = \rho_{EF}$, $\Sigma EA/N = \rho_{EA}$, $\Sigma E^2/N = \rho_{EE}$, and $\Sigma EV/N = \rho_{EV}$. Recall from above that the correlation between a variable and itself is 1.0, that is $\rho_{EE} = 1.0$. Furthermore, we assume the independent variables in a given equation are uncorrelated with the residual variable; that is, we assume $\rho_{EV} = 0$.

Substituting these results into the last equation above, we have:

$$\begin{aligned} \rho'_{EF} &= p_{FA}\,\rho_{EA} + p_{FE} \\ &= p_{FE} + p_{FA}\,\rho_{EA} \end{aligned}$$

In words, the implied correlation between education and the frequenting of bars is due to a direct path from education to the frequenting of bars, p_{FE}, plus the product of the path from age to frequenting bars, times the correlation between education and age, that is, $p_{FA}\,\rho_{EA}$.

A similar set of steps can be used to examine how each of the correlations between the independent variables and the dependent variable in the *second* equation can be implied by the hypothesized causal structure. For *each* of the independent variables we follow these steps:

1. Multiply the dependent variable by the independent variable.
2. Multiply each of the independent variables on the right-hand side of the equation by the independent variable.
3. Take sums of both sides of the equation, distributing the sum across all terms on the right-hand side of the equation.
4. Divide both sides of the equation by N in order to form correlations between the independent variable and all variables in the equation.
5. Simplify the result, taking into account two assumptions:
   a. A variable correlated with itself is 1.0.
   b. The correlation between an independent variable and the residual variable is zero.
6. Repeat steps 1–5 for each independent variable in the equation.

As an exercise, you should be able to prove the following results:

$$\rho'_{FD} = p_{DF} + p_{DA}\rho_{FA} + p_{DE}\rho_{FE}$$

$$\rho'_{AD} = p_{DA} + p_{DF}\rho_{AF} + p_{DE}\rho_{AE}$$

$$\rho'_{ED} = p_{DE} + p_{DF}\rho_{EF} + p_{DA}\rho_{EA}$$

The five equations we have derived above (two for the first equation and three for the second) can be summarized by what is called the **fundamental theorem of path analysis**. It states:

fundamental theorem of path analysis—an equation stating that the correlation between variables i and j implied by a hypothesized causal model is the sum of all products involving both a path from variable q to variable i, and the correlation between variable q and variable j. The sum of these products is formed over all Q variables that have direct paths to variable i

---

$$\rho'_{ij} = \sum_{q=1}^{Q} p_{iq}\rho_{qj}$$

The fundamental theorem of path analysis states that the correlation between variables i and j *implied* by the hypothesized causal model is the sum of all products involving both a path from variable q to variable i, and the correlation between variable q and variable j. The sum of these products is formed over all Q variables that have direct paths to variable i. Convince yourself that the fundamental theorem of path analysis could have been used to generate each of the five equations above.

Notice that each of the three equations above contains one or more correlations involving F. These correlations were decomposed into path components in analyzing the first equation. In particular, as we saw above:

$$\rho'_{AF} = p_{FA} + p_{FE}\rho_{AE}$$

$$\rho'_{EF} = p_{FE} + p_{FA}\rho_{EA}$$

If we substitute these two quantities into the equation for ρ'_{FD} above, for ρ_{FA} and ρ_{FE} we obtain:

$$\begin{aligned}\rho'_{FD} &= p_{DF} + p_{DA}(p_{FA} + p_{FE}\rho_{AE}) + p_{DE}(p_{FE} + p_{FA}\rho_{EA}) \\ &= p_{DF} + p_{DA}p_{FA} + p_{DA}p_{FE}\rho_{AE} + p_{DE}p_{FE} \\ &\quad + p_{DE}p_{FA}\rho_{EA}\end{aligned}$$

Similarly,

$$\begin{aligned}\rho'_{AD} &= p_{DA} + p_{DF}(p_{FA} + p_{FE}\rho_{AE}) + p_{DE}\rho_{AE} \\ &= p_{DA} + p_{DF}p_{FA} + p_{DF}p_{FE}\rho_{AE} + p_{DE}\rho_{AE}\end{aligned}$$

and:

$$\rho'_{ED} = p_{DE} + p_{DF}(p_{FE} + p_{FA}\rho_{EA}) + p_{DA}\rho_{EA}$$
$$= p_{DE} + p_{DF}p_{FE} + p_{DF}p_{FA}\rho_{EA} + p_{DA}\rho_{EA}$$

We recognize that these results are somewhat complex. Bear in mind that we have expressed the implied correlations between the independent variables and the dependent variable in both equations as a function of the hypothesized causal parameters. Notice, too, that the right side of each of the final versions of the equations contains only path coefficients and the correlation between age and education (ρ_{AE}). As a general principle, the final version of the decomposition of a correlation should contain only path coefficients and correlations among the predetermined variables.

12.3.5 Decomposing Implied Correlations by Tracing Paths

We can also obtain the decompositions for the five implied correlations by *tracing paths* in the diagram itself. In stating the following set of rules, which summarizes how to trace paths to obtain decompositions, we assume that variable *j* is causally prior to variable *i*. The steps are:

1. Beginning with dependent variable *i*, trace backward along the arrow that comes from variable *j*, if one exists. This is the simple *direct path coefficient*, p_{ij}. To its value should be added all the compound paths found, as noted in the following steps.
2. If other arrows come to variable *i* from third variables, *q*, trace all the connections between *i* and *j* which involve each *q*, multiplying the values of the path coefficients for these compound paths. In general, two kinds of compound linkages will occur:
   a. Variable *q* sends arrows to both *i* and *j* (either directly or through still other intervening variables). In this case, trace backward along the paths from *i* to *q*, then forward along the paths from *q* to *j*, multiplying coefficient values as you go. If more than one distinct compound pathway exists for a given *q*, treat each separately.

   b. Variable j sends an arrow to variable q, which in turn sends an arrow to variable i (either in two steps or through yet other intervening variables). In this case, simply trace backward from i, through q to j, multiplying path values as you go. If more than one distinct compound pathway back to j exists, treat each separately.

3. The following rules must be observed during tracing:
   a. You may trace backward along a series of arrows (from arrow head to arrow tail) for as many links as necessary to reach variable q. But once the direction has been changed in order to trace forward from q to j (from arrow tail to arrow head, as allowed in rule 2*a*), no subsequent reversals of direction are allowed.
   b. A particular double-headed curved arrow (for the correlation between predetermined variables) can be traversed only once during the tracing of a given compound path. And only one double-headed arrow can be traversed during any given compound linkage. Note that a traverse of a double-headed arrow always results in a change of direction, from backward to forward tracing. A traverse of a double-headed arrow results in a multiplication of the compound path by that correlation coefficient.
   c. All the legitimate compound paths presented in the path diagram must be traced and their values multiplied to determine the magnitude and sign of the compound effects.
4. When all direct and compound path values have been calculated, add them together to obtain the correlation between i and j (ρ'_{ij}) implied by the causal model.

To illustrate the use of tracing procedures, we will show how to decompose the correlation between drinking *(D)* and education *(E)*, ρ_{DE}. In Figure 12.4 (above), we can see that, using rule 1, the direct path between education *(E)* and drinking *(D)* is:

$$p_{DE}$$

Using rule 2, we see that there are two q variables, age *(A)* and frequenting bars *(F)*, with direct paths to drinking. Therefore these must provide indirect links with education. Under rules 2*a* and 3*b*, we trace backward from drinking to age (p_{DA}), and traverse the double-headed arrow to reach education (ρ_{EA}). This compound path will be added to the equation for decomposing ρ'_{DE} into direct and indirect effects:

$$p_{DA}\rho_{EA}$$

Notice that we *cannot* trace a compound path via age, such as $p_{DA}p_{FA}p_{FE}$, because it violates rule 3*a* about changing direction more than once. However, a second compound path connecting drinking to education through age *is* permissible:

$$p_{DF}p_{FA}\rho_{EA}$$

Since no other compound paths via age occur, we next turn to the indirect connections involving frequenting bars. One of these, also involving age, has just been noted. The only remaining compound path allowed by rule 2b is:

$$p_{DF}p_{FE}$$

Now, putting all these direct and indirect paths together and reordering terms, we arrive at the final dissection of the implied correlation between drinking and education:

$$\rho'_{DE} = p_{DE} + p_{DF}p_{FE} + p_{DF}p_{FA}\rho_{EA} + p_{DA}\rho_{EA}$$

which is the same result obtained with algebraic methods at the end of Section 12.3.4. Notice that the basic theorem did not permit any indirect pathways involving the paths from the residual factors, V and W. As stated above, one assumption of path analysis is that residual variables are uncorrelated with the independent

TABLE 12.1

Decompositions of Correlations in Path Model Shown in Figure 12.4

	Education (E)	***Age (A)***	***Frequenting Bars (F)***
Age (A)	ρ_{EA}		
Frequenting bars (F)	$p_{FE} + p_{FA}\rho_{EA}$	$p_{FA} + p_{FE}\rho_{EA}$	
Drinking (D)	$p_{DE} + p_{DF}p_{FE} + p_{DA}\rho_{EA} + p_{DF}p_{FA}\rho_{EA}$	$p_{DA} + p_{DF}p_{FA}p_{DE}\rho_{AE} + p_{DF}p_{FE}\rho_{AE}$	$p_{DF} + p_{DA}p_{FA} + p_{DE}p_{FE} + p_{DA}p_{FE}\rho_{AE} + p_{DE}p_{FA}\rho_{EA}$

variables in the model, and hence no compound paths involve these residuals.

Table 12.1 shows how the entire correlation matrix can be rewritten in terms of path equations. (The results have been rearranged slightly compared to those derived earlier.) As a check of your understanding of the basic path theorem, you should derive these equations yourself, using the tracing rules presented above, and compare them to the table. Since the same results can be obtained using either algebra or the tracing method, the technique you choose to employ is immaterial.

12.3.6. An Example Using Path Analysis: Estimating the Drinking Behavior Model

We now have all the tools we need to estimate the model shown in Figure 12.4. To do this we will use data from the 1978 General Social Survey, coding respondents' ages in years and education in number of years of schooling completed. Frequenting bars is measured by a seven-category response to the question: "How often do you go to a bar or tavern?" The categories, coded from 7 to 1, are almost every day, once or twice a week, several times a month, once a month, several times a year, once a year, or never. Drinking behavior is coded from responses to two items: "Do you ever have occasion to use any alcoholic beverages, such as liquor, wine, or beer, or are you a total abstainer?" A user is asked: "Do you sometimes drink more than you think you should?" Abstainers are scored 1, moderate drinkers 2, and those who sometimes drink excessively 3. The observed pairwise correlations among the variables are shown in Table 12.2.

TABLE 12.2
Correlation Matrix of Variables in Path Model

	Education (E)	*Age (A)*	*Frequenting Bars (F)*	*Drinking (D)*
Age *(A)*	−0.310			
Frequenting bars *(F)*	0.145	−0.340		
Drinking *(D)*	0.248	−0.283	0.536	
*N**	1,526	1,525	1,528	1,521

**Ns* differ because of pairwise deletion of missing data.
Source: 1978 General Social Survey.

TABLE 12.3
Regression Analyses for Frequenting Bars and Drinking Behavior Variables

Independent Variables	*Frequenting Bars (F)*	*Drinking (D)*
Age *(A)*	−0.326*	−0.068*
Education *(E)*	0.044	0.156*
Frequenting bars *(F)*	—	0.490*
Coefficient of determination (R^2)	0.116*	0.320*

*Significant for $\alpha = .01$.

Note: Regression analyses based on pairwise deletion procedures.

The results of the regression analysis are shown in Table 12.3. Since we are interested only in the standardized regression coefficients, the raw data regression coefficients are not tabled. For a recursive model, the path coefficients are simply the beta weights, and a residual path is simply the square root of 1 minus the appropriate coefficient of determination. The path coefficients for this example are also shown on the path diagram in Figure 12.5.

Since path coefficients are standardized values, they reflect causal effects in terms of standard deviation (Z-score) units. For example, $p_{DF} = 0.490$ means that a 1 standard deviation increase in frequenting bars and taverns leads to about a one-half standard deviation increase in the reported level of drinking. And since all path coefficients are standardized, comparison between the direct effects of causal variables is straightforward, as it is in multiple regression. Thus we see that the direct causal impacts of age and of education on drinking are opposite in direction, with older

FIGURE 12.5
Path Diagram with Numerical Estimates of Path Coefficients, Drinking Behavior Model

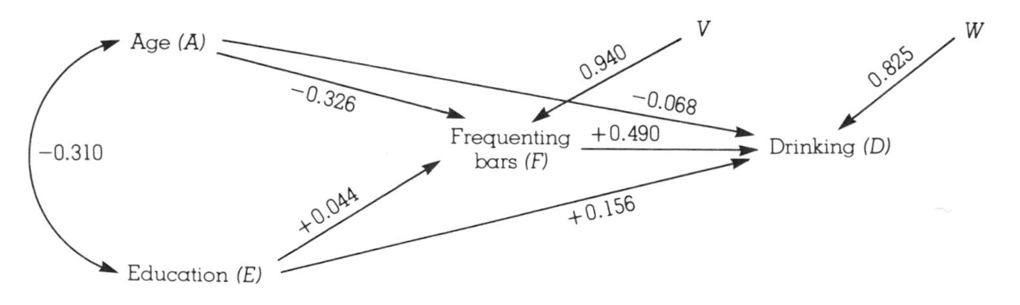

people drinking less and better educated people drinking more (contrary to expectation, as shown in Figure 12.3). We also find that both variables have much smaller effects on drinking than going to bars does. The diagram also shows that older persons frequent bars less often ($p_{FA} = -0.326$), but education has little impact on visiting bars.

Path coefficients also permit the calculation of indirect causal effects through the multiplication of path values of compound paths connecting two variables via intervening variables. As we suggested in Section 12.3.4, the sample estimates are:

$$r'_{DA} = \hat{p}_{DA} + \hat{p}_{DF}\hat{p}_{FA} + \hat{p}_{DF}\hat{p}_{FE}r_{AE} + \hat{p}_{DE}r_{AE}$$

The first term on the right is, of course, the estimated *direct effect* ($\hat{p}_{DA} = -0.068$). The second term is the estimated indirect effect. The final two terms are called **correlated effects,** since they depend on the correlation of A with E. The *indirect effect* in Figure 12.5 is $\hat{p}_{DF}\hat{p}_{FA}$—the effect of age on drinking via frequenting bars. Note that the estimated indirect effect of age lies in the same direction as the direct effect; $(-0.326)(0.490) = -0.159$, but it is more than twice the magnitude of the estimated direct effect in the hypothesized model.

correlated effect—a component in the decomposition of a correlation coefficient that is due to a correlation among predetermined variables

By adding together the direct and indirect causal effects and comparing the sum to the observed correlation in Table 12.2, we can see how much of the covariation is due to the correlated effects involving education. Thus, the observed age-drinking correlation is -0.283, the direct effect is -0.068, and the indirect effect is -0.159, which leaves -0.056 of the observed correlation due to their dependency on education. That is:

$$\begin{aligned}\hat{p}_{DF}\hat{p}_{FE}r_{AE} + \hat{p}_{DE}r_{AE} &= (0.490)(0.044)(-0.310) \\ &\quad + (0.156)(-0.310) \\ &= -0.056\end{aligned}$$

As an exercise, you should determine how much of the observed correlation between drinking and education is due to a direct effect, an indirect effect, and common correlation on age.

By squaring the path coefficients from the residual variables we can discover how much variance in the dependent variables remains unexplained by the hypothesized causal process. Both values are large: 88.4% of the variance in frequenting bars and 68.1% of the self-reported drinking levels cannot be explained by the causal structure. Clearly, this simple example does not come close to containing all the important social and psychological causes of these two behavioral measures. If we were to pursue this research, we would want to specify more elaborate models, including additional possible sources of drinking behavior.

> Fully recursive path models—in which all possible one-way arrows between variables are present—will always exactly reproduce the observed correlations when using the basic path theorem. Since the causal sequence among variables can be arbitrarily reordered, the empirical estimates of path coefficients generally provide no definitive answer to the question of whether a causal model is valid.

For example, if the location of frequenting bars and drinking behavior were switched in Figure 12.3, we could still derive path coefficients for each arrow from the observed correlations. Or, even more drastically, we could switch the places between the two predetermined variables and the two dependent measures and still generate path coefficients that would add up to the observed correlations. Clearly, the credibility of a path model cannot be based on statistical criteria alone.

> A causal model specification must have nonstatistical bases for its justification. In this text we have stressed the importance of theoretical understanding of social behavior in guiding empirical research. In organizing a causal model for path analysis, all the researcher's knowledge of social theory, past research, and logical deduction must be brought to bear.

In specifying the sequence of variables, understanding their temporal order is often indispensable. Our understanding that people's ages and formal schooling are set early in life, for example, enabled us to treat these variables as causally antecedent to the two contemporary indicators of alcoholic beverage use (frequency of visiting bars and amount of drinking). Unless causal analysis is firmly grounded in basic principles of social behavior, the results of path model analysis can be no firmer than the foundations of a house built on sand.

12.3.7. An Example of a Chain Path Model: School Busing Attitude

We have stated that fully recursive path models offer no statistical basis for their own rejection. Now we suggest that path models which are not fully recursive—that is, those in which some possible arrows are not present—do offer limited bases for deciding whether a specified model fits the data. When some possible causal paths are hypothesized not to operate, the implied correlations (r'_{ij}) from the path model need not equal the observed correlations (r_{ij}). When such discrepancies occur, the analyst may conclude that the model was incorrectly specified as a representation of the causal process, unless the discrepancy is small enough to be caused by sampling fluctuation.

To illustrate, we will use data from the 1972–1974–1976 Survey Research Center's panel study of American voters. A sample of voters was reinterviewed at two-year intervals about various issues and activities. One item asked respondents to rate their position on a seven-point scale from favoring school busing for racial integration to opposing busing. Among the simplest models of attitude causation through time is the chain model, where responses at time $t + 1$ depend solely on responses at the immediately preceding time, t. (This property is called the *Markovian principle,* which maintains that history prior to time t has no causal impact.)

In path diagrammatic terms, a **chain path model** for school busing attitude is shown in Figure 12.6. We assume that attitude toward busing in 1976 (Y_3) is caused by attitude toward busing in 1974 (Y_2) but not by attitude toward busing in 1972 (Y_1). And attitude toward busing in 1974 (Y_2) is assumed to be caused only by busing in 1972.

chain path model— a causal model in which variables measured on the same sample at three or more times are depicted as the causes of their own subsequent values.

Figure 12.6 implies two structural equations. They are:

$$Y_2 = p_{21} Y_1 + p_{2V} V$$
$$Y_3 = p_{32} Y_2 + p_{3U} U$$

Since these are two recursive equations, they can be estimated by the two beta coefficients obtained by regressing Y_2 on Y_1 and by regressing Y_3 on Y_2:

FIGURE 12.6
Causal Diagram of Presumed Causal Relationships among Attitudes toward Busing, 1972, 1974, and 1976, with Estimated Path Coefficients

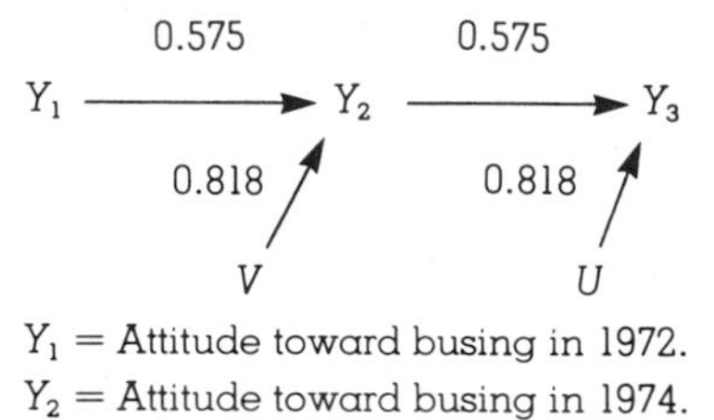

Y_1 = Attitude toward busing in 1972.
Y_2 = Attitude toward busing in 1974.
Y_3 = Attitude toward busing in 1976.

Source: Survey Research Center, University of Michigan, panel study of American voters.

$$\hat{p}_{21} = \beta^*_{21}$$
$$\hat{p}_{2V} = \sqrt{1 - R^2_{2\cdot 1}}$$
$$\hat{p}_{32} = \beta^*_{32}$$
$$\hat{p}_{3U} = \sqrt{1 - R^2_{3\cdot 2}}$$

The intercorrelations among the three variables are shown in Table 12.4. Since beta weights are zero-order correlations in the case of a single independent variable, we have $\hat{p}_{21} = 0.575$, $\hat{p}_{2V} = 0.818$, $\hat{p}_{32} = 0.575$, and $\hat{p}_{3U} = 0.818$.

TABLE 12.4
Intercorrelations among Attitudes toward Busing in 1972, 1974, and 1976

	Y_1	Y_2	Y_3
Y_1: Attitude in 1972	—	.575	.535
Y_2: Attitude in 1974		—	.575
Y_3: Attitude in 1976			—

Source: Survey Research Center, University of Michigan's panel study of American voters.

Now we want to test whether the correlation implied between Y_1 and Y_3 equals the observed correlation (0.535). To determine the implied correlation we can follow the methodology employed in Section 12.3.4. We begin with:

$$Y_3 = p_{32}Y_2 + p_{3U}U$$

Then we multiply this equation by Y_1, yielding:

$$Y_1Y_3 = p_{32}Y_1Y_2 + p_{3U}Y_1U$$

We sum both sides, distribute the summation, and divide by N:

$$\frac{\Sigma Y_1 Y_3}{N} = \frac{p_{32}\Sigma Y_1 Y_2}{N} + \frac{p_{3U}\Sigma Y_1 U}{N}$$

Since we assume $\rho_{Y_1U} = 0$, we have: $\rho'_{13} = p_{32}\rho_{21} = p_{32}\,p_{21}$, since $p_{21} = \beta^*_{21} = \rho_{21}$. *That is, the correlation between Y_1 and Y_3 implied by the causal model equals the product of the two paths, p_{32} and p_{21}.* Now convince yourself that you get the same results by tracing paths between Y_3 and Y_1.

In the SRC data,

$$r'_{13} = (0.575)\,(0.575) = 0.331$$

But as seen in Table 12.4, the actual observed correlation is 0.535. Hence, the large discrepancy between observed and implied correlations, $r_{13} - r'_{13} = 0.204$, suggests **misspecification** of the causal model.

misspecification— a condition in which a structural equation or path model includes incorrect variables or excludes correct variables

A logical alternative model for these three variables is shown in Figure 12.7. In it we have assumed a lagged causal effect from

FIGURE 12.7
Causal Diagram of an Alternative Set of Causal Relationships among Attitudes toward Busing, 1972, 1974, and 1976

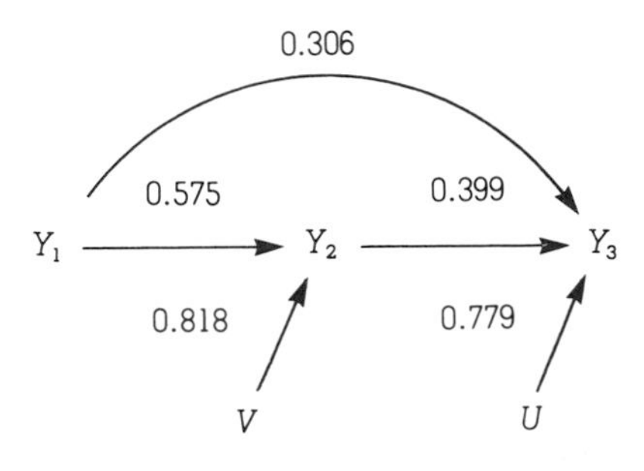

1972 to 1976 attitude. (A lagged causal effect is the same as a direct effect.) The two structural equations in this case are:

$$Y_2 = p_{21} Y_1 + p_{2V} V$$

$$Y_3 = p_{31} Y_1 + p_{32} Y_2 + p_{3U} U$$

The estimates of the path coefficients are:

$$\hat{p}_{21} = \beta^*_{21}$$

$$\hat{p}_{2V} = \sqrt{1 - R^2_{2\cdot 1}}$$

$$\hat{p}_{31} = \beta^*_{31}$$

$$\hat{p}_{32} = \beta^*_{32}$$

$$\hat{p}_{3U} = \sqrt{1 - R^2_{3\cdot 21}}$$

Using the data from Table 12.4, these estimates are as shown in Figure 12.7. While the estimates look plausible, this model must fit the observed correlations exactly, since all three possible one-way paths are present.

This alternative is only one of several possible causal models which could account for the pattern of correlations over time. Other models would include additional independent variables, with correlated residual variables and measurement errors. The specification and solution of such complex models requires advanced statistical techniques beyond the level of this text. We have, however, shown that the simple causal chain must be rejected as the explanation of the observed pattern of covariation.

In this chapter we have introduced the bare essentials of path analysis. We hope we have convinced you that causal inferences can be drawn from nonexperimental data if you have strong theoretical propositions. When you cannot experiment, path models can be very useful for estimating presumed causal processes. As we suggested above, however, to avoid nonsensical results you must pay close attention to meeting the assumptions of the model.

Review of Key Concepts

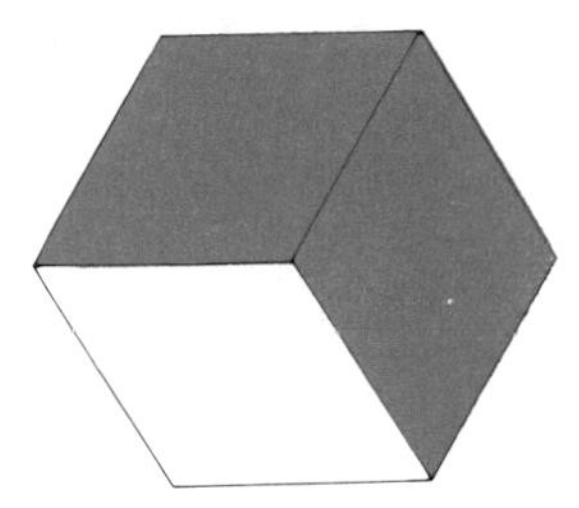

This list of the key concepts introduced in Chapter 12 is in the order of appearance in the text. Combined with the definitions in the margins, it will help you review the material and can serve as a self-test for mastery of the concepts.

covariation
time order
nonspuriousness
causal diagram
direct effect
indirect effect
predetermined variables
residual variables
probabilistic (stochastic)
deterministic
recursive model
nonrecursive model
epistemic relationships
path analysis
structural equations
path coefficient
decomposition
fundamental theorem of path analysis
correlated effects
chain path model
misspecification

PROBLEMS

General Problems

1. State why this statement cannot easily be recast as a causal proposition: "Everything else being equal, those who are more politically liberal are also more sexually permissive."
2. Rewrite this statement as a causal proposition: "The more expectations of economic well-being outstrip a society's capacity to fulfill such expectations, the more likely the society is to experience turmoil in the form of riots, strikes, and civil strife."
3. Diagram the causal process implied by these hypotheses:
   a. An increase in variable X increases variable Y.
   b. A decrease in variable Y increases variable Z.
   c. An increase in Z decreases W.
   d. An increase in W increases Q.
   e. When X decreases, Z increases and W increases.

4. Diagram the causal process implied by these hypotheses:
   a. The higher a person's rank in a group, the greater that person's centrality.
   b. The more central a person is to a group, the greater that person's conformity to group norms.
   c. The more central a person is to a group, the more influence that person has on group decisions.
   d. The greater the conformity to group norms, the more influence a person has on group decisions.

5. In the following path equation, H, J, and K are observable variables, and U is an unmeasured variable:

$$H = p_{HJ}J - p_{HK}K + p_{HU}U$$

Derive the formula for the correlation between H and J in terms of path coefficients and correlations.

6. In the following path equations, A, B, C, and D are observables, and U and V are unobservables:

$$D = p_{DC}C + p_{DA}A + p_{DV}V$$
$$C = p_{CB}B + p_{CA}A + P_{CU}U$$

   a. Derive the formula for the correlation between B and D in terms of path coefficients and correlations.
   b. What are the direct effects of B on D?

7. Here is a path diagram in which H is grade point average, J is teacher's encouragement, K is parental social status, and L is student's IQ:

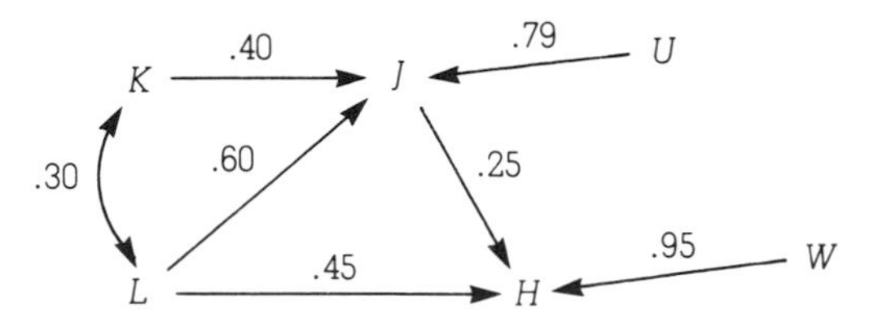

   a. Does K or L have the larger *indirect* effect on H?
   b. How much variance in J is explained by the linear combination of K and L?

8. Consider the following path diagram:

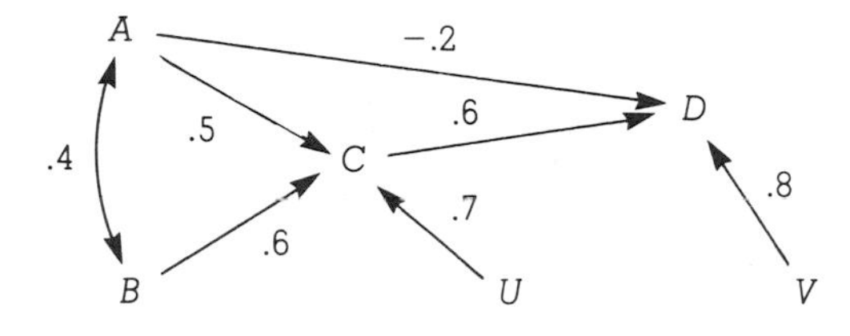

a. Does A or B have larger indirect effects on D?
b. How much variance in D is explained by A, B, and C?

9. In a panel study of sexual morality, an index of sexual traditionalism is measured on the same respondents at three intervals spaced evenly across a decade. The correlation between the first and second measurements is 0.667, and the correlation between the second and third measurements is 0.750. What is the implied correlation between the first and third measurements of sexual traditionalism if a simple causal chain is operating?

10. In a chain model where:

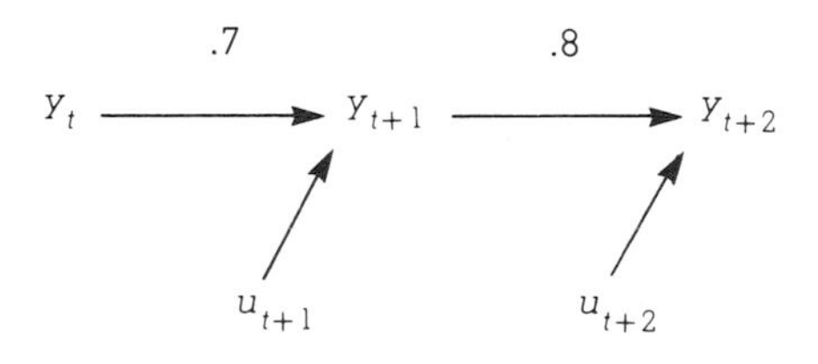

What is $r'_{t,t+2}$?

11. In the following causal diagram, A is age at which respondent was married, B is the social status of the family, C is current financial security, and D is marital happiness:

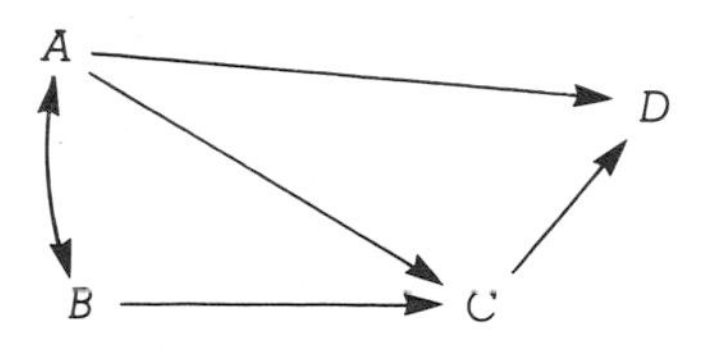

In symbolic terms, write an equation for the correlation between marital happiness *(D)* and social status *(B)*, designating the portion that arises from causal connections and the portion due to mutual dependency.

12. Consider the following causal diagram:

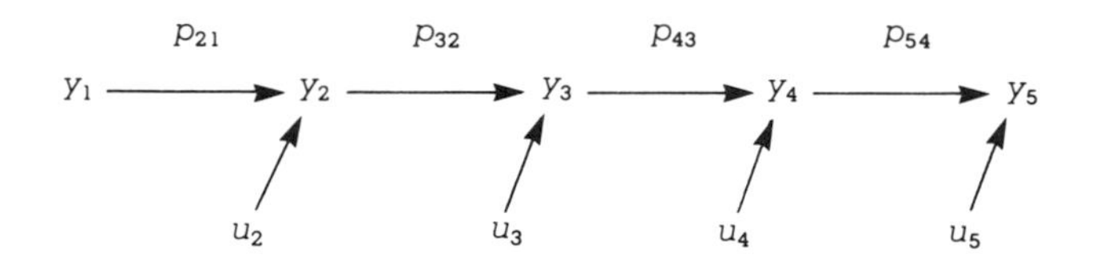

Assume all $p_{ij} = 0.5$, and generate the matrix of correlations implied among the five variables.

13. *a.* In Problem 11, what is the correlation between *B* and *D* if the following values are found: $r_{AB} = +0.44$; $p_{CB} = +0.60$; $p_{CA} = -0.10$; $p_{DC} = +0.35$; and $p_{DA} = +0.50$.

    b. What is the correlation between *A* and *D?*

14. Consider the following causal diagram:

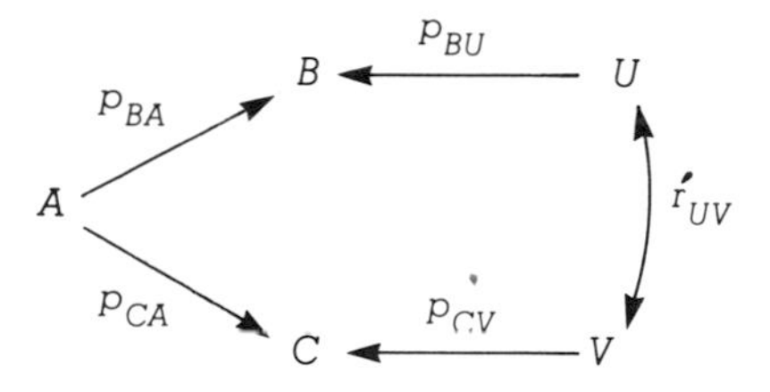

a. Prove that $r'_{UV} = r_{BC \cdot A}$; i.e., the partial correlation between *U* and *V*, controlling for *A*.

b. If $p_{BA} = 0.40$, $p_{CA} = 0.30$ and $r_{BC} = 0.15$, what does r'_{UV} equal?

Problems Requiring the 1980 General Social Survey Data

15. Some political sociologists have argued that the lower-class way of life produces individuals with rigid and intolerant approaches to politics. Test this idea by using an index of intoler-

ance (INTOL) toward those opposed to churches and religion created by summing responses to SPKATH (should be allowed to make speeches), COLATH (should be allowed to teach in a college or university), and LIBATH (book written by speaker should be removed from public library). The path model to be estimated (for males) is:

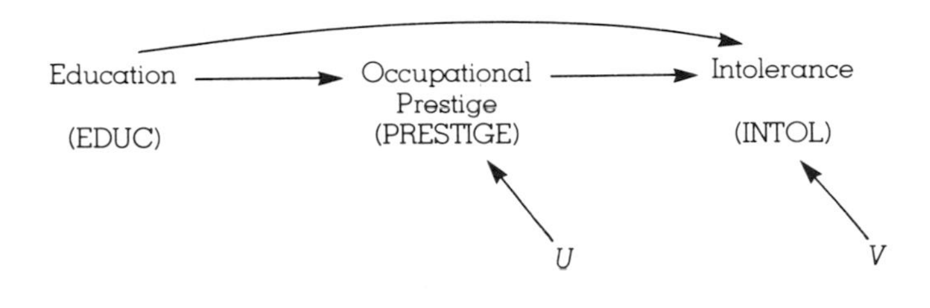

Interpret the results.

16. Estimate the following path model as an explanation of personal happiness:

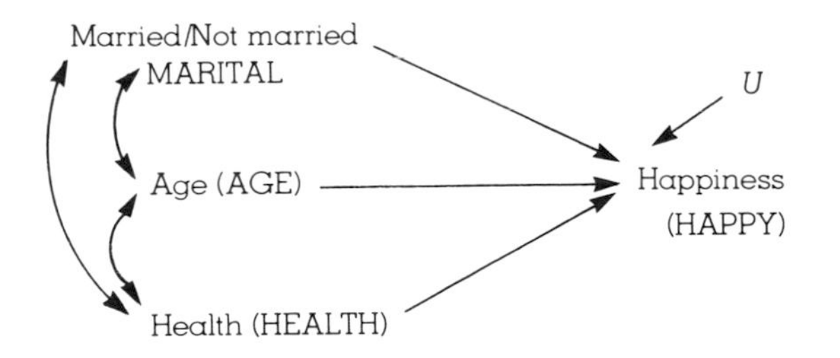

In estimating this model, use the following recodes: HEALTH (1=4, 2=3, 3=2, 4=1); HAPPY (1=3, 3=1); and MARITAL (2, 3, 4, 5 = 0).

17. Estimate the path coefficients of this causal model of status attainment among working men. Select cases where SEX = 1 and WRKSTAT equals or is less than 4.

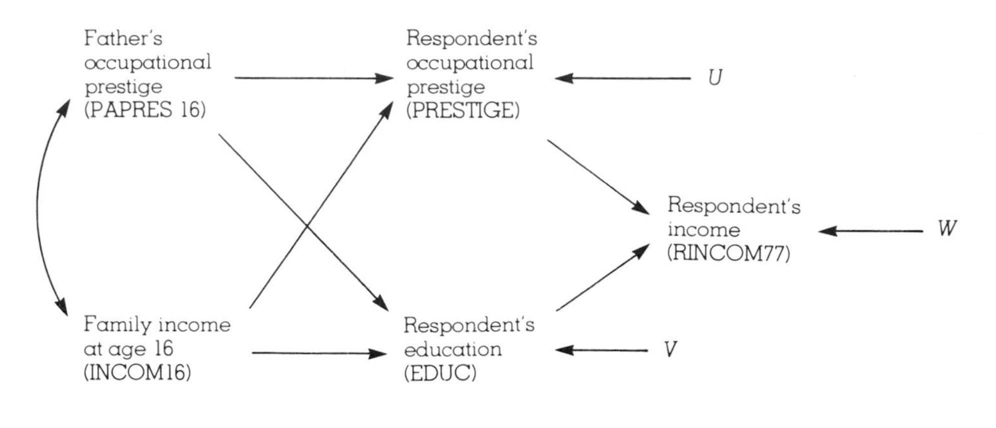

Interpret the results.

Problems Requiring the 63-Cities Data

18. The zero-order correlation between percentage of blacks in the city population (BLACKPCT) and police expenditures (POLICEXP) per capita in the 63-cities data is +0.57. Estimate a path model in which the crime rate is an intervening variable, and find how much the direct path from black percentage to police expenditures is reduced.

19. Estimate the path coefficients for the following model:

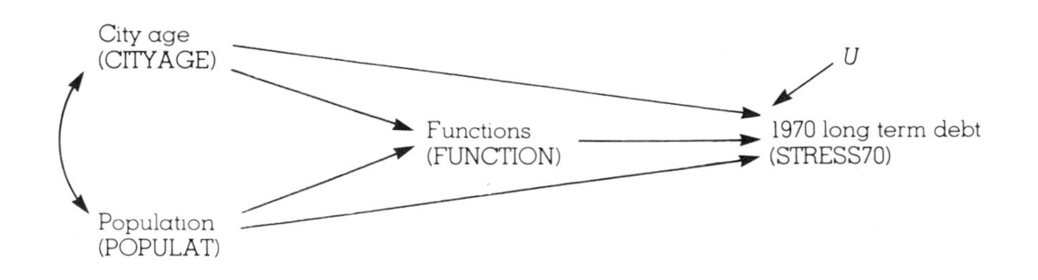

20. Estimate the path coefficients for the following model:

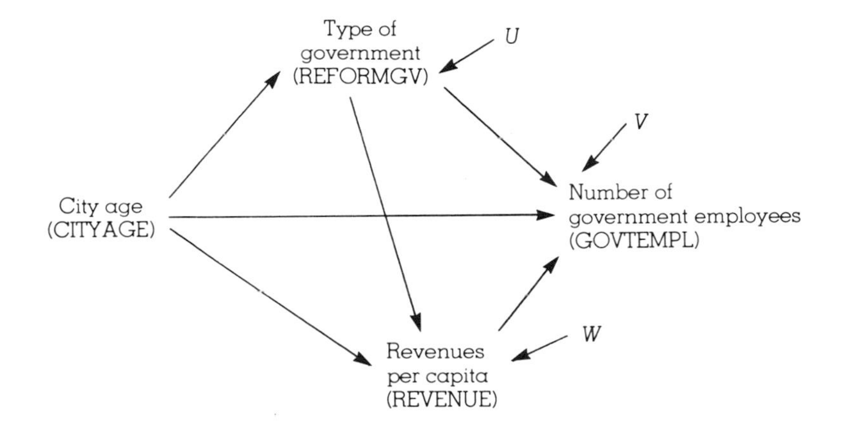

Appendix A

The Use of Summations

1. Variables and Subscripts

In this text we use the letters X, Y, and Z to stand for *variables.* Variables have outcomes that can be kept track of through the use of *subscripts.* If we have N individuals in our sample, then X_i denotes the particular value or outcome observed for individuals, i. For example, if we have four persons in a sample, then the four outcomes associated with them are represented by X_1, X_2, X_3 and X_4.

2. Sums

Many of the statistical techniques used in this text depend on the *sum* of observations across the N individuals in the sample. The Greek symbol sigma (Σ) is used to stand for the sum of the values that immediately follow the summation sign. An index value below Σ indicates the lowest value the summation will take, and an index value above Σ indicates the highest value the summation will take. Therefore:

$$\sum_{i=1}^{N} X_i$$

is read as "the sum of the N outcomes of X_i from X_1 to X_N," or:

$$\sum_{i=1}^{N} X_i = X_1 + X_2 + X_3 + \ldots + X_N$$

Suppose we observe four individuals (i.e., $N = 4$), and the four outcomes are $X_1 = 2$, $X_2 = 6$, $X_3 = 0$, and $X_4 = 3$. Then:

$$\sum_{i=1}^{4} X_i = 2 + 6 + 0 + 3 = 11$$

The simplest use of the summation is in computing the *mean,* which is simply the average value across a set of observations. As you probably know, an average of this sort is computed by adding up all the outcomes and then dividing the sum by the number of observations. That is, the mean is simply:

$$\frac{\sum_{i=1}^{N} X_i}{N}$$

or, in the example above, the mean is:

$$\frac{\sum_{i=1}^{4} X_i}{4} = \frac{11}{4} = 2.75$$

After you become familiar with the use of Σ and it becomes clear from the context that we are summing across all observations, we may use either $\sum_i X_i$ or ΣX_i instead of the longer $\sum_{i=1}^{N} X_i$ notation.

Sometimes variables are represented by more than a single subscript. This will be done for two reasons. The first is that we sometimes wish to represent not only an individual, *i,* but also a group, *j,* to which the person belongs (e.g., sex or religious identification). In this case the notation is X_{ij}. If the last observation in group j is notated by n_j, then the sum across the n_j individuals in group j is:

$$\sum_{i=1}^{n_j} X_{ij}$$

Written out, this is:

$$\sum_{i=1}^{n_j} X_{ij} = X_{1j} + X_{2j} + X_{3j} + \ldots + X_{n_j j}$$

If we wish to sum across all J groups of n_j individuals, this is symbolized by:

$$\sum_{j=1}^{J} \sum_{i=1}^{n_j} X_{ij}$$

and when written out, it is:

$$\sum_{j=1}^{J}\sum_{i=1}^{n_j} X_{ij} = (X_{11} + X_{21} + \ldots + X_{n_1 1}) + (X_{12} + X_{22} + \ldots + X_{n_2 2}) + (X_{1J} + X_{2J} + \ldots + X_{n_J J})$$

As an example, suppose we have three political groups where 1 = Republican, 2 = Democrat, and 3 = other, and there are four Republicans, three Democrats, and two others in the groups. We observe the following outcomes:

	i = 1	2	3	4
j = 1	3	4	2	3
2	2	0	1	
3	2	4		

Now if we want to sum the values of the Republicans (where j = 1), we have:

$$\sum_{i=1}^{4} X_{i1} = 3 + 4 + 2 + 3 = 12$$

The sum of the others is:

$$\sum_{i=1}^{2} X_{i3} = 2 + 4 = 6$$

Or, summing *all* observations across groups:

$$\sum_{j=1}^{3}\sum_{i=1}^{n_j} = (3 + 4 + 2 + 3) + (2 + 0 + 1) + (2 + 4) = 12 + 3 + 6 = 21$$

Where there is no ambiguity about the groups and individuals being summed across, we will also use $\sum_j\sum_i X_{ij}$ or $\Sigma\Sigma X_{ij}$, instead of the more cumbersome $\sum_{j=1}^{J}\sum_{i=1}^{n_j}$.

The second use of double subscripts is applied when we want to distinguish the same individual on two different variables. For example, we may have two variables, X_1 and X_2. In this case X_{1i} and X_{2i} symbolize the ith individual's outcomes on the two variables, and:

$$\sum_{i=1}^{N} X_{1i} = X_{11} + X_{12} + \ldots + X_{1N}$$

We use this notation in Chapter 11.

3. Rules of Summation

There are a few simple rules of summation that you should learn. If you do, you should have no difficulty with the few derivations presented in this book.

Rule 1: The sum over a constant for N observations equals N times the constant. That is, if a is a constant, then:

$$\sum_{i=1}^{N} a = Na$$

This may not seem intuitively obvious, but an example should make the rule clear. Suppose we have $N = 4$ observations, and each observation equals 5. Then $a = 5$ and

$$\sum_{i=1}^{4} a = (5 + 5 + 5 + 5) = (4)(5) = 20$$

We can also extend this rule, as shown below.

Rule 2: If each observation is multiplied times a constant, the sum of the constant times the observations equals the constant times the sum of the observations. That is,

$$\sum_{i=1}^{N} aX_i = a \sum_{i=1}^{N} X_i$$

For example, consider $a = 4$ and $X_1 = 2$, $X_2 = 6$, and $X_3 = 1$. Then:

$$\sum_{i=1}^{3} 4X_i = 4 \cdot 2 + 4 \cdot 6 + 4 \cdot 1 = 4(2 + 6 + 1)$$

$$= 4 \sum_{i=1}^{3} X_i = 36$$

This rule can also be applied to double sums. That is:

$$\sum_{j=1}^{J} \sum_{i=1}^{n_j} aX_{ij} = a \sum_{j=1}^{J} \sum_{i=1}^{n_j} X_{ij}$$

Rule 3: If the only operation to be carried out before a summation is itself a sum, the summation can be distributed.

This rule sounds more complex than it is. Consider the following example.

$$\begin{aligned}\sum_{i=1}^{3} (X_i + 2) &= (X_1 + 2) + (X_2 + 2) + (X_3 + 2) \\ &= (X_1 + X_2 + X_3) + (2 + 2 + 2) \\ &= \sum_{i=1}^{3} X_i + \sum_{i=1}^{3} 2 \\ &= \sum_{i=1}^{3} X_i + (3)(2) \\ &= \sum_{i=1}^{3} X_i + 6\end{aligned}$$

A more general expression of this rule is:

$$\begin{aligned}\sum_{i=1}^{N} (X_i \pm a) &= \sum_{i=1}^{N} X_i \pm \sum_{i=1}^{N} a \\ &= \sum_{i=1}^{N} X_i \pm Na\end{aligned}$$

This last step, that $\sum_{i=1}^{N} a = Na$, follows from Rule 1. Note, however, that:

$$\sum_{i=1}^{N} (X_i + a)^2 \neq \sum_{i=1}^{N} X_i^2 + \Sigma a^2$$

We can only distribute the summation sign when the term within the parentheses is itself a simple sum or difference.

If we expand the term $(X + a)^2$, we can then distribute the sum, that is:

$$\Sigma(X_i + a)^2 = \Sigma(X_i^2 + 2aX_i + a^2)$$
$$= \Sigma X_i^2 + \Sigma 2aX_i + \Sigma a^2$$

Now it follows from Rules 1 and 2 that we can simplify this expression even further, to:

$$\Sigma X_i^2 + 2a\Sigma X_i + Na^2$$

4. Sums of Two or More Variables

Sometimes we will want to examine sums of two or more variables at once. Suppose we ask what the sum of a product of two variables across N observations is.

Rule 4: If each observation has a score on two variables, X_i and Y_i, then:

$$\sum_{i=1}^{N} X_i Y_i = X_1Y_1 + X_2Y_2 + \ldots + X_NY_N$$

Suppose there are three persons for whom we have observations on two variables, X and Y. The observations are:

i	X_i	Y_i
1	2	1
2	4	−2
3	2	−3

Then

$$\sum_{i=1}^{3} X_i Y_i = (2)(1) + (4)(-2) + (2)(-3) = -12$$

Convince yourself, using this last example, that in general:

$$\sum_{i=1}^{N} X_i Y_i \neq \Sigma X_i \Sigma Y_i$$

Thus, we *cannot* distribute a summation sign across products. But Rule 2 *does* apply to products. That is:

$$\Sigma aX_iY_i = a\Sigma X_iY_i$$

This can be seen using the data from the example immediately above. Let $a = 3$; then:

$$\sum_{i=1}^{3} 3X_iY_i = 3(2)(1) + 3(4)(-2) + 3(2)(-3)$$
$$= 3((2)(1) + (4)(-2) + (2)(-3)) = (3)(-12) = -36$$
$$= 3 \sum_{i=1}^{3} X_iY_i$$

Rule 5: The sum of two or more variables equals the sum of the sums of the variables. That is,

$$\sum_i (X_{1i} + X_{2i} + \ldots + X_{ki}) = \sum_i X_{1i} + \sum_i X_{2i} + \ldots + \sum_i X_{ki}$$

Where:

$X_1, X_2, \ldots X_k = k$ different variables.

A special case of this rule is:

$$\sum_i (X_i + Y_i) = \sum_i X_i + \sum_i Y_i$$

Again using the data from the example above, we can show this rule as follows:

$$\sum_{i=1}^{3} (X_i + Y_i) = (2 + 1) + (4 + (-2)) + (2 + (-3))$$
$$= (2 + 4 + 2) + (1 + (-2) + (-3)) = 4$$
$$= \sum_{i=1}^{3} X_i + \sum_{i=1}^{3} Y_i$$

Rule 6: For constants a and b,

$$\sum_i (aX_i + bY_i) = a\sum_i X_i + b\sum_i Y_i$$

Rule 6 is really derivative from Rules 2 and 5. Rule 5 states that we can distribute the summation sign across sums of variables, and Rule 2 states that we can pull a constant out in front of a sum.

Again using the example data above, let $a = 2$ and $b = 4$. Then:

$$\sum_{i=1}^{3} (2X_i + 4Y_i) = ((2)(2) + (4)(1)) + ((2)(4) + (4)(-2)) + ((2)(2) + (4)(-3))$$

$$= 2(2 + 4 + 2) + 4(1 - 2 - 3)$$

$$= 0$$

$$= 2 \sum_{i=1}^{3} X_i + 4 \sum_{i=1}^{3} Y_i$$

With this set of rules you should be able to follow the algebra used in this text. Our one piece of advice is not to be overwhelmed by sums. When in doubt about an equivalence, for example, try writing out the sums. They usually are not as complex as they may seem.

Critical Values of Chi Square

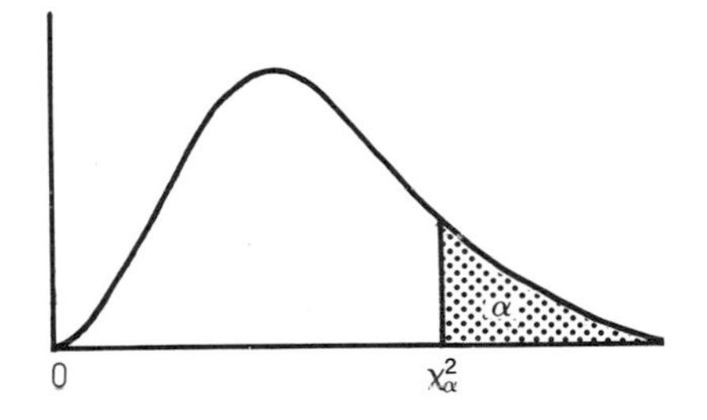

df	Level of Significance (α)					
	.100	.050	.025	.010	.005	0.001
1	2.7055	3.8414	5.0238	6.6349	7.8794	10.828
2	4.6051	5.9914	7.3777	9.2103	10.5966	13.816
3	6.2513	7.8147	9.3484	11.3449	12.8381	16.266
4	7.7794	9.4877	11.1433	13.2767	14.8602	18.467
5	9.2363	11.0705	12.8325	15.0863	16.7496	20.515
6	10.6446	12.5916	14.4494	16.8119	18.5476	22.458
7	12.0170	14.0671	16.0128	18.4753	20.2777	24.322
8	13.3616	15.5073	17.5346	20.0902	21.9550	26.125
9	14.6837	16.9190	19.0228	21.6660	23.5893	27.877
10	15.9871	18.3070	20.4831	23.2093	25.1882	29.588
11	17.2750	19.6751	21.9200	24.7250	26.7569	31.264
12	18.5494	21.0261	23.3367	26.2170	28.2995	32.909
13	19.8119	22.3621	24.7356	27.6883	29.8194	34.528
14	21.0642	23.6848	26.1190	29.1413	31.3193	36.123
15	22.3072	24.9958	27.4884	30.5779	32.8013	37.697
16	23.5418	26.2962	28.8454	31.9999	34.2672	39.252
17	24.7690	27.5871	30.1910	33.4087	35.7185	40.790
18	25.9894	28.8693	31.5264	34.8058	37.1564	42.312
19	27.2036	30.1435	32.8523	36.1908	38.5822	43.820
20	28.4120	31.4104	34.1696	37.5662	39.9968	45.315
21	29.6151	32.6705	35.4789	38.9321	41.4010	46.797
22	30.8133	33.9244	36.7807	40.2894	42.7956	48.268
23	32.0069	35.1725	38.0757	41.6384	44.1813	49.728
24	33.1963	36.4151	39.3641	42.9798	45.5585	51.179

continued

Source: Abridged from Table IV of Fisher and Yates: *Statistical Tables for Biological, Agricultural and Medical Research,* published by Longman Group Ltd. London (1974) 6th edition. (Previously published by Oliver & Boyd Ltd. Edinburgh) and by permission of the authors and publishers.

Critical Values of Chi Square (Cont.)

	Level of Significance (α)					
df	.100	.050	.025	.010	.005	0.001
25	34.3816	37.6525	40.6465	44.3141	46.9278	52.620
26	35.5631	38.8852	41.9232	45.6417	48.2899	54.052
27	36.7412	40.1133	43.1944	46.9680	49.6449	55.476
28	37.9159	41.3372	44.4607	48.2782	50.9933	56.892
29	39.0875	42.5569	45.7222	49.5879	52.3356	58.302
30	40.2560	43.7729	46.9792	50.8922	53.6720	59.703
40	51.8050	55.7585	59.3417	63.6907	66.7659	73.402
50	63.1671	67.5048	71.4202	76.1539	79.4900	86.661
60	74.3970	79.0819	83.2976	88.3794	91.9517	99.607
70	85.5271	90.5312	95.0231	100.425	104.215	112.317
80	96.5782	101.879	106.629	112.329	116.321	124.839
90	107.565	113.145	118.136	124.116	128.299	137.208
100	118.498	124.342	129.561	135.807	140.169	149.449

Appendix C

Area under the Normal Curve

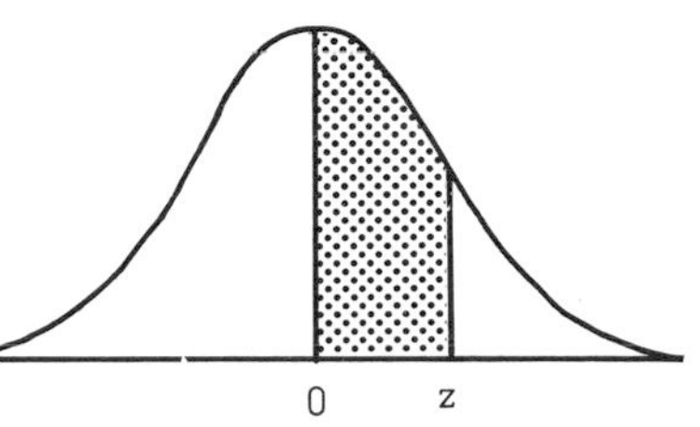

z	.00	.01	.02	.03	.04	.05	.06	.07	.08	.09
0.0	.0000	.0040	.0080	.0120	.0160	.0199	.0239	.0279	.0319	.0359
0.1	.0398	.0438	.0478	.0517	.0557	.0596	.0636	.0675	.0714	.0753
0.2	.0793	.0832	.0871	.0910	.0948	.0987	.1026	.1064	.1103	.1141
0.3	.1179	.1217	.1255	.1293	.1331	.1368	.1406	.1443	.1480	.1517
0.4	.1554	.1591	.1628	.1664	.1700	.1736	.1772	.1808	.1844	.1879
0.5	.1915	.1950	.1985	.2019	.2054	.2088	.2123	.2157	.2190	.2224
0.6	.2257	.2291	.2324	.2357	.2389	.2422	.2454	.2486	.2517	.2549
0.7	.2580	.2611	.2642	.2673	.2704	.2734	.2764	.2794	.2823	.2852
0.8	.2881	.2910	.2939	.2967	.2995	.3023	.3051	.3078	.3106	.3133
0.9	.3159	.3186	.3212	.3238	.3264	.3289	.3315	.3340	.3365	.3389
1.0	.3413	.3438	.3461	.3485	.3508	.3531	.3554	.3577	.3599	.3621
1.1	.3643	.3665	.3686	.3708	.3729	.3749	.3770	.3790	.3810	.3830
1.2	.3849	.3869	.3888	.3907	.3925	.3944	.3962	.3980	.3997	.4015
1.3	.4032	.4049	.4066	.4082	.4099	.4115	.4131	.4147	.4162	.4177
1.4	.4192	.4207	.4222	.4236	.4251	.4265	.4279	.4292	.4306	.4319
1.5	.4332	.4345	.4357	.4370	.4382	.4394	.4406	.4418	.4429	.4441
1.6	.4452	.4463	.4474	.4484	.4495	.4505	.4515	.4525	.4535	.4545
1.7	.4554	.4564	.4573	.4582	.4591	.4599	.4608	.4616	.4625	.4633
1.8	.4641	.4649	.4656	.4664	.4671	.4678	.4686	.4693	.4699	.4706
1.9	.4713	.4719	.4726	.4732	.4738	.4744	.4750	.4756	.4761	.4767
2.0	.4772	.4778	.4783	.4788	.4793	.4798	.4803	.4808	.4812	.4817
2.1	.4821	.4826	.4830	.4834	.4838	.4842	.4846	.4850	.4854	.4857
2.2	.4861	.4864	.4868	.4871	.4875	.4878	.4881	.4884	.4887	.4890
2.3	.4893	.4896	.4898	.4901	.4904	.4906	.4909	.4911	.4913	.4916
2.4	.4918	.4920	.4922	.4925	.4927	.4929	.4931	.4932	.4934	.4936
2.5	.4938	.4940	.4941	.4943	.4945	.4946	.4948	.4949	.4951	.4952
2.6	.4953	.4955	.4956	.4957	.4959	.4960	.4961	.4962	.4963	.4964
2.7	.4965	.4966	.4967	.4968	.4969	.4970	.4971	.4972	.4973	.4974
2.8	.4974	.4975	.4976	.4977	.4977	.4978	.4979	.4979	.4980	.4981
2.9	.4981	.4982	.4982	.4983	.4984	.4984	.4985	.4985	.4986	.4986
3.0	.4987	.4987	.4987	.4988	.4988	.4989	.4989	.4989	.4990	.4990

Source: Abridged from Table I of *Statistical Tables and Formulas,* by A. Hald (New York: John Wiley & Sons, Inc., 1952). Reproduced by permission of A. Hald and the publishers, John Wiley & Sons, Inc.

Appendix D

Student's *t* Distribution

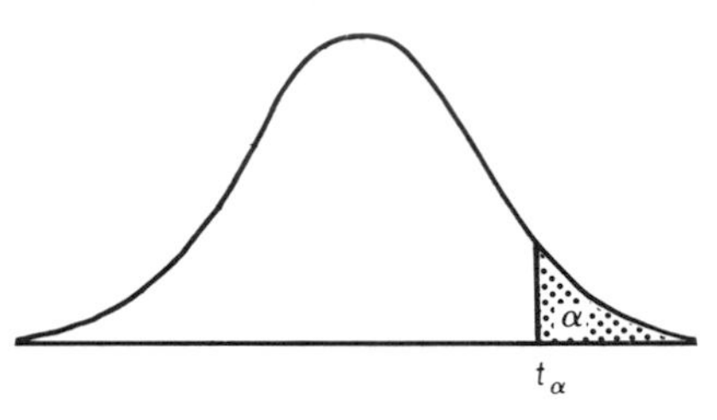

df	Level of Significance (α)					
	.10	.05	.025	.01	.005	.0005
1	3.078	6.314	12.706	31.821	63.657	636.619
2	1.886	2.920	4.303	6.965	9.925	31.598
3	1.638	2.353	3.182	4.541	5.841	12.941
4	1.533	2.132	2.776	3.747	4.604	8.610
5	1.476	2.015	2.571	3.365	4.032	6.859
6	1.440	1.943	2.447	3.143	3.707	5.959
7	1.415	1.895	2.365	2.998	3.499	5.405
8	1.397	1.860	2.306	2.896	3.355	5.041
9	1.383	1.833	2.262	2.821	3.250	4.781
10	1.372	1.812	2.228	2.764	3.169	4.587
11	1.363	1.796	2.201	2.718	3.106	4.437
12	1.356	1.782	2.179	2.681	3.055	4.318
13	1.350	1.771	2.160	2.650	3.012	4.221
14	1.345	1.761	2.145	2.624	2.977	4.140
15	1.341	1.753	2.131	2.602	2.947	4.073
16	1.337	1.746	2.120	2.583	2.921	4.015
17	1.333	1.740	2.110	2.567	2.898	3.965
18	1.330	1.734	2.101	2.552	2.878	3.922
19	1.328	1.729	2.093	2.539	2.861	3.883
20	1.325	1.725	2.086	2.528	2.845	3.850
21	1.323	1.721	2.080	2.518	2.831	3.819
22	1.321	1.717	2.074	2.508	2.819	3.792
23	1.319	1.714	2.069	2.500	2.807	3.767
24	1.318	1.711	2.064	2.492	2.797	3.745
25	1.316	1.708	2.060	2.485	2.787	3.725
26	1.315	1.706	2.056	2.479	2.779	3.707
27	1.314	1.703	2.052	2.473	2.771	3.690
28	1.313	1.701	2.048	2.467	2.763	3.674
29	1.311	1.699	2.045	2.462	2.756	3.659
30	1.310	1.697	2.042	2.457	2.750	3.646
40	1.303	1.684	2.021	2.423	2.704	3.551
60	1.296	1.671	2.000	2.390	2.660	3.460
120	1.289	1.658	1.980	2.358	2.617	3.373
∞	1.282	1.645	1.960	2.326	2.576	3.291

Source: Adapted from Table III of Fisher & Yates: *Statistical Tables for Biological, Agricultural and Medical Research,* published by Longman Group Ltd. London (1974) 6th edition. (Previously published by Oliver & Boyd Ltd. Edinburgh) and by permission of the authors and publishers.

Appendix E

F Distribution

The *F* distribution table consists of three parts, for $\alpha = .05$, $\alpha = .01$, and $\alpha = .001$. These tables appear on the next three pages.

$\alpha = .05$

ν_2 \ ν_1	1	2	3	4	5	6	7	8	9	10	12	15	20	24	30	40	60	120	∞
1	161.4	199.5	215.7	224.6	230.2	234.0	236.8	238.9	240.5	241.9	243.9	245.9	248.0	249.1	250.1	251.1	252.2	253.3	254.3
2	18.51	19.00	19.16	19.25	19.30	19.33	19.35	19.37	19.38	19.40	19.41	19.43	19.45	19.45	19.46	19.47	19.48	19.49	19.50
3	10.13	9.55	9.28	9.12	9.01	8.94	8.89	8.85	8.81	8.79	8.74	8.70	8.66	8.64	8.62	8.59	8.57	8.55	8.53
4	7.71	6.94	6.59	6.39	6.26	6.16	6.09	6.04	6.00	5.96	5.91	5.86	5.80	5.77	5.75	5.72	5.69	5.66	5.63
5	6.61	5.79	5.41	5.19	5.05	4.95	4.88	4.82	4.77	4.74	4.68	4.62	4.56	4.53	4.50	4.46	4.43	4.40	4.36
6	5.99	5.14	4.76	4.53	4.39	4.28	4.21	4.15	4.10	4.06	4.00	3.94	3.87	3.84	3.81	3.77	3.74	3.70	3.67
7	5.59	4.74	4.35	4.12	3.97	3.87	3.79	3.73	3.68	3.64	3.57	3.51	3.44	3.41	3.38	3.34	3.30	3.27	3.23
8	5.32	4.46	4.07	3.84	3.69	3.58	3.50	3.44	3.39	3.35	3.28	3.22	3.15	3.12	3.08	3.04	3.01	2.97	2.93
9	5.12	4.26	3.86	3.63	3.48	3.37	3.29	3.23	3.18	3.14	3.07	3.01	2.94	2.90	2.86	2.83	2.79	2.75	2.71
10	4.96	4.10	3.71	3.48	3.33	3.22	3.14	3.07	3.02	2.98	2.91	2.85	2.77	2.74	2.70	2.66	2.62	2.58	2.54
11	4.84	3.98	3.59	3.36	3.20	3.09	3.01	2.95	2.90	2.85	2.79	2.72	2.65	2.61	2.57	2.53	2.49	2.45	2.40
12	4.75	3.89	3.49	3.26	3.11	3.00	2.91	2.85	2.80	2.75	2.69	2.62	2.54	2.51	2.47	2.43	2.38	2.34	2.30
13	4.67	3.81	3.41	3.18	3.03	2.92	2.83	2.77	2.71	2.67	2.60	2.53	2.46	2.42	2.38	2.34	2.30	2.25	2.21
14	4.60	3.74	3.34	3.11	2.96	2.85	2.76	2.70	2.65	2.60	2.53	2.46	2.39	2.35	2.31	2.27	2.22	2.18	2.13
15	4.54	3.68	3.29	3.06	2.90	2.79	2.71	2.64	2.59	2.54	2.48	2.40	2.33	2.29	2.25	2.20	2.16	2.11	2.07
16	4.49	3.63	3.24	3.01	2.85	2.74	2.66	2.59	2.54	2.49	2.42	2.35	2.28	2.24	2.19	2.15	2.11	2.06	2.01
17	4.45	3.59	3.20	2.96	2.81	2.70	2.61	2.55	2.49	2.45	2.38	2.31	2.23	2.19	2.15	2.10	2.06	2.01	1.96
18	4.41	3.55	3.16	2.93	2.77	2.66	2.58	2.51	2.46	2.41	2.34	2.27	2.19	2.15	2.11	2.06	2.02	1.97	1.92
19	4.38	3.52	3.13	2.90	2.74	2.63	2.54	2.48	2.42	2.38	2.31	2.23	2.16	2.11	2.07	2.03	1.98	1.93	1.88
20	4.35	3.49	3.10	2.87	2.71	2.60	2.51	2.45	2.39	2.35	2.28	2.20	2.12	2.08	2.04	1.99	1.95	1.90	1.84
21	4.32	3.47	3.07	2.84	2.68	2.57	2.49	2.42	2.37	2.32	2.25	2.18	2.10	2.05	2.01	1.96	1.92	1.87	1.81
22	4.30	3.44	3.05	2.82	2.66	2.55	2.46	2.40	2.34	2.30	2.23	2.15	2.07	2.03	1.98	1.94	1.89	1.84	1.78
23	4.28	3.42	3.03	2.80	2.64	2.53	2.44	2.37	2.32	2.27	2.20	2.13	2.05	2.01	1.96	1.91	1.86	1.81	1.76
24	4.26	3.40	3.01	2.78	2.62	2.51	2.42	2.36	2.30	2.25	2.18	2.11	2.03	1.98	1.94	1.89	1.84	1.79	1.73
25	4.24	3.39	2.99	2.76	2.60	2.49	2.40	2.34	2.28	2.24	2.16	2.09	2.01	1.96	1.92	1.87	1.82	1.77	1.71
26	4.23	3.37	2.98	2.74	2.59	2.47	2.39	2.32	2.27	2.22	2.15	2.07	1.99	1.95	1.90	1.85	1.80	1.75	1.69
27	4.21	3.35	2.96	2.73	2.57	2.46	2.37	2.31	2.25	2.20	2.13	2.06	1.97	1.93	1.88	1.84	1.79	1.73	1.67
28	4.20	3.34	2.95	2.71	2.56	2.45	2.36	2.29	2.24	2.19	2.12	2.04	1.96	1.91	1.87	1.82	1.77	1.71	1.65
29	4.18	3.33	2.93	2.70	2.55	2.43	2.35	2.28	2.22	2.18	2.10	2.03	1.94	1.90	1.85	1.81	1.75	1.70	1.64
30	4.17	3.32	2.92	2.69	2.53	2.42	2.33	2.27	2.21	2.16	2.09	2.01	1.93	1.89	1.84	1.79	1.74	1.68	1.62
40	4.08	3.23	2.84	2.61	2.45	2.34	2.25	2.18	2.12	2.08	2.00	1.92	1.84	1.79	1.74	1.69	1.64	1.58	1.51
60	4.00	3.15	2.76	2.53	2.37	2.25	2.17	2.10	2.04	1.99	1.92	1.84	1.75	1.70	1.65	1.59	1.53	1.47	1.39
120	3.92	3.07	2.68	2.45	2.29	2.17	2.09	2.02	1.96	1.91	1.83	1.75	1.66	1.61	1.55	1.50	1.43	1.35	1.25
∞	3.84	3.00	2.60	2.37	2.21	2.10	2.01	1.94	1.88	1.83	1.75	1.67	1.57	1.52	1.46	1.39	1.32	1.22	1.00

Source: Adapted from E. S. Pearson and H. O. Hartley *Biometrika Tables for Statisticians.* (2nd Ed.). Cambridge: Cambridge University Press, 1962.

F Distribution (Cont.)

$\alpha = .01$

ν_2 \ ν_1	1	2	3	4	5	6	7	8	9	10	12	15	20	24	30	40	60	120	∞
1	4052	4999.5	5403	5625	5764	5859	5928	5982	6022	6056	6106	6157	6209	6235	6261	6287	6313	6339	6366
2	98.50	99.00	99.17	99.25	99.30	99.33	99.36	99.37	99.39	99.40	99.42	99.43	99.45	99.46	99.47	99.47	99.48	99.49	99.50
3	34.12	30.82	29.46	28.71	28.24	27.91	27.67	27.49	27.35	27.23	27.05	26.87	26.69	26.60	26.50	26.41	26.32	26.22	26.13
4	21.20	18.00	16.69	15.98	15.52	15.21	14.98	14.80	14.66	14.55	14.37	14.20	14.02	13.93	13.84	13.75	13.65	13.56	13.46
5	16.26	13.27	12.06	11.39	10.97	10.67	10.46	10.29	10.16	10.05	9.89	9.72	9.55	9.47	9.38	9.29	9.20	9.11	9.02
6	13.75	10.92	9.78	9.15	8.75	8.47	8.26	8.10	7.98	7.87	7.72	7.56	7.40	7.31	7.23	7.14	7.06	6.97	6.88
7	12.25	9.55	8.45	7.85	7.46	7.19	6.99	6.84	6.72	6.62	6.47	6.31	6.16	6.07	5.99	5.91	5.82	5.74	5.65
8	11.26	8.65	7.59	7.01	6.63	6.37	6.18	6.03	5.91	5.81	5.67	5.52	5.36	5.28	5.20	5.12	5.03	4.95	4.86
9	10.56	8.02	6.99	6.42	6.06	5.80	5.61	5.47	5.35	5.26	5.11	4.96	4.81	4.73	4.65	4.57	4.48	4.40	4.31
10	10.04	7.56	6.55	5.99	5.64	5.39	5.20	5.06	4.94	4.85	4.71	4.56	4.41	4.33	4.25	4.17	4.08	4.00	3.91
11	9.65	7.21	6.22	5.67	5.32	5.07	4.89	4.74	4.63	4.54	4.40	4.25	4.10	4.02	3.94	3.86	3.78	3.69	3.60
12	9.33	6.93	5.95	5.41	5.06	4.82	4.64	4.50	4.39	4.30	4.16	4.01	3.86	3.78	3.70	3.62	3.54	3.45	3.36
13	9.07	6.70	5.74	5.21	4.86	4.62	4.44	4.30	4.19	4.10	3.96	3.82	3.66	3.59	3.51	3.43	3.34	3.25	3.17
14	8.86	6.51	5.56	5.04	4.69	4.46	4.28	4.14	4.03	3.94	3.80	3.66	3.51	3.43	3.35	3.27	3.18	3.09	3.00
15	8.68	6.36	5.42	4.89	4.56	4.32	4.14	4.00	3.89	3.80	3.67	3.52	3.37	3.29	3.21	3.13	3.05	2.96	2.87
16	8.53	6.23	5.29	4.77	4.44	4.20	4.03	3.89	3.78	3.69	3.55	3.41	3.26	3.18	3.10	3.02	2.93	2.84	2.75
17	8.40	6.11	5.18	4.67	4.34	4.10	3.93	3.79	3.68	3.59	3.46	3.31	3.16	3.08	3.00	2.92	2.83	2.75	2.65
18	8.29	6.01	5.09	4.58	4.25	4.01	3.84	3.71	3.60	3.51	3.37	3.23	3.08	3.00	2.92	2.84	2.75	2.66	2.57
19	8.18	5.93	5.01	4.50	4.17	3.94	3.77	3.63	3.52	3.43	3.30	3.15	3.00	2.92	2.84	2.76	2.67	2.58	2.49
20	8.10	5.85	4.94	4.43	4.10	3.87	3.70	3.56	3.46	3.37	3.23	3.09	2.94	2.86	2.78	2.69	2.61	2.52	2.42
21	8.02	5.78	4.87	4.37	4.04	3.81	3.64	3.51	3.40	3.31	3.17	3.03	2.88	2.80	2.72	2.64	2.55	2.46	2.36
22	7.95	5.72	4.82	4.31	3.99	3.76	3.59	3.45	3.35	3.26	3.12	2.98	2.83	2.75	2.67	2.58	2.50	2.40	2.31
23	7.88	5.66	4.76	4.26	3.94	3.71	3.54	3.41	3.30	3.21	3.07	2.93	2.78	2.70	2.62	2.54	2.45	2.35	2.26
24	7.82	5.61	4.72	4.22	3.90	3.67	3.50	3.36	3.26	3.17	3.03	2.89	2.74	2.66	2.58	2.49	2.40	2.31	2.21
25	7.77	5.57	4.68	4.18	3.85	3.63	3.46	3.32	3.22	3.13	2.99	2.85	2.70	2.62	2.54	2.45	2.36	2.27	2.17
26	7.72	5.53	4.64	4.14	3.82	3.59	3.42	3.29	3.18	3.09	2.96	2.81	2.66	2.58	2.50	2.42	2.33	2.23	2.13
27	7.68	5.49	4.60	4.11	3.78	3.56	3.39	3.26	3.15	3.06	2.93	2.78	2.63	2.55	2.47	2.38	2.29	2.20	2.10
28	7.64	5.45	4.57	4.07	3.75	3.53	3.36	3.23	3.12	3.03	2.90	2.75	2.60	2.52	2.44	2.35	2.26	2.17	2.06
29	7.60	5.42	4.54	4.04	3.73	3.50	3.33	3.20	3.09	3.00	2.87	2.73	2.57	2.49	2.41	2.33	2.23	2.14	2.03
30	7.56	5.39	4.51	4.02	3.70	3.47	3.30	3.17	3.07	2.98	2.84	2.70	2.55	2.47	2.39	2.30	2.21	2.11	2.01
40	7.31	5.18	4.31	3.83	3.51	3.29	3.12	2.99	2.89	2.80	2.66	2.52	2.37	2.29	2.20	2.11	2.02	1.92	1.80
60	7.08	4.98	4.13	3.65	3.34	3.12	2.95	2.82	2.72	2.63	2.50	2.35	2.20	2.12	2.03	1.94	1.84	1.73	1.60
120	6.85	4.79	3.95	3.48	3.17	2.96	2.79	2.66	2.56	2.47	2.34	2.19	2.03	1.95	1.86	1.76	1.66	1.53	1.38
∞	6.63	4.61	3.78	3.32	3.02	2.80	2.64	2.51	2.41	2.32	2.18	2.04	1.88	1.79	1.70	1.59	1.47	1.32	1.00

***F* Distribution (Cont.)** $\alpha = .001$

ν_2 \ ν_1	1	2	3	4	5	6	7	8	9	10	12	15	20	24	30	40	60	120	∞
1	4053*	5000*	5404*	5625*	5764*	5859*	5929*	5981*	6023*	6056*	6107*	6158*	6209*	6235*	6261*	6287*	6313*	6340*	6366*
2	998.5	999.0	999.2	999.2	999.3	999.3	999.4	999.4	999.4	999.4	999.4	999.4	999.4	999.5	999.5	999.5	999.5	999.5	999.5
3	167.0	148.5	141.1	137.1	134.6	132.8	131.6	130.6	129.9	129.2	128.3	127.4	126.4	125.9	125.4	125.0	124.5	124.0	123.5
4	74.14	61.25	56.18	53.44	51.71	50.53	49.66	49.00	48.47	48.05	47.41	46.76	46.10	45.77	45.43	45.09	44.75	44.40	44.05
5	47.18	37.12	33.20	31.09	29.75	28.84	28.16	27.64	27.24	26.92	26.42	25.91	25.39	25.14	24.87	24.60	24.33	24.06	23.79
6	35.51	27.00	23.70	21.92	20.81	20.03	19.46	19.03	18.69	18.41	17.99	17.56	17.12	16.89	16.67	16.44	16.21	15.99	15.75
7	29.25	21.69	18.77	17.19	16.21	15.52	15.02	14.63	14.33	14.08	13.71	13.32	12.93	12.73	12.53	12.33	12.12	11.91	11.70
8	25.42	18.49	15.83	14.39	13.49	12.86	12.40	12.04	11.77	11.54	11.19	10.84	10.48	10.30	10.11	9.92	9.73	9.53	9.33
9	22.86	16.39	13.90	12.56	11.71	11.13	10.70	10.37	10.11	9.89	9.57	9.24	8.90	8.72	8.55	8.37	8.19	8.00	7.81
10	21.04	14.91	12.55	11.28	10.48	9.92	9.52	9.20	8.96	8.75	8.45	8.13	7.80	7.64	7.47	7.30	7.12	6.94	6.76
11	19.69	13.81	11.56	10.35	9.58	9.05	8.66	8.35	8.12	7.92	7.63	7.32	7.01	6.85	6.68	6.52	6.35	6.17	6.00
12	18.64	12.97	10.80	9.63	8.89	8.38	8.00	7.71	7.48	7.29	7.00	6.71	6.40	6.25	6.09	5.93	5.76	5.59	5.42
13	17.81	12.31	10.21	9.07	8.35	7.86	7.49	7.21	6.98	6.80	6.52	6.23	5.93	5.78	5.63	5.47	5.30	5.14	4.97
14	17.14	11.78	9.73	8.62	7.92	7.43	7.08	6.80	6.58	6.40	6.13	5.85	5.56	5.41	5.25	5.10	4.94	4.77	4.60
15	16.59	11.34	9.34	8.25	7.57	7.09	6.74	6.47	6.26	6.08	5.81	5.54	5.25	5.10	4.95	4.80	4.64	4.47	4.31
16	16.12	10.97	9.00	7.94	7.27	6.81	6.46	6.19	5.98	5.81	5.55	5.27	4.99	4.85	4.70	4.54	4.39	4.23	4.06
17	15.72	10.66	8.73	7.68	7.02	6.56	6.22	5.96	5.75	5.58	5.32	5.05	4.78	4.63	4.48	4.33	4.18	4.02	3.85
18	15.38	10.39	8.49	7.46	6.81	6.35	6.02	5.76	5.56	5.39	5.13	4.87	4.59	4.45	4.30	4.15	4.00	3.84	3.67
19	15.08	10.16	8.28	7.26	6.62	6.18	5.85	5.59	5.39	5.22	4.97	4.70	4.43	4.29	4.14	3.99	3.84	3.68	3.51
20	14.82	9.95	8.10	7.10	6.46	6.02	5.69	5.44	5.24	5.08	4.82	4.56	4.29	4.15	4.00	3.86	3.70	3.54	3.38
21	14.59	9.77	7.94	6.95	6.32	5.88	5.56	5.31	5.11	4.95	4.70	4.44	4.17	4.03	3.88	3.74	3.58	3.42	3.26
22	14.38	9.61	7.80	6.81	6.19	5.76	5.44	5.19	4.99	4.83	4.58	4.33	4.06	3.92	3.78	3.63	3.48	3.32	3.15
23	14.19	9.47	7.67	6.69	6.08	5.65	5.33	5.09	4.89	4.73	4.48	4.23	3.96	3.82	3.68	3.53	3.38	3.22	3.05
24	14.03	9.34	7.55	6.59	5.98	5.55	5.23	4.99	4.80	4.64	4.39	4.14	3.87	3.74	3.59	3.45	3.29	3.14	2.97
25	13.88	9.22	7.45	6.49	5.88	5.46	5.15	4.91	4.71	4.56	4.31	4.06	3.79	3.66	3.52	3.37	3.22	3.06	2.89
26	13.74	9.12	7.36	6.41	5.80	5.38	5.07	4.83	4.64	4.48	4.24	3.99	3.72	3.59	3.44	3.30	3.15	2.99	2.82
27	13.61	9.02	7.27	6.33	5.73	5.31	5.00	4.76	4.57	4.41	4.17	3.92	3.66	3.52	3.38	3.23	3.08	2.92	2.75
28	13.50	8.93	7.19	6.25	5.66	5.24	4.93	4.69	4.50	4.35	4.11	3.86	3.60	3.46	3.32	3.18	3.02	2.86	2.69
29	13.39	8.85	7.12	6.19	5.59	5.18	4.87	4.64	4.45	4.29	4.05	3.80	3.54	3.41	3.27	3.12	2.97	2.81	2.64
30	13.29	8.77	7.05	6.12	5.53	5.12	4.82	4.58	4.39	4.24	4.00	3.75	3.49	3.36	3.22	3.07	2.92	2.76	2.59
40	12.61	8.25	6.60	5.70	5.13	4.73	4.44	4.21	4.02	3.87	3.64	3.40	3.15	3.01	2.87	2.73	2.57	2.41	2.23
60	11.97	7.76	6.17	5.31	4.76	4.37	4.09	3.87	3.69	3.54	3.31	3.08	2.83	2.69	2.55	2.41	2.25	2.08	1.89
120	11.38	7.32	5.79	4.95	4.42	4.04	3.77	3.55	3.38	3.24	3.02	2.78	2.53	2.40	2.26	2.11	1.95	1.76	1.54
∞	10.83	6.91	5.42	4.62	4.10	3.74	3.47	3.27	3.10	2.96	2.74	2.51	2.27	2.13	1.99	1.84	1.66	1.45	1.00

* Multiply these entries by 100.

Appendix F

Fisher's *r* to *Z* Transformation

r	Z	*r*	Z	*r*	Z	*r*	Z	*r*	Z
.000	.000	.200	.203	.400	.424	.600	.693	.800	1.099
.005	.005	.205	.208	.405	.430	.605	.701	.805	1.113
.010	.010	.210	.213	.410	.436	.610	.709	.810	1.127
.015	.015	.215	.218	.415	.442	.615	.717	.815	1.142
.020	.020	.220	.224	.420	.448	.620	.725	.820	1.157
.025	.025	.225	.229	.425	.454	.625	.733	.825	1.172
.030	.030	.230	.234	.430	.460	.630	.741	.830	1.188
.035	.035	.235	.239	.435	.466	.635	.750	.835	1.204
.040	.040	.240	.245	.440	.472	.640	.758	.840	1.221
.045	.045	.245	.250	.445	.478	.645	.767	.845	1.238
.050	.050	.250	.255	.450	.485	.650	.775	.850	1.256
.055	.055	.255	.261	.455	.491	.655	.784	.855	1.274
.060	.060	.260	.266	.460	.497	.660	.793	.860	1.293
.065	.065	.265	.271	.465	.504	.665	.802	.865	1.313
.070	.070	.270	.277	.470	.510	.670	.811	.870	1.333
.075	.075	.275	.282	.475	.517	.675	.820	.875	1.354
.080	.080	.280	.288	.480	.523	.680	.829	.880	1.376
.085	.085	.285	.293	.485	.530	.685	.838	.885	1.398
.090	.090	.290	.299	.490	.536	.690	.848	.890	1.422
.095	.095	.295	.304	.495	.543	.695	.858	.895	1.447
.100	.100	.300	.310	.500	.549	.700	.867	.900	1.472
.105	.105	.305	.315	.505	.556	.705	.877	.905	1.499
.110	.110	.310	.321	.510	.563	.710	.887	.910	1.528
.115	.116	.315	.326	.515	.570	.715	.897	.915	1.557
.120	.121	.320	.332	.520	.576	.720	.908	.920	1.589
.125	.126	.325	.337	.525	.583	.725	.918	.925	1.623
.130	.131	.330	.343	.530	.590	.730	.929	.930	1.658
.135	.136	.335	.348	.535	.597	.735	.940	.935	1.697
.140	.141	.340	.354	.540	.604	.740	.950	.940	1.738
.145	.146	.345	.360	.545	.611	.745	.962	.945	1.783
.150	.151	.350	.365	.550	.618	.750	.973	.950	1.832
.155	.156	.355	.371	.555	.626	.755	.984	.955	1.886
.160	.161	.360	.377	.560	.633	.760	.996	.960	1.946
.165	.167	.365	.383	.565	.640	.765	1.008	.965	2.014
.170	.172	.370	.388	.570	.648	.770	1.020	.970	2.092
.175	.177	.375	.394	.575	.655	.775	1.033	.975	2.185
.180	.182	.380	.400	.580	.662	.780	1.045	.980	2.298
.185	.187	.385	.406	.585	.670	.785	1.058	.985	2.443
.190	.192	.390	.412	.590	.678	.790	1.071	.990	2.647
.195	.198	.395	.418	.595	.685	.795	1.085	.995	2.994

SPSS File for the 63-Cities Data Set

```
RUN NAME          CREATE AN SPSS FILE FOR 63 CITIES STUDY
FILE NAME         CITIES
VARIABLE LIST     IDNUMBER, REGION, CITYAGE, FUNCTION, POPULAT, INCOMEPC, MANUFPCT,
                  WCPCT, BLACKPCT, POPCHANG, HOUSEPCT, BIRTHS, FEMLABOR, CRIMRATE,
                  RETAIL, REVENUE, EXPENDIT, GOVTEMPL, POLICEXP, REFORMGV, STRESS70,
                  STRESS74, DECINDEX
INPUT MEDIUM      CARDS
N OF CASES        63
INPUT FORMAT      FIXED (2(F2.0,1X), F3.0, 1X, F1.0, 1X, F4.0, 1X, F3.0, 1X, 3(F2.0,1X), F3.0, 1X,
                  4(F2.0,1X), F5.0, 1X, 2(F3.0,1X), 2(F2.0,1X), F1.0, 1X, F3.0, 1X, F4.0, 1X, F3.2)
MISSING VALUES    FUNCTION (0)/FEMLABOR (0)/DECINDEX (9.99)
VAR LABELS        IDNUMBER CITY IDENTIFICATION NUMBER/
                  REGION CENSUS REGION OF CITY LOCATION/
                  CITYAGE YEARS SINCE CITY REACHED 10,000 POPULATION/
                  FUNCTION NUMBER OF MUNICIPAL FUNCTIONS PERFORMED, 1962/
                  POPULAT NUMBER OF RESIDENTS, IN 1,000S/
                  INCOMEPC INCOME PER CAPITA, IN $10S/
                  MANUFPCT PER CENT OF LABOR FORCE IN MANUFACTURING/
                  WCPCT PER CENT OF LABOR FORCE IN NONMANUAL OCCUPATIONS/
                  BLACKPCT PER CENT BLACK/
                  POPCHANG PER CENT POPULATION CHANGE, 1960 TO 1970/
                  HOUSEPCT PER CENT OF HOUSING UNITS BUILT BEFORE 1950/
                  BIRTHS BIRTHRATE PER 1,000 POPULATION IN 1968/
                  FEMLABOR PER CENT OF LABOR FORCE FEMALE/
                  CRIMRATE NUMBER OF SERIOUS CRIMES PER 1,000 POPULATION/
                  RETAIL RETAIL SALES IN $ MILLIONS/
                  REVENUE GENERAL MUNICIPAL REVENUES PER CAPITA/
                  EXPENDIT GENERAL MUNICIPAL EXPENDITURES PER CAPITA/
                  GOVTEMPL LOCAL GOVERNMENT EMPLOYEES PER 1,000 POPULATION/
                  POLICEXP POLICE EXPENDITURES PER CAPITA/
                  REFORMGV INDEX OF REFORM GOVERNMENT STRUCTURE/
                  STRESS70 LONGTERM DEBT PER CAPITA, 1970/
                  STRESS74 LONGTERM DEBT PER CAPITA, 1974/
                  DECINDEX DECENTRALIZED DECISION-MAKING INDEX, 1967/
VALUE LABELS      REGION (1) NORTHEAST (2) SOUTH (3) MIDWEST (4) FAR WEST/
                  REFORMGV (3) MANAGR, NONPARTY, AT-LARG (2) ANY TWO OF THESE
                  (1) ANY ONE OF THESE (0) NONE OF THESE/
                  IDNUMBER (0) AKRON, OHIO (1) ALBANY, N.Y. (2) AMARILLO, TEXAS
                  (3) ATLANTA, GEO. (4) BERKELEY, CALIF. (5) BIRMINGHAM, ALA.
                  (6) BLOOMINTON, MINN. (7) BOSTON, MASS. (8) BUFFALO, N.Y.
                  (9) CAMBRIDGE, MASS. (10) CHARLOTTE, N.C. (11) CLIFTON, N.J.
                  (12) DULUTH, MINN. (13) EUCLID, OHIO (14) FT. WORTH, TEXAS
                  (15) FULLERTON, CALIF. (16) GARY, IND. (17) HAMILTON, OHIO
                  (18) HAMMOND, IND. (19) INDIANAPOLIS, IND. (20) IRVINGTON, N.J.
                  (21) JACKSONVILLE, FLA. (22) LONG BEACH, CALIF. (23) MALDEN, MASS.
                  (24) MANCHESTER, N.H. (25) MEMPHIS, TENN. (26) MILWAUKEE, WISC.
                  (27) MINNEAPOLIS, MINN. (28) NEWARK, N.J. (29) PALO ALTO, CALIF.
                  (30) PASADENA, CALIF. (31) PHOENIX, ARIZ. (32) PITTSBURGH, PA.
                  (33) ST. LOUIS, MO. (34) ST. PAUL, MINN. (35) ST. PETERSBURG, FLA.
```

SPSS File for 63-Cities Data (Cont.)

```
                    (36) SALT LAKE, UTAH (37) SAN FRANCISCO, CALIF. (38) SANTA ANA, CALIF.
                    (39) SAN JOSE, CALIF. (40) SANTA MONICA, CALIF. (41) SCHENECTADY, N.Y.
                    (42) SEATTLE, WASH. (43) SOUTH BEND, IND. (44) TAMPA, FLA. (45) TYLER, TEXAS
                    (46) UTICA, N.Y. (47) WACO, TEXAS (48) WARREN, MICH.
                    (49) WATERBURY, CONN. (50) WAUKEGON, ILL. (51) NEW YORK CITY, N.Y.
                    (52) CHICAGO, ILL. (53) LOS ANGELES, CALIF. (54) PHILADELPHIA, PA.
                    (55) DETROIT, MICH. (56) HOUSTON, TEXAS (57) BALTIMORE, MD.
                    (58) DALLAS, TEXAS (59) SAN DIEGO, CALIF. (60) SAN ANTONIO, TEXAS
                    (61) WASHINGTON, D.C. (62) CLEVELAND, OHIO
FREQUENCIES         GENERAL=IDNUMBER

READ INPUT DATA
 0  3  86 4 275 327 38 45 18  —5 70 19 37 48   497 161 200 10 20 0 161  292 750
 1  1 146 5 116 347 11 59 12 —11 84 21 45 29   261 294 316 26 21 0 115  462 663
 2  2  46 4 127 301 12 54  5  —8 38 19 38 31   277 107  98 10 14 3 286  228 333
 3  2 106 4 497 316 17 50 51   2 49 21 45 55  1346 198 236 15 22 1 538 1175 650
 4  4  66 4 117 395 10 70 24   5 72 12 44 55   216 171 164 10 29 3  35   48 592
 5  2  86 4 301 257 22 47 42 —12 60 22 42 44   705 141 136 12 19 2 397  852 588
 6  3  16 3  82 375 23 64  0  62  9 21 38 22   148 123 134  4 10 2 372  576 445
 7  1 176 9 641 309 18 55 16  —8 85 18 45 60  1470 536 554 38 49 2 338  573 725
 8  1 126 4 463 288 32 42 20 —13 93 18 41 40   796 343 367 29 37 0 347  558 867
 9  1 116 8 100 390 17 67  7  —7 87 16 46 75   266 421 451 29 31 3 198  276 500
10  2  66 3 241 327 19 57 30  20 36 21 43 54   566 147 156 11 23 3 336  402 625
11  1  56 5  82 410 40 53  0   0 66 12 40 15   116 232 221 16 21 3 162  154 590
12  3  86 3 101 299 16 51  1  —6 79 18 39 21   209 157 152 15 13 1 124  211 525
13  3  26 3  72 407 44 57  0  14 44 16 39  9    95 138 127  9 18 2 150  409 693
14  2  86 4 393 324 28 50 20  10 49 21 39 40   824 111 124 10 18 3 296  483 675
15  4  16 2  86 435 34 61  1  53 15 16 37 33   174 117 110  6 21 2  68   58 645
16  3  56 3 175 281 49 34 53  —2 65 22 37 65   317 116 115  9 18 0 202  173 675
17  3  76 3  68 299 39 44  7  —6 68 19 37 30   131 119 115 10 17 3 182  304 600
18  3  66 3 108 333 46 39  4  —3 65 16 33 41   209 133 122  9 19 0 194  332 775
19  3 106 3 743 347 28 52 18  57 53 14 40 34  1155  93  91  8 15 0 202  385 900
20  1  56 7  60 362 34 49  4   1 75 12 43 30    98 277 288 25 22 1  96  163 767
21  2  76 4 518 285 12 54 22 163 39 11 37 49   515 119 132 12 16 0 443  855 625
22  4  56 4 359 396 24 56  5   4 56 17 36 44   704 247 260 13 34 3  82  227 475
23  1  86 8  56 324 28 51  1  —3 82 18 41 15    85 345 330 22 21 1 111  123 850
24  1 116 6  88 294 34 44  0  —1 75 18 41 15   199 253 283 22 15 0 320  470 497
25  2 116 7 624 279 21 51 39  25 41 18 41 35   973 288 294 39 21 1 588  776 638
26  3 126 5 717 318 35 45 15  —3 66 20 42 28  1275 215 219 14 37 1 311  321 775
27  3  96 5 434 348 21 55  4 —10 78 18 46 54   994 195 177 12 21 1 183  306 800
28  1 126 7 382 249 37 34 54  —6 81 24 42 83   644 425 451 41 60 1 262  417 913
29  4  26 3  56 526 22 77  3   7 39 13 41 44   161 169 265 13 26 3 279  202 650
30  4  66 4 113 459 18 62 16  —3 61 16 42 63   392 226 206 15 27 3 162  302 550
31  4  56 3 582 325 20 54  5  32 26 18 39 51   968 142 130  9 22 3 295  397 775
32  1 136 4 520 307 21 50 20 —14 84 15 40 55  1187 169 192 14 42 1 165  222 775
33  3 126 8 622 273 28 42 41 —17 85 20 45 74  1199 242 225 22 45 0 254  278 800
34  3 106 6 310 340 25 54  3  —1 72 19 43 49   599 174 214  9 19 1 447  602 850
35  2  46 5 216 318 10 54 15  19 36 11 43 36   395 135 121 17 16 2 235  245 675
```

SPSS File for 63-Cities Data (Cont.)

36	4	96	4	176	326	13	59	1	—7	70	23	42	59	461	136	111	10	18	0	97	124	713
37	4	116	8	716	423	12	62	13	—3	82	16	43	80	1664	674	603	29	42	1	432	537	775
38	4	46	2	157	309	28	45	4	55	28	23	37	36	359	99	103	6	19	3	31	27	650
39	4	76	4	444	339	30	56	2	119	21	23	36	33	705	143	157	6	21	3	207	204	563
40	4	46	3	88	464	21	65	5	6	47	16	44	63	251	183	177	12	29	3	132	89	633
41	1	76	3	78	325	27	55	4	—5	89	21	41	15	185	152	149	12	22	1	59	84	575
42	4	86	4	531	405	19	59	7	—5	66	17	41	59	1360	210	209	19	32	2	530	873	750
43	3	86	2	126	330	32	52	14	—5	70	17	39	44	277	104	111	10	16	0	151	189	700
44	2	66	4	278	278	17	49	20	1	47	18	39	50	614	148	139	15	24	2	303	290	825
45	2	36	5	58	314	22	54	21	13	47	18	40	30	139	88	88	10	13	3	263	382	767
46	1	116	4	92	286	30	48	6	—9	84	18	42	8	166	158	243	13	17	0	378	435	938
47	2	76	4	95	263	17	53	20	—3	52	18	42	40	191	125	199	11	25	3	363	359	325
48	3	16	3	179	366	48	47	0	101	19	23	32	29	278	111	100	5	19	2	139	227	550
49	1	86	7	108	328	43	42	10	1	72	18	42	31	208	351	357	25	26	1	301	408	875
50	3	46	3	65	366	36	50	13	17	53	23	41	34	166	127	115	5	21	0	148	142	767
51	1	186	9	7896	370	21	58	21	2	74	17	0	66	12073	839	838	47	62	0	918	1283	999
52	3	126	6	3367	340	32	48	33	—5	78	20	0	38	6424	183	205	13	47	0	248	352	999
53	4	86	5	2816	395	24	57	18	13	52	18	0	62	5292	189	169	15	39	1	452	796	999
54	1	186	7	1949	302	28	48	34	—3	79	18	41	23	2986	276	299	18	44	0	481	609	999
55	3	126	7	1511	320	36	41	44	—9	84	19	39	84	2546	272	234	17	37	2	381	479	999
56	2	86	4	1233	338	18	54	26	31	35	21	38	49	2252	112	112	8	16	1	365	465	999
57	2	146	8	906	288	26	44	46	—4	75	18	42	69	1539	618	639	41	56	0	609	575	999
58	2	86	4	844	370	20	56	25	24	35	21	42	60	1738	143	158	13	22	2	348	604	999
59	4	66	4	697	352	18	59	8	22	37	17	32	33	1013	137	137	8	20	3	176	184	999
60	2	96	5	654	242	12	51	8	11	46	22	38	42	980	84	88	13	14	3	272	345	999
61	2	146	0	757	384	5	58	71	—1	68	20	48	78	1603	908	910	64	75	0	459	1123	999
62	3	116	3	751	282	38	37	38	—14	86	20	40	59	1333	194	198	21	42	1	371	363	999

FINISH

Glossary of Terms

A

ANOVA. See *analysis of variance.*

A posteriori comparison. See *post hoc comparison.*

A priori comparison. See *planned comparison.*

Alpha. The probability level for rejecting a null hypothesis when it is true, conventionally set at .05 or lower. Symbolized α.

Alpha error. See *Type I error; false rejection error.*

Alternative hypothesis. A secondary hypothesis about the value of a population parameter that often mirrors the research or operational hypothesis; the hypothesis that is accepted when the null hypothesis is rejected. Symbolized H_1.

Analysis of covariance. A multiple regression equation including one or more dummy variables with a continuous independent variable and no interaction terms.

Analysis of variance. A statistical test of the difference of means for two or more groups. Symbolized ANOVA.

Association. See *covariation; measures of association.*

Asymmetric measure of association. A measure of covariation where a conceptual and computational distinction between the independent and dependent variable is required, such as lambda and Somers's d_{yx}.

Asymmetry. A property of a distribution that exhibits skewness.

Average. See *mean; central tendency.*

Average deviation. The mean of the absolute values of the difference between a set of continuous measures and their mean. Symbolized AD.

B

Bar chart. A type of graph for discrete variables in which the numbers or percentages of cases in each outcome are displayed.

Beta. The probability level for failing to reject a null hypothesis when it is false. Symbolized β.

Beta coefficient. A standardized regression coefficient indicating the amount of net change, in standard deviation units, of the dependent variable for an independent variable change of 1 standard deviation. Symbolized β^*.

Beta error. See *Type II error; false acceptance error.*

Beta weight. See *beta coefficient.*

Between sum of squares. A value obtained by subtracting the grand mean from each group mean, squaring these differences for all individuals, and summing them; used in the F test for analysis of variance. Symbolized SS_{BETWEEN}.

Bias. An inaccurate estimate of a population parameter as represented by a sample statistic.

Bivariate association. Covariation between two variables; measures of bivariate association include lambda, gamma, Yule's Q, tau b, tau c, Spearman's rho, phi, eta, the correlation coefficient, and the bivariate regression coefficient.

Bivariate linear relationship. Covariation between two variables which can be represented by a straight line.

Bivariate regression coefficient. A parameter estimate of a bivariate regression equation that measures the amount of increase or decrease in the dependent variable for a one-unit difference in the independent variable.

Box-and-whisker diagram. A type of graph for discrete and continuous variables in which boxes and lines represent central tendency and variability.

C

Causal chain model. See *chain path model.*

Causal diagram. A visual representation of the cause-and-effect relationships between variables, using keyword names and directed arrows.

Causality. A condition which is present when the occurrence of one event or situation is sufficient reason to expect the production of another event or situation. To establish causality, the criteria of covariation, time order, and nonspuriousness must be met.

Cell. The intersection of a row and a column in a crosstabulation of two or more variables. Numerical values contained within cells may be cell frequencies, cell proportions, or cell percentages.

Central limit theorem. A mathematical theorem stating that if repeated random samples of size N are selected from a normally distributed population with mean $= \mu$ and standard deviation $= \sigma$, then the means of the samples will be normally distributed with mean $= \mu$ and standard error $= \sigma/\sqrt{N}$ as N gets large.

Central tendency. A value that describes the typical outcome of a distribution of scores.

Chain path model. A causal model in which variables measured on the same sample at three or more times are depicted as the causes of their own subsequent values.

Chebycheff's inequality. A theorem which states that regardless of the shape of a distribution, the probability of an observation being k standard deviations above (or below) the population mean is less than or equal to $1/k^2$.

Chi-square distribution. A family of distributions, each of which has different degrees of freedom, on which the chi-square test statistic is based.

Chi-square test. A test of statistical significance based on a comparison of the observed cell frequencies of a joint contingency table with frequencies that would be expected under the null hypothesis of no relationship. Symbolized χ^2.

Coefficient of determination. A PRE statistic for linear regression that expresses the amount of variation in the dependent variable explained or accounted for by the independent variable(s) in a regression equation. Symbolized $R^2_{Y \cdot X}$.

Coefficient of nondetermination. A statistic that expresses the amount of variation in a dependent variable that is left *un*explained by the independent variable(s) in a regression equation. Symbolized $1 - R^2_{Y \cdot X}$.

Coefficient of relative variation. A measure of relative dispersion obtained by dividing the standard deviation by the mean. Symbolized CRV.

Column marginals. The frequency distribution of the variable shown across the columns of a crosstabulation.

Concept. Person, object, relationship, or event which is the referent of a social theory.

Concordant pairs. In a crosstabulation of two orderable discrete variables, the number of pairs having the same rank order of inequality on both variables. Symbolized n_s.

Conditional correlation coefficients. Correlation coefficients calculated between two crosstabulated continuous variables within each category of a third variable.

Conditional mean. The expected average score on the dependent variable, Y, for a given value of the independent variable, X.

Confidence interval. A range of values constructed around a point estimate (for example, around a sample mean or around the difference between two sample means). A $1 - \alpha$ percent confidence interval permits an inference that, over repeated samples,

$1 - \alpha$ percent of the intervals will contain the population parameter between the upper and lower confidence limits.

Confidence intervals for mean differences. A confidence interval constructed around the point estimate of the difference between two population means.

Confidence limits. The extreme upper and lower values of a confidence interval.

Consistency. A desirable quality of an estimator achieved when the estimated value approximates the population parameter more closely as the sample size *(N)* gets larger.

Consistent estimator. An estimator of a population parameter that approximates the parameter more closely as N gets large.

Constant. A value that does not change.

Constructs. Unobserved concepts social scientists may use to explain observations.

Contingency coefficient. A measure of association for nonorderable discrete variables based on chi square.

Contingency table. See *crosstabulation.*

Continuous probability distribution. A probability distribution for a continuous variable, with no interruptions or spaces between the outcomes of the variable.

Continuous variable. A variable that, in theory, can take on all possible numerical values in a given interval.

Contrast. A set of weighted population means that sum to zero, used in making post hoc comparisons of treatment groups.

Control variable. A third or additional variable that is held constant in an analysis to observe its effect on the original covariation between two or more variables.

Controlling. See *holding constant.*

Correlated effect. A component in the decomposition of a correlation coefficient that is due to a correlation among predetermined variables.

Correlation coefficient. A measure of association between two continuous variables that estimates the direction and strength of linear relationship. Pearson's product-moment correlation coefficient is symbolized r_{XY} in the sample.

Correlation difference test. A statistical test to determine whether two correlation coefficients differ in the population.

Covariance. The sum of the product of deviations of the Xs and Ys about their respective means, divided by $N - 1$ in the sample or N in the population. Symbolized s_{XY} in the sample.

Covariates. The continuous variables in an analysis of covariance.

Covariation. Joint variation, or association, between a pair of variables.

Cramer's V. A measure of association for discrete variables based on chi square. Symbolized V.

Critical region. See *region of rejection.*

Critical value. The minimum value of a test statistic that is necessary to reject the null hypothesis at a given probability level. Symbolized c.v.

Cronbach's alpha. A measure of internal reliability for multi-item summed indexes.

Cross classification. See *crosstabulation.*

Crosstabulation. A tabular display of the joint frequency distribution of two discrete variables that has r rows and c columns.

Cumulative frequency. For a given score or outcome of a variable, the total number of cases in a distribution at or below that value.

Cumulative frequency distribution. A distribution of scores showing the number of cases at or below each outcome of the variable being displayed in the distribution.

Cumulative percentage. For a given score or outcome of a variable, the percentage of cases in a distribution at or below that value.

Cumulative percentage distribution. A distribution of scores showing the percentage of cases at or below each outcome of the variable being displayed in the distribution.

Curvilinear relationship. Covariation between two variables that is best fit by a curved rather than a straight line.

D

Data. The numbers or scores assigned to outcomes of variables in a given sample or population of observations.

Deciles. The outcomes or scores that divide a set of observations into 10 groups of equal size.

Decomposition. The division of a correlation coefficient into its component parts, involving direct effects, indirect effects, and dependence on common causes.

Deduction. The process of deriving a conclusion about relationships among concepts by logical reasoning about their connections to common concepts.

Degrees of freedom. The number of values free to vary when computing a statistic. Symbolized *df.*

Dependent variable. A variable that has a consequent, or affected, role in relation to an independent variable.

Descriptive statistics. Numbers that describe features of a set of observations; examples include percentages, modes, variances, and correlations.

Deterministic. A causal relationship in which change in one variable always produces a constant change in another variable.

Diagram. A visual representation of a set of data.

Direct effect. A connecting path in a causal model between two variables without an intervening third variable.

Direction of association. For continuous or orderable discrete variables, a pattern of association in which variables either positively covary (high scores coincide with the high scores on both measures and low scores coincide with the low scores on both measures) or negatively covary (high scores on one variable coincide with low scores on another variable, and vice versa).

Discordant pairs. In a crosstabulation of two orderable discrete variables, the number of pairs having reverse rank order of inequality on both variables. Symbolized n_d.

Discrete probability distribution. A probability distribution for a discrete variable.

Discrete variable. A variable that classifies persons, objects, or events according to the kind or quality of their attributes.

Dispersion. See *variation.*

Dummy variable. A variable coded 1 to indicate the presence of an attribute and coded 0 to indicate its absence.

E

Effect. The impact on a dependent variable of being in a certain treatment group. Symbolized α_j.

Efficient estimator. The estimator of a population parameter among all possible estimators that has the smallest sampling variance.

Elaboration. A process of analyzing bivariate crosstabulations within the categories of a third variable to examine the effects of the third variable on the original bivariate relationship.

Empirical probability distribution. A probability distribution for a set of empirical observations.

Epistemic relationships. The relationship between abstract theoretical (unobserved) concepts and their corresponding operational (observed) measurements.

Equiprobable distribution. A probability distribution in which each outcome has an equal chance of occurrence.

Error sum of squares. A numerical value obtained in ANOVA or linear regression by subtracting the between sum of squares or the regression sum of squares, respectively, from the total sum of squares. Symbolized SS_{ERROR}. See *within sum of squares.*

Error term. The difference between an observed score and a score predicted by the model.

Eta squared. A measure of nonlinear covariation between a discrete and a continuous variable; the ratio of SS_{BETWEEN} to SS_{TOTAL}. Symbolized η^2.

Exhaustiveness. A property of a classification system that provides sufficient categories so that *all* observations can be located in some category.

Expected frequency. In a chi-square test, the value that cell frequencies are expected to take, given the hypothesis under study (ordinarily, the null hypothesis).

Expected value. The number that best describes the typical observation in a population or sampling distribution; the expected value equals the mean of the distribution. Symbolized $E(Y)$.

Explanation. Covariation between two variables due to an intervening third variable.

F

F distribution. A theoretical probability distribution for one of a family of F ratios, having $J - 1$ and $N - J$ *df* in the numerator and denominator, respectively.

F ratio. A test statistic formed by the ratio of two mean-square estimates of the population error variance. Symbolized F.

False acceptance error. See *Type II error.*

False rejection error. See *Type I error.*

First-order table. A subtable containing the crosstabulation or covariation between two variables, given a single outcome of a third, control variable.

Frequency. The number of cases in a given outcome of a variable. Symbolized f.

Frequency distribution. A table of the outcomes of a variable and the number of times each outcome is observed in a sample.

Fundamental theorem of path analysis. An equation stating that the correlation between variables i and j implied by a hypothesized causal model is the sum of all products involving both a path from variable q to variable i, and the correlation between variable q and variable j. The sum of these products is formed over all Q variables that have direct paths to variable i.

G

Gamma. A symmetric measure of association for orderable discrete variables that takes into account only the number of untied pairs. Symbolized γ.

Goodness-of-fit test. The chi-square statistic applied to a single discrete variable, with $K - 1$ degree of freedom, where expected frequencies are generated on some theoretical basis.

Grand mean. In the analysis of variance, the mean of all observations. Symbolized μ.

Graph. See *diagram.*

Grouped data. Data that have been collapsed into a smaller number of categories.

Grouping error. Error that arises from use of the midpoint of grouped frequency distributions to represent all scores in the measurement class.

H

Histogram. A type of diagram that uses bars to represent the frequency, proportion, or percentage of cases associated with each outcome or interval of outcomes of a variable.

Holding constant. A statistical technique for observing the effects of a control variable on the original covariation between two variables.

Homoscedasticity. A condition in which the variances of two or more population distributions are equal.

Horizontal axis. In graphing, the horizontal line along which values of an independent variable, usually labeled X, are located; also called the *abscissa.*

Hypothesis. See *operational hypothesis.*

Hypothesis testing. A branch of statistics in which hypotheses are tested to determine whether the observed sample data have been generated by chance from a population in which the null hypothesis is true.

I

Independent random sample. A sample drawn according to random selection procedures, in which the choice of one observation for a sample does not affect the probability of another observation being chosen for a different sample.

Independent variable. A variable that has an antecedent, or causal, role in relation to a dependent variable.

Index. A variable that is a summed composite of other variables that are assumed to reflect some underlying construct.

Index of diversity. A measure of variation for a discrete variable that indicates the likelihood two observations drawn at random from a population are from different categories of the variable. Symbolized D.

Index of qualitative variation. A measure of variation for discrete variables; a standardized version of the index of diversity. Symbolized IQV.

Indicators. Observable measures of underlying unobservable theoretical constructs.

Indirect effect. A compound path connecting two variables in a causal model through an intervening third variable.

Inference. The process of making generalizations or drawing conclusions about the attributes of a population from evidence contained in a sample.

Inferential statistics. Numbers that represent generalizations, or inferences, drawn about some characteristic of a population, based on evidence from a sample of observations from the population.

Interaction effect. A difference in the relationship between two variables within categories of a control variable.

Intercept. A constant value in a regression equation that shows the point at which the regression line crosses the Y axis when values of X equal zero. Symbolized a.

Interpretation. See *explanation.*

Interquartile range. A measure of dispersion indicating the difference in scores between the 25th and 75th percentiles. Symbolized IQR.

Interval level of measurement. Level of measurement or scale that assigns numbers to observations that reflect a constant unit length between categories.

J

Joint contingency table. See *crosstabulation.*

Joint frequency distribution. See *crosstabulation.*

L

Lambda. An asymmetric measure of association for nonorderable discrete variables based on prediction from modes. Symbolized λ.

Least squares. The sum of squared deviations of a set of scores about the mean that is a minimum.

Level of measurement. A classification of measurement scales according to the amount of information recorded about observations; this information generates four types of scales: interval, nominal, ordinal, and ratio.

Level of significance. See *probability level.*

Linear relationship. Covariation in which the value of the dependent variable is proportional to the value of the independent variable.

Listwise deletion. The removal of all cases from a multiple regression analysis that have missing values on *any* of the variables.

Lower confidence limit. The lowest value of a confidence interval.

M

Marginal distributions. The frequency distributions of each of two crosstabulated variables.

Matrix of correlations. A row by column tabular display of correlation coefficients for pairs of variables.

Mean. A measure of central tendency for continuous variables calculated as the sum of all scores in a distribution divided by the number of scores; the arithmetic average. Symbolized $\bar{Y}$ in the sample.

Mean difference hypothesis test. A statistical test of a hypothesis about the difference between two population means.

Mean difference test. A statistical test of whether two means differ in the population.

Mean of a probability distribution. The expected value of a population of scores. Symbolized μ_Y.

Mean square. Estimate of variance used in the analysis of variance.

Mean square between. A value in ANOVA obtained by dividing the between sum of squares by its degrees of freedom. Symbolized $MS_{BETWEEN}$.

Mean square error. A value in linear regression or ANOVA obtained by dividing the error sum of squares (or within sum of squares) by its degrees of freedom. Symbolized MS_{ERROR} or MS_{WITHIN}.

Mean square regression. A value in linear regression obtained by dividing the regression sum of squares by its degrees of freedom. Symbolized $MS_{REGRESSION}$.

Mean square within. See *mean square error.*

Measurement. The assignment of numbers to observations according to a set of rules.

Measurement class. A range of scores on a variable into which observations can be grouped.

Measurement interval. See *measurement class.*

Measures of association. Statistics that show the direction and/or magnitude of a relationship between variables.

Median. The value or score that exactly divides an ordered frequency distribution into equal halves; the outcome associated with the 50th percentile.

Midpoint. A number exactly halfway between the true upper and lower limits of a measurement class or interval, obtained by adding the upper to the lower limit and dividing the total by 2.

Misspecification. A condition in which a structural equation or path model includes incorrect variables or excludes correct variables.

Mode. The value of the response category in a frequency distribution that has the largest number or percentage of cases.

Multicollinearity. A condition of high or near-perfect correlation among the independent variables in a multiple regression equation that leads to unstable estimates of regression coefficients and large standard errors.

Multiple causation. The view that social behavior is caused by more than one factor.

Multiple correlation coefficient. The coefficient for a multiple regression equation that, when squared, equals the ratio of the sum of squares due to regression to the total sum of squares. Symbolized $R_{Y.X}$ for two independent variables.

Multiple regression analysis. A statistical technique for estimating the relationship between a continuous dependent variable and two or more continuous or discrete independent variables.

Multiple regression coefficient. A measure of association showing the amount of increase or decrease in a continuous dependent variable for a one-unit difference in the independent variable, controlling for the other independent variables in the equation.

Multivariate contingency analysis. Statistical techniques for analyzing relationships among three or more discrete variables.

Mutual exclusiveness. A property of a classification system that places each observation in one and only one category of a variable.

N

Negative relationships. A pattern of covariation in which high values of one variable are associated with low values of another variable, and vice versa.

Negative skew. A property of frequency distribution in which the larger frequencies are found toward the positive end and the smaller frequences toward the negative end.

Nominal level of measurement. A level of measurement or scale that assigns a name or number to observations in purely arbitrary sequence.

Nonindependent sample. A sample drawn according tó procedures in which the choice of one observation for a sample determines the probability of another observation being chosen for a different sample.

Nonlinear regression. Regression analysis performed on variables that do not bear a linear relationship to one another.

Nonorderable discrete variable. A discrete measure in which the sequence of categories cannot be meaningfully ordered.

Nonrecursive model. A model in which causal influences may occur in both directions between dependent variables.

Nonspuriousness. Covariation between two variables that is causal and not due to the effects of a third variable.

Normal curve. See *normal distribution.*

Normal distribution. A smooth, bell-shaped theoretical probability distribution for continuous variables that can be generated from a formula. (All bell-shaped distributions are not normal distributions.)

Normal equations. Algebraic equations used in the estimation of linear regression coefficient values.

Null hypothesis. A statement that no relationship exists between variables in a population. Symbolized H_0.

Null hypothesis about a single mean. A null hypothesis that the population mean is equal to or unequal to a specific value.

Number of cases. See *sample size.*

Numbers. Numerical symbols assigned to empirical observations.

O

Observations. The outcomes of specific empirical cases under investigation.

Observed frequencies. The frequencies actually observed in the data in the chi-square test of independence and goodness-of-fit test.

One-tailed hypothesis test. A hypothesis test in which the alternative is stated in such a way that the probability of making a Type I error is entirely in one tail of a probability distribution.

Operation. Any method for observing and recording those aspects of persons, objects, or events that are relevant to testing a hypothesis.

Operational hypothesis. A proposition in which observable, concrete referents or terms are restated to replace abstract concepts.

Order of table. The number of control variables used in forming partial tables or subtables from a zero-order crosstabulation. See *first-order table; zero-order table.*

Orderable discrete variable. A discrete measure in which the categories are arranged from smallest to largest, or vice versa.

Ordinal level of measurement. A level of measurement or scale that assigns numbers to observations in sequence, from lesser to greater amounts of the measured attribute.

Ordinary least squares. A method for obtaining estimates of regression equation coefficients that minimizes the error sum of squares. Symbolized OLS.

Outcome. A response category of a variable.

Outliers. Extreme values of observed variables in a scatterplot that can distort estimates of regression coefficients.

P

PRE measure of association. See *proportional reduction in error.*

p value. The probability of observing a test statistic under the assumption that the null hypothesis is true.

Pairs. The set of all possible pairs of observations in a sample, used in several measures of association for discrete orderable variables. See *tied cases; untied cases; concordant pairs; discordant pairs.*

Pairwise deletion. In multiple regression analysis, the removal of a case from the calculation of a correlation coefficient only if it has missing values for one of the variables.

Parameter. See *population parameter.*

Part correlation. When squared, a measure of the proportion of variance in a dependent variable that an independent variable can explain after controlling for the other independent variable in a multiple regression equation.

Partial correlation coefficient. A measure of association for continuous variables that shows the magnitude and direction of covariation between two variables that remain after the effects of a control variable have been held constant. Symbolized $r_{YX \cdot W}$ in the sample.

Partial regression coefficient. See *multiple regression coefficient.*

Partial table. A subtable containing the crosstabulation between two variables, given an outcome of a control variable, to observe its effects. A partial table is given for each outcome of the control variable.

Path analysis. A method for analyzing correlations that yields empirical estimates of effects in a hypothesized causal system.

Path coefficient. A numerical estimate of the causal relationships between two variables in a path analysis.

Pearson's product-moment correlation coefficient. See *correlation coefficient.*

Percentage. A number created by multiplying a proportion by 100.

Percentage frequency distribution. A distribution of relative frequencies or proportions in which each entry has been multiplied by 100.

Percentaging rule. A rule that provides that in crosstabulation, percentages should be computed within categories of the independent variable.

Percentile. The outcome or score below which a given percentage of observations fall.

Phi. A symmetric measure of association for 2 $\times$ 2 crosstabulations, equivalent to the correlation coefficient. Symbolized ϕ.

Phi adjusted. A symmetric measure of association for 2 $\times$ 2 crosstabulations, in which phi is divided by phi maximum to take into account the largest covariation possible, given the marginals. Symbolized ϕ_{adj}.

Phi coefficient. See *phi.*

Phi maximum. The largest value that phi can attain for a given 2 $\times$ 2 crosstabulation, used in adjusting phi for its marginals. Symbolized ϕ_{max}.

Planned comparison. Hypothesis test of differences between and among population means carried out before doing an analysis of variance.

Point estimate. A sample statistic used to estimate a population parameter, such as a sample mean used to estimate the population mean.

Point estimate for mean differences. The difference between the sample means used to estimate the difference between two population means.

Polygon. A diagram constructed by connecting the midpoints of a histogram with a straight line.

Population. A set of persons, objects, or events

having at least one common attribute to which the researcher wishes to generalize on the basis of a representative sample of observations.

Population parameter. A descriptive characteristic of a population, such as a mean, variance, or correlation; usually designated by a Greek letter.

Population regression equation. A regression equation for a population rather than a sample.

Positive relationship. A pattern of covariation in which high values of one variable are associated with high values of another variable, and low values of both variables are associated.

Positive skew. An asymmetrical frequency distribution characteristic whereby, in a graphic display, larger frequencies are found toward the negative end of the diagram and smaller frequencies toward the positive end.

Post hoc comparison. Hypothesis test of the differences among population means carried out following an analysis of variance.

Power of a test. The probability of rejecting the null hypothesis when it is false and thus should be rejected.

Predetermined variable. A variable in a path model whose causes are not specified.

Prediction equation. A regression equation without the error term, useful for predicting the score on the dependent variable from the independent variable(s).

Probabilistic. A causal relationship in which change in one variable produces change in another variable, with a certain probability of occurrence.

Probability distribution. A set of outcomes, each of which has an associated probability of occurrence.

Probability level. The probability selected for rejection of a null hypothesis, which is the likelihood of making a Type I error. Symbolized α.

Probability theory. That branch of mathematics concerned with the laws of the chance occurrence of events.

Proportion. A number formed by dividing the cases associated with an outcome of a variable by the total number of cases. Symbolized p. See *relative frequency.*

Proportional reduction in error. A characteristic of some measures of association that allows the calculation of reduction in errors in predicting the dependent variable, given knowledge of its relationship to an independent variable. Symbolized PRE.

Proposition. A statement about the relationship between abstract concepts.

Q

Q. See *Yule's Q.*

Quantile. A division of observations into groups with known proportions in each group. See *quartiles, quintiles, deciles,* and *percentile.*

Quartiles. The outcomes or scores that divide a set of observations into four groups of equal size.

Quintiles. The outcomes or scores that divide a set of observations into five groups of equal size.

R

R squared. See *coefficient of determination.*

r-to-Z transformation. A natural logarithm transformation in the value of the correlation coefficient to a Z score, to test the probability of observing r under the null hypothesis.

Random assignment. In an experiment, the assignment of subjects to treatment levels on a chance basis.

Random sample. A sample whose cases or elements are selected at random from a population.

Random sampling. A procedure for selecting a set of representative observations from a population, in which each observation has an equal chance of being selected for the sample.

Range. A measure of dispersion based on the difference between the largest and smallest outcomes in a distribution.

Rank. The position occupied by an observation when scores on some variable have been ordered from smallest to largest (or vice versa).

Ranked data. Orderable discrete measures in which each observation is assigned a number from 1 to *N*, to reflect its standing relative to the other observations.

Ratio level of measurement. A level of measurement or scale that assigns numbers to observations to reflect the existence of a true, or absolute, zero point.

Recursive model. A model in which all the causal influences are assumed to be asymmetric (one-way).

Region of rejection. An area in the tail(s) of a sampling distribution for a test statistic determined by the probability level chosen for rejection of the null hypothesis.

Regression equation. See *regression model.*

Regression line. A line that is the best fit to the points in a scatterplot, computed by ordinary least squares regression.

Regression model. An equation for the linear relationship between a continuous dependent variable and one or more independent variables, plus an error term.

Regression sum of squares. A number obtained in linear regression by subtracting the mean of a set of scores from the value predicted by linear regression, squaring, and summing these values. Symbolized $SS_{\text{REGRESSION}}$.

Regression toward the mean. A condition demonstrated when the predicted scores on the dependent variable show less variability about the mean than the observed scores do, due to the imperfect correlation between two variables.

Relationship. A connection between two concepts or variables, of either a covariational or causal nature.

Relative frequency. A number formed by dividing the cases associated with an outcome of a variable by the total number of cases. See *proportion.*

Reliability. The degree to which different operations of the same concept yield the same results.

Representativeness. The degree to which characteristics of a sample accurately stand for the population from which observations were selected.

Research hypothesis. See *operational hypothesis.*

Residual term. See *error term.*

Residual variable. An unmeasured variable in a path model that is posited as causing the unexplained part of an observed variable.

Rounding. Expressing digits in more convenient and interpretable units, such as tens, hundreds, or thousands, by applying an explicit rule.

Row marginals. The frequency distribution of the variable shown across the rows of a crosstabulation.

S

Sample. A subset of cases or elements selected from a population.

Sample size. The number of cases or observations selected from a population for a specific sample. Symbolized *N.*

Sample statistic. See *statistic.*

Sampling distribution. A theoretical frequency distribution of a test statistic under the assumption that all possible random samples of a given size have been drawn from some population.

Sampling distribution of sample means. The population distribution of all possible means for samples of size *N* selected from a population.

Scatterplot. A type of diagram that displays the

covariation of two continuous variables as a set of points on a Cartesian coordinate system.

Scope conditions. The times, places, or units of analysis for which the propositions of a social theory are expected to be valid.

Scores. Numerical values assigned to empirical observations.

Significance. See *probability level.*

Significance testing with proportions. Using statistical tests to determine whether the observed difference between sample proportions could occur by chance in the populations from which the samples were selected.

Simple random sampling. A procedure for selecting cases from a population in which each observation has an equal chance of being selected.

Skewed distribution. A frequency distribution that is asymmetric with regard to its dispersion. See *positive skew; negative skew.*

Skewness. A property of a frequency distribution that refers to the degree of asymmetry. See *positive skew; negative skew.*

Social theory. A set of two or more propositions in which concepts referring to certain social phenomena are assumed to be causally related.

Somers's d_{yx}. An asymmetric measure of association for two orderable discrete variables that takes into account the numbers of untied pairs and of pairs tied only on the dependent variable. Symbolized d_{yx}.

Spearman's rho. An asymmetric measure of association for two ranked variables. Symbolized ρ_s.

Spuriousness. Covariation between two variables due only to the effect of a third variable.

Standard deviation. The square root of the variance. Symbolized s_Y for a sample.

Standard error. The standard deviation of a sampling distribution of means. Symbolized $\sigma_{\bar{Y}}$.

Standard scores. See *Z scores.*

Standardized regression coefficient. See *beta coefficient.*

Statistic. A numerical characteristic of a sample, usually designated by an italic English letter.

Statistics. The branch of mathematics concerned with describing and inferring something about data.

Statistical independence. A condition of no relationship between variables in a population.

Statistical significance. Significance of a relationship in a statistical sense, as indicated by rejection of a null hypothesis at a particular level. Because results can be due to a large sample size, statistical significance does not necessarily reveal practical importance. See *substantive significance.*

Statistical significance test. A test of inference that conclusions based on a sample of observations also hold true for the population from which the sample was selected.

Statistical table. A numerical display which either summarizes data or presents the results of a data analysis.

Status variable. A variable whose outcomes cannot be manipulated.

Stem-and-leaf diagram. A type of diagram that displays the observed values and frequency counts of a frequency distribution.

Stochastic. See *probabilistic.*

Structural equation. A mathematical equation representing the structure of hypothesized relationships among variables in a social theory.

Substantive significance. The practical importance of a research finding for theory, policy, or explanation, apart from its statistical significance. See *statistical significance.*

Subtable. See *partial table.*

Sum of squares. See *between sum of squares; error sum of squares; regression sum of squares; total sum of squares; within sum of squares.*

Suspending judgment. A position taken by a researcher when the results of a statistical test permit neither clear rejection nor clear acceptance of the null or alternative hypotheses.

Symmetric measure of association. A measure of covariation where the distinction between independent and dependent variables is not required.

Symmetry. A property of a distribution that exhibits no skewness.

T

t distribution. One of a family of test statistics used with small samples selected from a normally distributed population or, for large samples, drawn from a population with any shape.

t scores. See *t variables.*

t test. A test of significance for continuous variables where the population variance is unknown and the sample is assumed to have been drawn from a normally distributed population.

t variables. A transformation of the scores of a continuous frequency distribution derived by subtracting the mean and dividing by the estimated standard error.

Tally. A count of the frequency of outcomes observed for a variable or the frequency of joint outcomes of several variables.

Tau b. A symmetric measure of association for two orderable discrete variables with the same number of categories that takes into account the number of tied cases as well as the number of untied cases. Symbolized τ_b.

Tau c. A symmetric measure of association for two orderable discrete variables with unequal numbers of categories that takes into account only the number of untied pairs. Symbolized τ_c.

Test statistic. A number used to evaluate a statistical hypothesis about a population, calculated from data on a sample selected from a population.

Test of significance. A statistical method to determine the probability of an observed finding in a sample, given a null hypothesis about a population parameter.

Theoretical probability distribution. A probability distribution for a set of theoretical observations.

Theory. See *social theory.*

Third variable. See *control variable.*

Tied cases. In a crosstabulation of two orderable discrete variables, the number of pairs of cases with at least one row or column in common. Symbolized n_s and n_d.

Time order. The condition that changes in a purported independent variable must precede in time changes in the dependent measure, when a causal relationship between the two is assumed.

Total sum of squares. A number obtained by subtracting the scores of a distribution from their mean, squaring, and summing these values. Symbolized SS_{TOTAL}.

Treatment level. A term in experimental research to indicate the experimental group to which a subject has been assigned.

True limits. The exact upper and lower limits of a measurement class or interval into which rounded values are grouped.

Two-tailed hypothesis test. A hypothesis test in which the region of rejection falls equally within both tails of the sampling distribution.

Type I error. A statistical decision error that occurs when a true null hypothesis is rejected; its probability is α.

Type II error. A statistical decision error that occurs when a false null hypothesis is not rejected; its probability is β.

U

Unbiased estimator. An estimator of a population parameter whose expected value equals the parameter.

Unit of analysis. The general level of social phenomena that is the object of observation (e.g., individual, nation).

Universe. See *population.*

Untied cases. In a crosstabulation of two orderable discrete variables, the sum of the number of concordant and discordant pairs. Symbolized $n_s + n_d$.

Upper confidence limit. The highest value of a confidence interval.

V

Validity. The degree to which an operation results in a measure that accurately reflects the concept it is intended to measure.

Variable. A characteristic or attribute of persons, objects, or events that differs in value across such persons, objects, or events.

Variance. A measure of dispersion for continuous variables indicating an average of squared deviations of scores about the mean. Symbolized s_Y^2 in a sample.

Variance of a probability distribution. The expected spread or dispersion of a population of scores. Symbolized σ_Y^2.

Variation. The spread or dispersion of a set of scores around some central value.

Venn diagram. A type of graph that uses overlapping shaded circles to demonstrate relationships or covariation among a set of variables.

Vertical axis. In graphing, the vertical line along which values of a dependent variable, usually labeled *Y*, are located; also called the *ordinate.*

W

Within sum of squares. A value obtained by sutracting each subgroup mean from each observed score, squaring, and summing. Symbolized SS_{WITHIN}.

Working hypothesis. See *operational hypothesis.*

Y

Y intercept. See *intercept.*

Yule's Q. A symmetric measure of association for 2 × 2 crosstabulations, equivalent to gamma. Symbolized *Q*.

Z

Z scores. A transformation of the scores of a continuous frequency distribution by subtracting the mean from each outcome and dividing by the standard deviation. Symbolized *Z*.

Zero-order table. A crosstabulation of two variables in which no additional variables have been controlled.

Z test. A test of significance for continuous variables where the sampling distribution is normally distributed and population variance is known.

List of Mathematical and Statistical Symbols

a	1. Frequency in the upper-left cell of a 2×2 crosstabulation. 2. Intercept term in a regression equation for sample data.
$a < b$	a is less than b.
$a > b$	a is greater than b.
$a \leq b$	a is less than or equal to b.
$a \geq b$	a is greater than or equal to b.
AD	Average deviation.
α	1. Greek letter, lowercase *alpha*. 2. Probability level for Type I error (false rejection error). 3. Cronbach's alpha, a measure of the internal consistency reliability of a set of items.
α_j	The effect of being in group j.
b	1. Frequency in the upper-right cell of a 2×2 crosstabulation. 2. Regression coefficient for sample data.
β	1. Greek letter, lowercase *beta*. 2. Probability of Type II error (false acceptance error). 3. Regression coefficient in a population.
β^*	Beta coefficient; beta weight; standardized regression coefficient for a sample.

c	Frequency in the lower-left cell of a 2×2 crosstabulation.
c_j	Contrast weight for the *j*th group in making multiple comparisons.
c_p	Cumulative frequency up to but not including the interval containing P_i.
$c\%$	Cumulative percentage.
cf	Cumulative frequency.
C	1. The contingency coefficient, a measure of association. 2. Number of columns in a crosstabulation.
CRV	Coefficient of relative variation.
χ^2	Chi square (Greek lowercase *chi,* squared)
χ^2_ν	Chi square with ν degrees of freedom.
c.v.	Critical value.
^	Caret; estimated value of a parameter or variable.
d	Frequency in the lower-right cell of a 2×2 crosstabulation.
d_i	Deviation of *i*th observation from the mean.
d_{yx}	Somers's *d,* a measure of association.
D	Index of diversity.
D_i	Score of the *i*th decile.
D_j	Dummy variable for outcome or group *j.*
D_jX_i	Product term for dummy variable *j* multiplied by continuous variable X_i.
df	Degrees of freedom.
e	A mathematical constant approximately equal to 2.7183.
e_i	Error term for the *i*th case in regression equation.
e_{ij}	Error term for *i*th case in *j*th group in the analysis of variance.
$E(Y)$	Expected value of *Y.*
$E[g(Y)]$	Expected value of function *g* of variable *Y.*
η	Greek letter, lowercase *eta.*
η^2	Eta squared, the correlation ratio.
f	Frequency, count, or tally.
f_i	Frequency of cases of type Y_i.
f_p	Frequency in the interval containing the *i*th percentile.
$\hat{f}_{ij}$	Expected frequency in row *i* and column *j* under

null hypothesis of independence in a crosstabulation.

F — F ratio.

F_{ν_1,ν_2} — F ratio with ν_1 and ν_2 degrees of freedom.

G — Value of gamma in a sample.

GSS — General Social Survey.

γ — 1. Greek letter, lowercase *gamma*.
2. A measure of association in a population.

H_0 — Null hypothesis.

H_1 — Alternative hypothesis.

H1 — A numbered hypothesis derived from the first proposition of a social theory.

i — A generic subscript value that may be replaced by actual numeric values.

IQR — Interquartile range.

IQV — Index of qualitative variation.

$\neq$ — Inequality sign ("does not equal").

∞ — Infinity.

j — A generic subscript value that may be replaced by actual numeric values.

J — Number of subgroups or treatment groups in the analysis of variance.

k — Number of independent variables in a multiple regression equation.

K — Number of outcomes or categories in the distribution of a variable.

K_i — Score of the ith quintile.

L_p — True lower limit of the interval containing the ith percentile.

λ — 1. Greek letter, lowercase *lambda*.
2. A measure of association in a population.

m — The smaller of the number of rows or columns in a crosstabulation.

Min(i,j) — The lower of two numbers, i or j.

μ — Greek letter, lowercase *mu*.

μ_Y — Population mean of variable Y.

$\mu_{\bar{Y}}$ — Mean of the sampling distribution of means for variable Y.

μ_{Y_0} — Population mean for variable Y under the null hypothesis.

μ_{Y_1}	Population mean for variable Y under the alternative hypothesis.
$\mu_{(\bar{Y}_2 - \bar{Y}_1)}$	Difference between the means of variables Y_1 and Y_2 in a population.
n_j	Number of observations in jth group.
n_d	Number of discordant untied pairs in computing association with discrete variables.
n_s	Number of concordant untied pairs in computing association with discrete variables.
N	Total number of observations in a sample.
N_i	Number of cases in subsample i.
$N \rightarrow \infty$	N (sample size) goes to infinity.
ν	1. Greek letter, lowercase *nu*. 2. Degrees of freedom for a χ^2 or t test of significance.
ν_1, ν_2	Number of degrees of freedom in an F test.
p	1. Proportion. 2. Probability of an event.
p_i	1. Proportion of cases in the ith category. 2. ith percentile written as a proportion.
p_{ij}	Path coefficient for direct effect of variable j on variable i.
$p(Y_i)$	Probability of outcome i for variable Y.
$p(\lvert Z \rvert \geq k)$	Probability that the absolute value of Z is equal to or greater than some number, k.
$p(a \leq Y_i \leq b)$	Probability that outcome i of variable Y lies between values a and b, inclusive.
P_i	Score of the ith percentile.
P1	A numbered proposition from a social theory.
ϕ	1. Greek letter, lowercase *phi*. 2. Phi, a measure of association in a 2×2 crosstabulation that equals the correlation coefficient.
ϕ_{adj}	Phi adjusted for marginal distributions.
ϕ_{max}	Maximum value of ϕ, given marginal distributions.
π	1. Greek letter, lowercase *pi*. 2. A mathematical constant approximately equal to 3.14159.
ψ	1. Greek letter, lowercase *psi*. 2. A contrast among J population means.
%	Percent.

%ile	Percentile.
q	Probability of events not defined by p; $q = 1 - p$.
Q	Yule's Q, a measure of association in a 2 × 2 crosstabulation.
Q_i	Score of the ith quartile.
r	Correlation coefficient in a sample.
$\bar{r}$	The mean intercorrelation among a set of k items in a sample.
r_s	Spearman's rho for a sample.
r_{XY}	Correlation coefficient between variables X and Y in a sample.
$r_{XY \cdot W}$	Partial correlation coefficient between variables X and Y, controlling for variable W.
$r_{Z_X Z_Y}$	Correlation coefficient between standardized scores of variables X and Y.
r'_{ij}	Implied correlation coefficient between variables i and j in a sample when decomposing correlations in path analysis.
$r^2_{YX_2 \cdot X_1}(1 - r^2_{X_1X_2})$	Part correlation squared.
R	Number of rows in a crosstabulation.
R^2	R squared, the sample coefficient of determination in a regression analysis.
R^2_i	Coefficient of determination for multiple regression equation i for a sample.
$R^2_{Y \cdot X}$	Coefficient of determination for the sample regression of variable Y on variable X.
$R^2_{Y \cdot X_1X_2}$	Coefficient of determination for the sample multiple regression of variable Y on variables X_1 and X_2.
$1 - R^2_{Y \cdot X}$	Coefficient of nondetermination.
ρ	Greek letter, lowercase *rho*.
ρ_s	Spearman's rho, a measure of association for a population.
ρ_{XY}	Correlation coefficient between variables X and Y in the population.
ρ'_{XY}	Implied correlation coefficient for variables X and Y in the population when decomposing correlations in path analysis.
ρ^2_i	Coefficient of determination for multiple regression equation i for a population.

$\rho^2_{Y \cdot X}$	Coefficient of determination for the regression of variable Y on X in the population.
$\rho^2_{Y \cdot X_1 X_2}$	Coefficient of determination for the multiple regression of variable Y on variables X_1 and X_2, for a population.
s_b	Estimated standard error for a regression coefficient.
s_p	Estimated standard error of sampling distribution of proportions.
$s_{(p_2 - p_1)}$	Estimated standard error of sampling distribution of the difference between two proportions.
s_Y	Sample standard deviation of variable Y.
s_{YX}	Covariance between variables X and Y.
s^2_Y	Variance of variable Y in a sample.
$s^2_{Z_Y}$	Variance of Y in standard-score (Z-score) form in a sample.
σ	Greek letter, lowercase *sigma*.
σ_b	Standard error for the regression coefficient, b.
σ_Q	Standard error for the sampling distribution of Yule's Q.
σ_Y	Standard deviation of variable Y in the population.
$\sigma_{\bar{Y}}$	Standard error of the sampling distribution of sample means for variable Y.
$\sigma_{(\bar{Y}_2 - \bar{Y}_1)}$	Standard error of the sampling distribution for difference between two means, Y_2 and Y_1.
$\sigma_{d_{yx}}$	Standard error for Somers's d_{yx}.
σ_{r_s}	Standard error for Spearman's rho.
σ_{τ_b}	Standard error of tau b.
σ^2_e	Error variance in a population.
σ^2_Y	Variance of variable Y in a population.
$\sigma^2_{\bar{Y}}$	Variance of the sampling distribution of sample means for variable Y.
σ^2_Z	Variance of a Z score.
$\sigma^2_{\chi^2}$	Variance of a chi-square variable.
σ^2_ψ	Variance of a contrast, ψ, in a population.
Σ	1. Greek letter, uppercase *sigma*. 2. Summation sign.
$\sum_{i=1}^{N} Y_i$	Sum of all scores on variable Y for observations from $i = 1$ to N.

$\sum_{j=1}^{J}\sum_{i=1}^{N} Y_{ij}$	Sum of all scores on variable Y for observations from $i = 1$ to N, for groups from $j = 1$ to J.
$\sqrt{}$	Square root, or radical sign.
t	Student's t.
t_b	Tau b for a sample.
t_c	Tau c for a sample.
t_{N-2}	t test with $N-2$ degrees of freedom.
t_α	Critical value of t for probability level α, one-tailed test.
$t_{\alpha/2}$	Critical value of t for probability level α, two-tailed test.
t_ν	t score with ν degrees of freedom.
T	Number of observations in a population.
T_c	Number of ties for the column variable in computing measures of association for discrete variables.
T_r	Number of ties for the row variable in computing measures of association for discrete variables.
τ	Greek letter, lowercase *tau*.
τ_b	Tau b, a measure of association in an $R \times C$ crosstabulation.
τ_c	Tau c, a measure of association in an $R \times C$ crosstabulation when $R = C$.
θ	1. Greek letter, lowercase *theta*. 2. A general notation to designate any population parameter.
2×2	Crosstabulation of two dichotomous variables, creating a table with four cells, *a, b, c,* and *d*.
V	Cramer's V, a measure of association.
W_i	Width of the interval containing P_i, the ith percentile score.
X_i	Score or value of ith observation on variable X.
X_{1i}	Score or value of ith observation on variable X_1.
Y_i	Score of the ith observation on variable Y.
$\hat{Y}_i$	Expected score at the ith observation on variable Y; used in regression analysis.
$\overline{Y}$	Mean of variable Y in a sample.
$\overline{Y}_j$	Mean of the jth group on variable Y in a sample.
$\lvert Y \rvert$	Absolute value of variable Y.
Z_i	Standard score or Z score of the ith observation.

$Z_{(p_2 - p_1)}$	Z score of the difference between two populations, p_2 and p_1.
Z_r	Correlation coefficient r transformed to a Z score.
Z_α	Critical value of Z score for probability level α, one-tailed test.
$Z_{\alpha/2}$	Critical value of Z score for probability level α, two-tailed test.
Z_0	Z score for the null hypothesis test statistic.
Z_1	Z score for the alternative hypothesis test statistic.

Answers to Problems

Answers to Chapter 1 Problems

1. These three propositions do not form a theory because they do not have a concept in common. In fact, the second statement is not a proposition but an operational definition of intergroup conflict.

2. (*a*) concepts are interpersonal attraction, intragroup conflict; relationship is inverse; (*b*) concept is intragroup conflict; no relationship; (*c*) concepts are group orientation and problem solution; relationship is positive covariation of task orientation and problem solving.

3. Several versions are possible. Examples are: "The greater the proportion of persons in a group who are close friends, the lower the frequency of thefts within the group"; "The more frequently people in a group visit one another socially, the fewer the psychosomatic symptoms the group members will exhibit."

4. Units of analysis appear to be societies or their economies, and scope conditions are those societies with no-growth or slow-growth economies that are undergoing a transition to self-sustaining growth during a 20–30-year period.

5. (*a*) local autonomy and democracy in the first proposition, and occupational identification, interest in union affairs, and participation in union affairs in the second; (*b*) in the first proposition, the union local; in the second, the individual worker.

6. Examples are: (*a*) total annual family income, (*b*) average of number of years of formal education of mother and father, (*c*) the occupational prestige score of the head of household, and (*d*) subjective

placement on a five-category scale, from "lower class" to "upper class."

7. For example, faculty members who belong to the local faculty club, participate in graduation and other ceremonies, and concentrate heavily on teaching would have a local orientation; those who change jobs frequently, participate in meetings of their professional societies, and publish numerous research articles would have a cosmopolitan orientation.

8. (*a*) variable; (*b*) variable; (*c*) constant; (*d*) variable; (*e*) constant; (*f*) variable; (*g*) constant; (*h*) constant.

9.

	Independent Variable	*Dependent Variable*
(*a*)	Television watching	Academic performance
(*b*)	Region of residence	Toleration of interpersonal violence
(*c*)	Gender	Earnings

Only region of residence and gender are status variables; the others can change for an individual.

10. "Proof" is too strong a conclusion, even though the relationship was found 50 times. Instead, the scientist should conclude that she could strongly reject the null hypothesis of no gender differences in sexual arousal by visual material. The possibility that some women will be more aroused than some men remains.

11. A valid measure is one whose observational procedures accurately capture the concept intended to be measured. A reliable measure points to consistency of results on repeated application. It is possible to have a reliable measure that is not valid: it could give the same results time after time but not measure the concept it was intended to measure.

12. (*a*) orderable discrete; (*b*) nonorderable discrete; (*c*) continuous; (*d*) continuous; (*e*) orderable discrete; (*f*) nonorderable discrete; (*g*) continuous; (*h*) orderable discrete.

13. (*a*) Both mutual exclusiveness (doctors can be coded in categories 1 and 3, and teachers in categories 2 and 3) and exhaustiveness (no categories for farmers, police officers, etc.) have been violated; (*b*) Increase the number of categories to cover other broad occupational groups (such as service workers, transportation, etc.), eliminate categories 1 and 2, and add a category for "no occupation."

14. (*a*) ratio; (*b*) nominal; (*c*) ratio; (*d*) ratio; (*e*) ordinal; (*f*) nominal; (*g*) ratio; (*h*) ordinal.

15. (*a*) values or beliefs; (*b*) variable; (*c*) operationalization; (*d*) reject;

(*e*) valid; (*f*) other or unclassifiable; (*g*) covariation or relationship; (*h*) inferences, population.

Answers to Chapter 2 Problems

1. *a.* Tally:

Activity	*Tally*	*Frequency*
Talking (T)	IIII IIII II	12
Talking and kissing (TK)	IIII I	6
Talking and holding hands (HT)	IIII	4
Kissing and holding hands (HK)	IIII	5
Other (O)	III	3
Total		30

b. and *c.* Table of relative frequencies and percentages:

Activity	*f*	*p*	*%*
T	12	0.400	40.0%
TK	6	0.200	20.0
HT	4	0.133	13.3
HK	5	0.167	16.7
O	3	0.100	10.0
Total	30	1.000	100.0%

d. Graph not shown.

2.

Region	*f*	*p*	*%*
Northeast	15	0.238	23.8%
Midwest	18	0.286	28.6
South	16	0.254	25.4
Far West	14	0.222	22.2
Total	63	1.000	100.0%

Graph not shown.

3.

Class	*f*	*p*	*%*
Senior	12	0.316	31.6%
Junior	17	0.447	44.7
Sophomore	6	0.158	15.8
Special	3	0.079	7.9
Total	38	1.000	100.0%

Histogram not shown.

4.

Age	f	p	%	Age	f	p	%
16	3	.048	4.8%	96	3	.048	4.8
26	2	.032	3.2	106	3	.048	4.8
36	1	.016	1.6	116	6	.095	9.5
46	5	.079	7.9	126	6	.095	9.5
56	5	.079	7.9	136	1	.016	1.6
66	6	.095	9.5	146	3	.048	4.8
76	5	.079	7.9	176	1	.016	1.6
86	11	.175	17.5	186	2	.032	3.2
				Total	63	1.001*	100.1%*

*Totals exceed 1.000 and 100.0% due to rounding
Graphs not shown.

5. Graphs not shown. A noticeable increase in single-adult households, little change in dual-adult households, and a drop in households with three adults.

6. 727, 389, 278, 44, and 41 (does not add to 1,480 due to rounding).

7. Graph not shown.

8.

No. of Reform Characteristics	f	c %
None	18	28.6%
One	16	54.0
Two	11	71.4
Three	18	100.0
Total	63	100.0%

Graph not shown.

9. (a) 1940, $p = 0.970$; 1950, $p = 0.960$; 1960, $p = 0.944$; 1970, $p = 0.927$; (b) Graph not shown. The proportion of whites among graduates gradually declined over 30 years. Alternatively, the proportion of nonwhites gradually increased.

10.

No. of Crimes	f	%
0–5,000	19	30.2
5,001–10,000	8	12.7
10,001–15,000	7	11.1
15,001–20,000	4	6.3
20,001–25,000	4	6.3
25,001–30,000	6	9.5
30,001–35,000	2	3.2
35,001–40,000	1	1.6
40,001–45,000	1	1.6
45,001–50,000	2	3.2
50,001–55,000	1	1.6
55,001–60,000	3	4.8
60,001 and over	5	7.9
Total	63	100.0%

By lumping together cities with 60,001 or more crimes, nonadjacent categories with a single frequency are compressed into what would have been a very flat distribution beyond the "60,001 or more" category. Graph not shown.

11. (*a*) 100,000; (*b*) 1 to 100,500, 100,500 to 200,500, 200,500 to 300,500, 300,500 to 400,500; (*c*) into the under 100,000 interval; (*d*) into the 101,000 to 200,000 interval.

12. *a*. Interval width is $5,000.

b.

True Limits	*Midpoints*
$ 5,500–$10,500	$8,000
$10,500–$15,500	$13,000
$15,500–$20,500	$18,000
$20,500–$25,500	$23,000

13.

True Limits	***Midpoints***	***Proportion***
0.5–9.5	4.5	0.021
9.5–19.5	14.5	0.050
19.5–29.5	24.5	0.079
29.5–39.5	34.5	0.119
39.5–49.5	44.5	0.530
49.5–59.5	54.5	0.086
59.5–69.5	64.5	0.064
69.5–79.5	74.5	0.023
79.5 or more	—	0.027
Total		1.000

14.

Interval	***True Limits***	***Midpoint***	***f***	***cf***	***%***	***c%***
1–200	0.5–200.5	100.5	14	14	22.2%	22.2%
201–400	200.5–400.5	300.5	37	51	58.7	80.9
401–600	400.5–600.5	500.5	4	55	6.3	87.3
601–800	600.5–800.5	700.5	4	59	6.3	93.7
801–1000	800.5–1000.5	900.5	2	61	3.2	96.8
1001–1200	1000.5–1200.5	1100.5	0	61	0	96.8
1201–1400	1200.5–1400.5	1300.5	1	62	1.6	98.4
1401–1600	1400.5–1600.5	1500.5	0	62	0	98.4
1601–1800	1600.5–1800.5	1700.5	1	63	1.6	100.0
Total			63	63	99.9%*	100.0%

*Total does not add to 100.0% due to rounding

15. Score for 30th percentile is 227; score for 90th percentile is 685.5.

16. Score for 80th percentile is 27.25.

17. 2nd, 5th, and 7th deciles; 1st, 2nd, and 3rd quartiles; 1st, 3rd, and 4th quintiles.

18. (*a*) 3.85; (*b*) 7.69; (*c*) 8.38.

19. The 50th percentile for neighbors is 3.74, and for friends it is 4.02,

suggesting that people are likely to spend a little more time with friends than with neighbors.

20. Graph not shown. The line for visiting with friends lies below that for neighbors at the three lowest levels of visiting, crosses at "about once a month," and is above the line for neighbors at the three higher levels.

21. Fiftieth percentile falls into the fifth category. Histogram not shown.

22.

Category	**HOMOSEX**	**XMARSEX**
Always wrong	73.3%	70.5%
Almost always wrong	6.0	15.9
Sometimes wrong	6.1	9.9
Not wrong at all	14.6	3.7
Total	100.0%	100.0%
N	1,397	1,444

Very similar distributions, although homosexuality is less condemned than extramarital sexual relations.

23. Fiftieth percentile for young men is 1.88; young women, 1.36; older men, 1.99; and older women, 1.61. This suggests age of first marriage is lower among those who were born after World War II, and the difference between men and women is slightly larger than it is for those born before the war. Of course, late marriers (those who first marry after age 36) have not had a chance to enter into the married population and raise the 50th percentile among the recent generation.

24. Polygon not shown. Distribution is skewed to the right, and its peak is at the extreme left (no organizational memberships).

25.

Category	**INCOM16**	**FINRELA**
Far below average	0.080	0.048
Below average	0.233	0.235
Average	0.516	0.525
Above average	0.158	0.169
Far above average	0.013	0.023
Total	1.000	1.000
N	1,455	1,454

There is virtually no difference in these distributions, except about 3% fewer rate their current family incomes as "far below average."

26.

No. of Functions	f	p	c%
0	0	0.000	0.0%
1	0	0.000	0.0
2	3	0.048	4.8
3	15	0.242	29.0
4	20	0.323	61.3
5	8	0.129	74.2
6	3	0.048	79.0
7	6	0.097	88.7
8	5	0.081	96.8
9	2	0.032	100.0
Total	62	1.000	100.0%

Histogram not shown.

27.

Birth Rate per 1,000	f	%
11	2	3.2%
12	3	4.8
13	1	1.6
14	1	1.6
15	1	1.6
16	7	11.1
17	5	7.9
18	16	25.4
19	5	7.9
20	5	7.9
21	8	12.7
22	3	4.8
23	5	7.9
24	1	1.6
Total	63	100.0%

$P_{25} = 16.65$; $P_{50} = 18.22$; $P_{75} = 20.66$.

28. The true interval limits are 500 above and 500 below every 100,000, i.e., 100,500–200,500; 200,500–300,500; etc. Midpoint values are 50,500; 150,500; 250,500; 350,500; etc.

Category	c%	Category	c%
0–100	25.4%	801–900	88.9
101–200	44.4	901–1,000	90.5
201–300	50.8	1,201–1,300	92.1
301–400	58.7	1,501–1,600	93.7
401–500	65.1	1,901–2,000	95.2
501–600	71.4	2,801–2,900	96.8
601–700	79.4	3,301–3,400	98.4
701–800	87.3	7,801–7,900	100.0

50th percentile is 288,000.

29. Detroit, 92nd; Euclid, 4th; Memphis, 87th; Washington, 100th; and Palo Alto, 17th.

30.

Percent Female	*f*	%	c%
32	2	3.3%	.033
33	1	1.7	.050
34	0	0.0	.050
35	0	0.0	.050
36	2	3.3	.083
37	6	10.0	.183
38	4	6.7	.250
39	7	11.7	.367
40	5	8.3	.450
41	9	15.0	.600
42	10	16.7	.767
43	5	8.3	.850
44	2	3.3	.883
45	4	6.7	.950
46	2	3.3	.483
47	0	0.0	.983
48	1	1.7	1.000
Total	60	100.0%	1.00

In the 45th percentile, 40% of the labor force is female.

Answers to Chapter 3 Problems

1. (*a*) 18 births per 1,000; (*b*) three modes: 18%, 24%, 48%; (*c*) no single modal value exists.
2. (*a*) $15.5 billion; (*b*) 23.5%; (*c*) 18.21 births per 1,000.
3. Mode = 4 (every day); median = 3.63.
4. (*a*) a slight positive skew; (*b*) a negative skew.
5. (*a*) $31.06 billion; (*b*) 24.5%; (*c*) 18.7 births per 1,000.
6. 2.78.
7. A positive skew.
8. (*a*) 6; (*b*) 5.5; (*c*) 5.14; (*d*) no skew.
9. (*a*) D = 0.643; (*b*) IQV = 0.803.
10. (*a*) 0.613; (*b*) 0.765.
11. (*a*) $109 billion; (*b*) 43%; (*c*) 7 births per 1,000.
12. $24,300.
13. IQR = 3.0.

14. AD = 9.51.

15. (*a*) 164.78; (*b*) 12.84.

16. (*a*) 37.34; (*b*) 14.13; (*c*) 1.96.

17. (*a*) +1.75; (*b*) −0.75; (*c*) +0.67; (*d*) −1.33; (*e*) +1.00; (*f*) −0.80.

18. (*a*) 16; (*b*) 9.34; (*c*) 21; (*d*) 10.5.

19. Lester is further from the mean, since the Z score for GPA = 1.65 is only −1.42.

20. (*a*) −0.59; (*b*) +1.82; (*c*) +0.69; (*d*) −0.72; (*e*) −0.30.

21. Ireland and Portugal, both above.

22. (*a*) 2; (*b*) 2.37; (*c*) 7; (*d*) 1.31; (*e*) 2.63; (*f*) 1.00; (*g*) 1.00; (*h*) $Z_1 = -1.63$.

23. CRV = 0.196 for both regions; hence, no difference in relative variation.

24. The distribution is skewed strongly to the right by the presence of the $30,000 income, which raises the mean to $8,000. This is well above the mode and median, at $5,000, which are more realistic central tendency statistics to represent "typical" incomes in this town.

25. In the two departments, the means and ranges are identical, but the standard deviation in the chemistry department ($s = 3.70$) is greater than it is in the biology department ($s = 2.97$), indicating a greater dispersion about the mean in the former. Biology junior faculty publication varies little about the median value of 6 (with one exception), whereas chemists' publication is spread widely across the range.

26. From 1972 to 1980, the mean ideal family size decreased by one-quarter child (from 2.88 to 2.63), and the dispersion about the mean also shrank (the standard deviation decreased from 1.20 to 1.00). However, the coefficients of relative variation are about the same (0.42 and 0.38, respectively), suggesting little difference in relative dispersion.

27. *a.*

	1972	*1975*	*1978*
Married	.719	.672	.627
Widowed	.086	.097	.100
Divorced	.040	.056	.088
Single	.024	.033	.031
Never married	.130	.142	.155
N	1,613	1,490	1,531

There appears to be a decrease in those who are married, an increase in those divorced, and a slight increase in those responding "never married."

b. $D_{72} = 0.457$, $IQV_{72} = 0.571$; $D_{75} = 0.515$, $IQV_{75} = 0.643$; $D_{78} = 0.564$, $IQV_{78} = 0.705$. The amount of variation appears to be increasing over time, probably because more individuals are opting for divorce or staying single.

28. *a.* $\overline{Y}_{72} = 0.791$, $s_{72} = 0.407$; $\overline{Y}_{75} = 0.837$, $s_{75} = 0.370$; $\overline{Y}_{78} = 0.832$, $s_{78} = 0.374$. Between 1972 and 1975 there appeared to be a small increase in support for abortion when a woman had been raped but virtually no change occurred between 1975 and 1978. There was a small decrease in variation between 1972 and 1975 but virtually no change between 1975 and 1978.

b. $p_{72} = .791$; $p_{75} = .837$; $p_{78} = .832$. When a dichotomy is coded 0 and 1, the mean equals the proportion of cases coded 1.

c. $D_{72} = 0.331$, $IQV_{72} = 0.661$; $D_{75} = .273$, $IQV_{75} = .546$; $D_{78} = .280$, $IQV_{78} = .559$. The conclusion about variation is the same —there appears to have been a small decrease between 1972 and 1975 but little change between 1975 and 1978.

29. *a.*

Discipline	***1952***	***1972***
Social sciences	20.0%	17.8%
Humanities	11.5	12.7
Natural sciences	49.4	44.0
Education and other	19.1	25.5
N	7,683	24,600

Yes. Relatively, there were slightly fewer in the social sciences and natural sciences and more in education and other in 1972.

b. $D_{52} = 0.666$ and $D_{72} = 0.694$. There appears to be slightly more variation in the 1972 distribution.

30. *a.* Modal cause of deaths for males is heart disease; for females, it is cancer.

b. $D_m = 0.586$ and $D_f = 0.637$, suggesting more variability in the causes of death in women among persons aged 45 to 64, compared to men.

31. (*a*) mean = 39.53, mode = 50.00, median = 39.82, range = 70.00, variance = 184.43, standard deviation = 13.58; (*b*) *Z* scores are (1) +3.13, (2)+2.69, (3) −0.55, (4)−0.18, and (5) −1.29.

32. (*a*) 0; (*b*) 1.13; (*c*) 16; (*d*) 1.59; (*e*) 3.26; (*f*) 1.81; (*g*) 1.14; (*h*) 1.88.

33. Confidence in the medical profession is much higher than confidence in Congress. Of respondents, 52.3% have a great deal of confidence in medicine, compared to only 9.3% with a great deal of

confidence in Congress. On a three-point scale with 1 = highest and 3 = lowest, the means are 1.54 for medicine and 2.25 for Congress.

34. Black men, CRV = 0.70; white men, CRV = 0.54; black women, CRV = 0.75; white women, CRV = 0.73. White men show the lowest dispersion relative to this mean income, but the other three groups have approximately the same relative variation.

35. Satisfaction with family life has a slightly higher mean (1.99) than satisfaction with friendships (2.09), on a seven-point scale with 1 = highest and 7 = lowest. The variances are 1.65 for family life and 1.35 for friendship. The corresponding CRVs are 0.65 and 0.55.

36. (*a*) 44; (*b*) 28; (*c*) 23.75; (*d*) 25.49; (*e*) 106.35; (*f*) 10.31.

37.

No. of Functions	*Z Score*
2	−1.46
3	−0.91
4	−0.36
5	+0.19
6	+0.73
7	+1.28
8	+1.83
9	+2.37

38. For example:

Category	*%*
0–11%	44.4%
11–20	25.4
21–30	11.1
31–40	6.3
41–50	6.3
51 and over	6.3
Total	99.8%

39.

Midwest	0.11
South	0.13
Northeast	0.13
Far West	0.16

40.

REFORMGV	***Standard Deviation***	***Mean***	***CRV***
None	2.09	4.88	0.43
Any one	1.78	5.38	0.33
Any two	2.01	4.36	0.46
All three	1.28	4.00	0.32

There is no apparent pattern of relative variation with increasing reform.

Answers to Chapter 4 Problems

1. Raw-Data Table:

	Major	*Nonmajor*	*Total*
Senior	25	5	30
Junior	15	10	25
Soph	10	15	25
Total	50	30	80

Percentage Table:

	Major	*Nonmajor*	*Total*
Senior	50.0	16.7	37.5
Junior	30.0	33.3	31.3
Soph	20.0	50.0	31.3
Total	100.0	100.0	100.0

Although more sociology majors than nonmajors enroll in introductory statistics, they tend to wait until their senior year, while a majority of nonmajors enroll as sophomores.

2.

	Religiosity		
Vote	***Fundamentalist***	***Liberal***	***Total***
Reagan	56.7	55.0	56.0
Carter	14.7	35.0	22.8
Nonvoter	28.7	10.0	21.2
Total	100.0%	100.0%	100.0%
N	150	100	250

Neither religious group was more likely to vote for Reagan; however, Carter voters were more than twice as prevalent among liberals, while nonvoting was about three times as great among fundamentalists.

3. 0.0077, or less than 1 chance in 100.

4. *a.* Young people are not more sexually permissive than older people.
   b. Marital status is unrelated to drinking behavior.
   c. Southern and non-Southern residents own the same number of firearms.
   d. Average house prices are the same in suburbs and the inner city.

5. H_0: Early-release and full-term prisoners do not differ in their rearrest rates.
   H_1: Early-release prisoners have a lower rate of rearrest than do full-term prisoners.

   If the H_0 is rejected, the chances are 1 in 100 that the two sets of prisoners actually do *not* differ in their rearrest rates. If H_0 is not rejected, the program may actually lower rearrest rates between the groups, even though the alternative is not accepted.

6. 9.38.

7.

Vote	Religiosity	
	Fundamentalist	*Liberal*
Reagan	84.0	56.0
Carter	34.2	22.8
Nonvoter	31.8	21.2

8. 11.03 and 20.77.

9. (*a*) 6; (*b*) 12; (*c*) 6.

10. (*a*) 20.09; (*b*) 3.84; (*c*) 46.797; (*d*) 15.51; (*e*) 34.81.

11. (*a*) 15.51; (*b*) 20.09; (*c*) 26.13.

12. Because the observed χ^2 is 2.14, and the critical value at $\alpha = .01$ is 6.63, the conclusion is that probation had no effect on recidivism.

13. Raw-Data Table:

	Undergrad	*Grad*	*Total*
Excellent	15	5	20
Good	45	15	60
Poor	90	30	120
Total	150	50	200

Percentage Table:

	Undergrad	*Grad*	*Total*
Excellent	10.0	10.0	10.0
Good	30.0	30.0	30.0
Poor	60.0	60.0	60.0
Total	100.0	100.0	100.0

Because the percentage distributions are the same for both categories of class level, chi square must be zero, and the null hypothesis of no difference of opinion can't be rejected.

14. For $df = 12$, the critical value for $\alpha = .025$ is 23.34; hence the chance of Type I error is almost exactly .025.

15. If you applied traditional α levels (e.g. .05 or .01) you could not reject the hypothesis of no difference between the number of birth defects among women taking the drug and those not taking it. But the critical factor is the probability of Type II error, since letting a truly harmful drug onto the market could cause a tragedy (recall the Thalidomide scandal). Information in this problem is insufficient to calculate probability of Type II error. Best decision is not to release the drug but to conduct more tests.

16. Independence Data:

	Traditional	*Open*	*Total*
High	10	10	20
Medium	10	10	20
Low	10	10	20
Total	30	30	60

$\chi^2 = 0.00$

Maximum Relationship Data:

	Traditional	*Open*	*Total*
High	20	0	20
Medium	10	10	20
Low	0	20	20
Total	30	30	60

$\chi^2 = 40.00$

17. $\chi^2 = 271.23$, $df = 2$, so a real change occurred.

18. $\chi^2 = 1.88$, $df = 1$, H_0 can be rejected only at $\alpha > .10$.

19. Chi-square goodness of fit is 2.70 for $df = 3$. Hence you cannot reject the null hypothesis of equal geographic distribution.

20. χ^2 goodness of fit is 33.33 for $df = 3$. Reject the hypothesis that the coins are fair (i.e., the observed distribution differs significantly from that which would be expected if the chance of a head or a tail is one-half).

21. (*a*) $\chi^2_{22} = 20.63$, not significant; (*b*) $\chi^2_{33} = 34.14$, not significant.

22. $\chi^2_1 = 18.63$, reject the null hypothesis. Smoking and drinking moderately covary, with about a 15% difference in smoking between drinkers and nondrinkers, and a 15% difference in drinking between smokers and nonsmokers.

23. $\chi^2_4 = 26.40$; reject the null hypothesis. The less confidence in Congress, the more likely respondents are to say their taxes are too high.

24. $\chi^2_1 = 8.72$; reject the null hypothesis. Persons favoring capital punishment are about 8% more likely to oppose gun laws than are persons against the death penalty.

25. $\chi^2_8 = 75.07$; reject the null hypothesis. There is a strong increase in belief in afterlife, from 60.3% of those who never attend church to 96.3% of those who attend church more than once a week.

26. $\chi^2_3 = 17.00$ for 1970 stress and $\chi^2_3 = 8.05$ for 1974 stress by region, both significant at $\alpha = .05$. High-stress cities were found substantially more frequently in the South than in the other three regions.

27. $\chi^2_6 = 18.27$; reject the null hypothesis. High decentralization is more common in nonreformed cities than in cities with more reform characteristics.

28. $\chi^2_9 = 20.41$; reject the null hypothesis. Reform characteristics are more common in the South and Far West, nonreform cities more frequently found in the Northeast and Midwest.

29. $\chi^2_4 = 24.81$; reject the null hypothesis. Cities in the highest crime-rate

category have much higher police expenditures than cities with medium or low crime rates.

30. $\chi^2_7 = 37.61$; reject the null hypothesis. The distribution of municipal functions is not equiprobable.

Answers to Chapter 5 Problems

1. (*a*) 1/365 = .0027; (*b*) 31/365 = .0849; (*c*) 1/7 = .1429.
2. (*a*) .111; (*b*) .476; (*c*) .896.
3. (*a*) 30.0; (*b*) 37.5.
4. (*a*) variance = 80, standard deviation = 8.94; (*b*) variance = 88.75, standard deviation = 9.42.
5. (*a*) .06; (*b*) .84; (*c*) .33; (*d*) .43.
6. Chebycheff's inequality shows that the probabilities are at least .25 and .16, respectively.
7. (*a*) .4418; (*b*) .4991; (*c*) .3389; (*d*) .0910.
8. (*a*) 1.88; (*b*) 1.04; (*c*) 2.06; (*d*) $\pm$ 2.33; (*e*) $\pm$ 1.28; (*f*) $\pm$ 3.30.
9. (*a*) .383; (*b*) .872; (*c*) .980; (*d*) .532.
10. (*a*) $\mu_{\bar{Y}} = 12.5$, $\sigma_{\bar{Y}} = .89$; (*b*) $\mu_{\bar{Y}} = 40$, $\sigma_{\bar{Y}} = 1.00$; (*c*) $\mu_{\bar{Y}} = 0$, $\sigma_{\bar{Y}} = .45$; (*d*) $\mu_{\bar{Y}} = 14$, $\sigma_{\bar{Y}} = 1.41$; (*e*) $\mu_{\bar{Y}} = 200$, $\sigma_{\bar{Y}} = 1.00$.
11. The means of all three sampling distributions are 4.5 hours. Standard errors are .40 for $N = 25$, .28 for $N = 50$, and .20 for $N = 100$.
12. *a*. Standard errors are (1) .56, (2) .32, and (3) .18.
    b. For $N = 100$, the standard error is .25, the Z score is 2.00, and the probability is .023.
13. (*a*) LCL = 45.6, UCL = 104.4; (*b*) LCL = 36.3, UCL = 113.7; (*c*) LCL = −.01, UCL = 7.21; (*d*) LCL = 8.74, UCL = 20.26.
14. Confidence interval is −0.20 to 13.52, which includes zero, so you cannot be 95% confident that the mean population productivity differs from zero.
15. H_0: $\mu_{CG} \leq 100$; H_1: $\mu_{CG} > 100$.
16. H_0: $\mu_{CG} = 100$; H_1: $\mu_{CG} = 110$.
17. $Z = +10.43$, so reject null hypothesis in Problem 15; and $Z = +10.43$, so reject the null hypothesis in Problem 16. But $Z =$

-4.47 for the alternative, so withhold judgment, since neither hypothesis is supported by the data.

18. (*a*) .49; (*b*) .495; (*c*) .495; (*d*) .475; (*e*) .495; (*f*) .495.

19. (*a*) 1.80; (*b*) 2.72; (*c*) 1.73; (*d*) ± 2.10; (*e*) 2.4; (*f*) ± 2.02; (*g*) ± 5.41; (*h*) ± 3.29.

20. (*a*) .055; (*b*) .05; (*c*) .035; (*d*) .055.

21. (*a*) -5.60; (*b*) $+4.71$; (*c*) -1.59.

22. 6.85 (7) subjects.

23. (*a*) $LCL_{95} = 81.83$, $UCL_{95} = 98.17$; (*b*) $LCL_{99} = 79.0$, $UCL_{99} = 101.0$.

24. With $df = 14$, the c.v. of t for $\alpha = .01$, one-tailed, is $+2.62$. The observed value is $t_{14} = +3.87$, so you reject the null hypothesis of no increase in reading readiness above the minimum level.

25. With $df = 26$, the observed value is $t = -1.15$, not sufficient to reject the null hypothesis that the population mean score is 8.20.

26.

No. of Children	*p*
0	.266
1	.161
2	.245
3	.145
4	.082
5	.044
6	.025
7	.012
8	.020
Total	1.010

Probability of four or more children is .183.

27. $Z = -1.248$ for 8 years or less. Area in tail of the normal distribution is .106.

28. $Z = -19.45$; reject null hypothesis at $\alpha = .05$, one-tailed. Blacks in the population feel too little is being done to improve the conditions of blacks.

29. $Z = 7.29$; reject null hypothesis at $\alpha = .001$. Population mean is greater than 2.5.

30. $Z = 0.54$; do not reject the hypothesis that the population mean is 20 hours or less.

31. $t_{62} = 3.75$; reject the null hypothesis.

32. $t_{63} = 2.49$; do not reject H_0 for a two-tailed test.

33. (*a*) 1.03; (*b*) 6.95; (*c*) 4.05; (*d*) 0.42.

34. $t_{62} = -2.05$; reject H_0.

35. (*a*) $LCL_{95} = 78.08$, $UCL_{95} = 98.04$; (*b*) $LCL_{99} = 74.79$, $UCL_{99} = 101.34$. A population mean of 75 years falls only inside the 99% confidence level.

Answers to Chapter 6 Problems

1. H_0: $\mu_N - \mu_L \geq 0$ and H_1: $\mu_N - \mu_L < 0$, where N = nonlocal decision-making groups, L = local groups. A one-tailed test is required.

2. H_0: $\mu_W - \mu_B \leq 0$ and H_1: $\mu_W - \mu_B > 0$, where W = whites, B = blacks. A one-tailed test is required.

3. (*a*) .85; (*b*) .33; (*c*) 1.20; (*d*) 1.10.

4. (*a*) $Z = -1.12$, do not reject H_0; (*b*) $Z = -3.54$, reject H_0; (*c*) $Z = 2.37$, do not reject H_0; (*d*) $Z = -6.45$, reject H_0.

5. $Z = 1.649$, do not reject H_0.

6. $Z = 5.31$, reject H_0.

7. $Z = \pm 2.33$.

8. $Z = -2.33$.

9. $df = 27$.

10. (*a*) $t = 1.78$; (*b*) $Z = \pm 1.96$; (*c*) $t = \pm 2.75$; (*d*) $t = 2.51$; (*e*) $t = \pm 4.02$.

11. $t_{11} = +2.68$, significant at $\alpha = .05$, one-tailed. Hypothesis is supported.

12. $LCL_{95} = 0.77$, $UCL_{95} = 7.85$; $LCL_{99} = -0.68$, $UCL_{99} = 9.30$.

13. (*a*) $LCL_{95} = -0.01$, $UCL_{95} = 0.07$; $LCL_{99} = -0.02$, $UCL_{99} = 0.08$; (*b*) $LCL_{95} = 0.05$, $UCL_{95} = 0.11$; $LCL_{99} = 0.04$, $UCL_{99} = 0.12$.

14. $t = 3.53$, c.v. = 2.53 for $df = 20$; reject null hypothesis.

15. $t = 2.03$, c.v. = 1.73 for $df = 18$; reject null hypothesis.

16. 95% confidence interval from -1.4 to 15.4; therefore, you cannot be 95% confident that the population difference is not zero.

17. 95% confidence interval from 14.1% to 35.9%; 99% confidence

interval from 10.7% to 39.3%. Since the interval does not contain zero, you can be 99% confident that upperclassmen are more in favor of abolition.

18. $t = 1.50$, while the c.v. $= \pm 1.96$. Therefore do not reject the null hypothesis.

19. Diagram not shown.

20. Diagram not shown.

21. $t_{838} = -7.64$. Ford voters are more conservative than Carter voters.

22. $t_{554} = -7.01$. Central-city residents are more afraid than rural residents.

23. $t_{1396} = -2.64$. Those who have received government aid are more likely to think too little is being spent on welfare.

24. $t_{1444} = -0.08$. No significant differences in TV viewing.

25. $t_{1415} = -3.33$. Whites are more likely than blacks to oppose interracial marriage.

26. $t_{61} = -1.75$. Cities having all three reform characteristics have *lower* per capita expenditures. Hypothesis is incorrect as stated.

27. $t_{61} = -2.23$. Cities having all three reform characteristics have *fewer* government employees per 1,000 residents. Hypothesis is incorrect as stated.

28. $t_{49} = -1.87$. South has less centralized decision making than other regions.

29. $t_{58} = -0.07$. No significant difference in birth rates.

30. $t_{61} = 6.90$. Per capita income is significantly higher in cities with larger percentages of white-collar workers.

Answers to Chapter 7 Problems

1. H_0: $\mu_{HiI,\,LoS} = \mu_{HiI,HiS} = \mu_{LoI,\,HiS} = \mu_{LoI,LoS}$; H_1: $\mu_{HiI,\,LoS} > \mu_{HiI,HiS} = \mu_{LoI,\,HiS} = \mu_{LoI,LoS}$.

2. H_0: $\mu_{teens} = \mu_{20s} = \mu_{30s}$; H_1: $\mu_{teens} > \mu_{20s} = \mu_{30s}$.

3. $\alpha_{CATHO} = 3.1$; $\alpha_{PROT} = 0.5$; $\alpha_{JEW} = -0.5$; $\alpha_{OTHER} = -2.4$.

4. $\alpha_{control} = 2.1$; $\alpha_{lo\ dose} = 1.9$; $\alpha_{hi\ dose} = -4.0$.

5. (*a*) $F_{3,56} = 2.76$; (*b*) $F_{5,18} = 2.77$; (*c*) $F_{2,27} = 5.49$; (*d*) $F_{6,33} = 3.47$; (*e*) $F_{11,188} = 2.35$.

6. (*a*) $F_{2,19} = 5.93$; (*b*) $F_{3,76} = 2.76$; (*c*) $F_{1,368} = 10.83$.
7. 7.0.
8. 11.33.
9. 15.43.
10. 28.92.
11. 2,712.60.
12. $SS_{TOTAL} = 119.733$; $SS_{BETWEEN} = 97.733$; and $SS_{WITHIN} = 22.000$. $MS_{BETWEEN} = 48.867$ and $MS_{WITHIN} = 1.833$. $F_{2,12} = 26.7$. Since c.v. = 3.89 for $\alpha = .05$, reject the null hypothesis.
13. 0.816
14.

Source	*SS*	*df*	*MS*	*F*
Between	76.55	3	25.517	6.38
Within	64.00	16	4.000	
Total	140.55	19		

Since the c.v. for 3 and 16 *df* is 3.24 and $F = 6.38$, reject the null hypothesis.

15. $\psi = (-\frac{1}{3})\mu_A + (-\frac{1}{3})\mu_B + (1)\mu_C + (-\frac{1}{3})\mu_D$, and $t = 4.086/1.033 = 3.960$. Since c.v. $= \sqrt{(3)(3.24)} = 3.118$, reject the null hypothesis and conclude that method C is superior to the other three.

16.

Source	*SS*	*df*	*MS*	*F*
Between	43.515	2	21.76	6.98
Within	65.443	21	3.12	
Total	108.958	23		

Since the c.v. for 2 and 21 degrees of freedom is 3.47, reject the null hypothesis.

17. $\hat{\psi} = (1)6.444 + (-\frac{1}{2})3.750 + (-\frac{1}{2})3.571 = 2.784$

$$\hat{\sigma}^2_{\hat{\psi}} = 3.12\left(\frac{1^2}{9} + \frac{(-\frac{1}{2})^2}{8} + \frac{(-\frac{1}{2})^2}{7}\right) = 0.556$$

$$t = \frac{2.784}{0.745} = 3.735.$$

Since c.v. $= \sqrt{2(3.47)} = 2.63$, the null hypothesis that reinforcement method A is not superior to B and C can be rejected.

18. 39.9%.

19. a.

Source	SS	df	MS	F
Between	315.68	6	52.61	2.21
Within	1427.32	60	23.79	
Total	1743.00	66		

b. Since c.v. = 3.12, do not reject the null hypothesis.
c. $\hat{\eta}^2 = .181$.

20.

Source	SS	df	MS	F
Between	25.0	1	25.00	3.24
Within	108.0	14	7.71	
Total	133.0	15		

Since, for $\alpha = .05$, c.v. = 4.60, the null hypothesis cannot be rejected.

21. $t_{14} = 1.800$ and $t_{14}^2 = 3.24$.

22. Diagram not shown.

23. (a) $F_{4,1461} = 16.42$, reject the null hypothesis; (b1) $t_{1461} = -2.447$, reject the null hypothesis; (b2) $t_{1461} = 6.35$, reject the null hypothesis; (c) Since age and marital status are probably related (those never married are younger, those widowed are older), the results could be due to age rather than marital status.

24. $F_{1,1450} = 13.27$; reject the null hypothesis. Clearly blacks are less satisfied with where they live than whites are.

25. $F_{3,532} = 2.95$, not significant. $\hat{\psi} = -.325$, $\hat{\sigma}_{\hat{\psi}} = 0.254$, $t_{532} = -1.28$, not significant. You cannot reject either of the nullhypotheses.

26. a. $F_{4,1380} = 14.41$; reject the null hypothesis.
b. The pattern of means suggests that the higher the family income when a 16-year-old, the higher the current family income. An exception is the category "far above average," which has only 19 respondents.
c. $\hat{\eta}^2 = .04$. 4% of the variance in current income can be explained by family income when the respondent was 16 years old.

27. (a) $F_{4,1389} = 62.86$, reject the null hypothesis; (b) The pattern of means suggests that means increase monotonically the higher the degree completed; (c) $\hat{\eta}^2 = .153$, or 15.3% of the variance in income can be explained by educational degree held.

28. (a) $F_{3,59} = 9.12$, reject the null hypothesis; (b) $t_{59} = 4.79$, reject the

null hypothesis. The West has more in nonmanual occupations than the other three regions.

29. (*a*) $F_{3,59} = 6.09$, reject the null hypothesis; (*b*) $t_{59} = 2.90$. Reject the null hypothesis. Even though the contrast is significant, the means indicate that the difference is due primarily to the age of the cities in the Northeast.

30. $F_{7,54} = 21.36$; reject the null hypothesis. The means suggest that, in general, as the number of functions increases, so do expenditures.

31. $F_{3,59} = 2.42$; you cannot reject the null hypothesis.

32. (*a*) $F_{9,53} = 45.77$; reject the null hypothesis, since the pattern of means clearly supports the research hypothesis; (*b*) $\hat{\eta}^2 = .886$, a very strong relationship.

Answers to Chapter 8 Problems

1. Scatterplot not shown. Relationship is that with larger income, fewer children.

2. (*a*) $\hat{Y}_i = 6.896 - 0.322X_i$; (*b*) 2.07 children; (*c*) -15.00.

3. *a.*

Person	e_i	Person	e_i
1	1.15	6	−0.68
2	−1.42	7	−0.96
3	0.54	8	0.68
4	−0.34	9	1.04
5	−1.35	10	1.39

b. $\Sigma e_i^2 = 10.47$.

4. (*a*) 3.5 per 100,000; (*b*) between 17.65 to 35.15 suicides per 100,000; (*c*) 2.74% unemployment.

5. An SAT of 450 or higher.

6. $\hat{Y}_i = -40 + 4X_i$.

7. $\hat{Y}_i = 16.86 + .075X_i$.

8. (*a*) 728; (*b*) 3.53; (*c*) 1,954.74; (*d*) 74.5.

9. (*a*) 54.000; (*b*) 43.524; (*c*) 10.48; (*d*) 0.806.

10. (*a*) 5,105.7; (*b*) 8,844.3; (*c*) 0.366.

11. (*a*) 0.64; (*b*) 0.36; (*c*) 0.80.

12. $r = 0.54$.

13. Given that $b_{YX} = r_{YX}\frac{s_Y}{s_X}$ and $b_{XY} = r_{XY}\frac{s_X}{s_Y}$ for $b_{YX} = b_{XY}$,

then $r_{YX}\frac{s_Y}{s_X} = r_{XY}\frac{s_X}{s_Y}$, or $\frac{s_Y}{s_X} = \frac{s_X}{s_Y}$. That is, $s_Y^2 = s_X^2$,

which is not always the case.

14. (*a*) 0.38; (*b*) 0.30; (*c*) −0.75; (*d*) −0.60; (*e*) −0.168; (*f*) 0.267.

15. $\beta^* = 0.70$.

16. (*a*) 0.585; (*b*) 0.536; (*c*) −0.333; (*d*) 0.800; (*e*) −0.269.

17.

Source	df	SS	MS	F
Regression	1	1,100	1,100	4.40*
Error	48	12,000	250	
Total	49	13,100		

*Significant for $\alpha = .05$

18. $F_{1,8} = 33.24$; reject H_0 for $\alpha = .001$.

19. (*a*) 4.22, do not reject H_0; (*b*) 44.22, reject H_0; (*c*) 13.45, reject H_0; (*d*) 6.00, do not reject H_0; (*e*) 7.73, do not reject H_0.

20. (*a*) 0.0559; (*b*) 5.76; (*c*) Reject H_0.

21. $LCL_{95} = 1.92$, $UCL_{95} = 7.64$; $LCL_{99} = 1.01$, $UCL_{99} = 8.55$. You can be 95% confident but not 99% confident that $\beta \neq 8.00$.

22. $R^2 = 0.069$; $F = 1.86$; $df = 1,25$. Not significantly different from zero for $\alpha = .05$.

23. (*a*) $t_{23} = 1.61$, do not reject H_0; (*b*) $t_{35} = 6.02$, reject H_0; (*c*) $t_{98} = 2.39$, reject H_0; (*d*) $t_{46} = 3.01$, reject H_0.

24. (*a*) $LCL_{95} = -0.54$, $UCL_{95} = 4.34$; (*b*) $LCL_{95} = 0.37$, $UCL_{95} = 0.74$; (*c*) $LCL_{95} = 0.07$, $UCL_{95} = 0.79$; (*d*) $LCL_{95} = 2.05$, $UCL_{95} = 10.35$.

25. (*a*) $Z = 0.61$, do not reject H_0; (*b*) $Z = 2.19$, reject H_0; (*c*) $Z = 2.66$, reject H_0; (*d*) $Z = -8.81$, reject H_0; (*e*) $Z = -6.88$, reject H_0.

26. $F_{1,59} = 5.28$; reject H_0.

27. $F_{4,78} = 2.79$; significant at $\alpha = .05$.

28. $t_{79} = 4.00$; significant at $\alpha = .05$.

29. (*a*) $t_{34} = -3.86$; (*b*) $t_{46} = 6.48$; (*c*) $t_{21} = 2.24$; (*d*) $t_{69} = -9.56$.

30. $Z = 0.96$; do not reject H_0.

31. *a.* $\hat{Y}_i = 3.999 - 0.095X_i$; $R^2 = 0.022$.

b. $\hat{Y}_i = 3.479 - 0.014X_i$; $R^2 = 0.009$.
c. $\hat{Y}_i = 3.821 - 0.093X_i$; $R^2 = 0.021$.
The three negative coefficients show that as socioeconomic variables increase, the number of children ever born to women now over 40 decreases. However, the coefficient for spouse's occupational prestige is not statistically significant at $\alpha = .05$, and the other two variables each explain only 2% of the variance.

32. $\hat{Y}_i = 2.688 - 0.00007X_i$; $R^2 = 0.0047$. As expected, an increase in city size leads to a decrease in satisfaction, significant at $\alpha = .01$.

33. For men: $\hat{Y}_i = 1.941 - 0.000011X_i$; $R^2 = 0.013$; $r = -0.115$.
For women: $\hat{Y}_i = 1.854 - 0.000015X_i$; $R^2 = 0.013$; $r = -.114$.
The negative coefficients for income effect on job satisfaction mean that the more money a person earns, the lower the *dis*satisfaction. For every $10,000, men's dissatisfaction scores drop by 0.11 points and women's scores drop by 0.15 points. Correlation coefficients are virtually identical, meaning no significant differences between genders in the size of the covariations.

34. For XMARSEX: $\hat{Y}_i = 1.810 - 0.086X_i$; $R^2 = 0.075$.
For HOMOSEX: $\hat{Y}_i = 2.075 - 0.113X_i$; $R^2 = .071$.
Both relationships significant at $\alpha = .001$.

35. $\hat{\eta}^2 = 0.0124$; $R^2 = 0.00017$; $F_{6,1444} = 2.98$; significant for $\alpha = .01$.

36. Scatter diagram not shown. $\hat{Y}_i = 241.01 - 1.39X_i$; $R^2 = 0.069$.

37. For every 10 years of CITYAGE, HOUSEPCT increases by 3.63%.

38. $r = -0.395$.

39. Non-Southern, $r = 0.02$, $N = 47$; Southern, $r = -0.03$, $N = 16$. No significant difference.

40. $\hat{Y}_i = -35.86 + 1.98X_i$; $R^2 = 0.121$. Hypothesis is supported at $\alpha = .01$ that cities with a larger percentage of women in the labor force have higher crime rates.

Answers to Chapter 9 Problems

1. (*a*) 0; (*b*) $\chi^2_4 = 27.24$; (*c*) $V = 0.138$, $C = 0.136$. Even though lambda is zero, χ^2, V, and C are not zero. Percentaging the table suggests that Catholics are slightly less likely to endorse abortion (76.3%) than Protestants (85.1%), Jews (100%), nones (91.2%) and others (92.9%).

2. (*a*) $\lambda = 0.652$, suggesting a strong relationship between current religious identification and the religion in which one was raised; (*b*) yes, since the c.v. = 39.25.

3. (*a*) Far West; (*b*) $\lambda = 0.25$; (*c*) $\chi^2_6 = 13.11$: Reject the null hypothesis, since c.v. = 12.59; (*d*) $V = 0.323$ and $C = 0.415$.

4. (*a*) those never married; (b) $\lambda = 0.056$; (*c*) yes, since the c.v. = 26.13; (*d*) $V = 0.132$ and $C = 0.183$.

5. (*a*) $\lambda = 0.0$; (*b*) since the c.v. = 26.30, conclude that region of residence and happiness are unrelated.

6. (*a*) $\chi^2_8 = 114.95$ (significant even for $\alpha = .001$), $V = 0.286$, and $C = 0.375$; (*b*) $\lambda = 0.004$; (*c*) percentaged table not shown; (*d*) While all religious groups favor local option over the other two response categories, Protestants and Catholics are more likely to favor required prayer than Jews and those of no religious identification, accounting for a weak relationship between the two variables.

7. *a.* As educational degree held increases, objection to bringing a black home for dinner decreases.
   b. $\gamma = 0.398$.
   c. Reject null hypothesis, since $Z = 5.86$ and c.v. = 2.33.

8. (*a*) $G = 0.412$; (*b*) 0.040; (*c*) reject the null hypothesis, since c.v. = 32.91.

9. $\tau_c = 0.1646$. $Z = 7.12$, far greater than the c.v. of 3.08, so reject the null hypothesis.

10. $\hat{d}_{yx} = 0.173$. $Z = 6.83$, far greater than the c.v. of 2.33, so reject the null hypothesis.

11. (*a*) $r_s = +0.464$; (*b*) $t_8 = +1.48$: Since c.v. = ± 2.31, do not reject the null hypothesis.

12. *a.* $r_s = 0.30$.
    b. $t_8 = 0.89$: Since c.v. = ± 2.31, do not reject the null hypothesis.
    c. The difference may be due to the fact that each sample contained only 10 cities. A better estimate of the relationship would be obtained using the data from *all* 63 cities.

13. (*a*) $\chi^2_1 = 28.80$; (*b*) $Q = -1.00$; (*c*) $\phi = -0.80$.

14. (*a*) $\chi^2_1 = 6.13$; (*b*) $Q = -1.00$; (*c*) $\phi = -0.312$, $\phi_{max} = 0.312$.

15. (*a*) $Q = 0.596$, $\hat{\sigma}_Q = 0.213$, and $Z = 2.800$: Since c.v. = 1.65, reject the null hypothesis; (*b*) $\phi = 0.272$.

16. (*a*) 0.096; (*b*) 0.002; (*c*) 0.185.

17. 200.

18. (*a*) 0.212; (*b*) 0.287.

19. (*a*) 0.40; (*b*) 0.371.

20. $\chi^2_{22} = 20.63$, $\lambda = 0.0$, $V = 0.084$, and $C = 0.118$. Since the c.v. for chi square with $df = 22$ is 40.29, the null hypothesis cannot be rejected. Astrological sign and happiness are not related.

21. $\chi^2_{33} = 34.14$, $\lambda = 0.001$, $V = 0.008$, and $C = 0.151$. Since the c.v. for chi square with $df = 33$ is 50.89, the null hypothesis cannot be rejected. Astrological sign is not related to health status, either.

22. $\chi^2_4 = 5.84$, $\lambda = 0$, $V = 0.063$, and $C = 0.063$. Since the c.v. for a chi square with $df = 4$ is 13.28, we cannot reject the null hypothesis. Catholics appear not to divorce any more or any less than other religious groups.

23. $G = 0.133$, $t_b = 0.091$, and $\hat{d}_{yx} = 0.081$. These results suggest a weak but statistically significant relationship. However, the direction is opposite from what might have been predicted. That is, the more urban one is, the more likely one is to believe that the government is spending too little on protecting the environment.

24. $G = 0.019$, $t_b = 0.014$, and $\hat{d}_{yx} = 0.013$. There appears to be no systematic relationship between political beliefs and attitude about spending levels for space exploration.

25. $G = 0.274$, $t_b = 0.186$, and $\hat{d}_{yx} = 0.167$. In this case, those who are more liberal also support increased spending to protect the environment.

26. $\chi^2_1 = 31.33$ and $\phi = 0.159$. Those who believe in life after death are less likely to favor abortion for a single woman who gets pregnant.

27. $\chi^2_1 = 9.58$ and $\phi = 0.089$. While the relationship is statistically significant, there is very little difference on attitude toward abortion when the child may have a serious defect as a function of belief in life after death.

28. $\chi^2_6 = 16.11$, $V = 0.358$, $C = 0.451$, and $\lambda = 0.167$. Since the c.v. for a χ^2 with 6 *df* is 12.59, we can reject the null hypothesis. The results suggest that the South has the lowest per capita income and the Far West the highest.

29. $\chi^2_6 = 13.11$, $V = 0.323$, $C = 0.415$, and $\lambda = 0.250$. Since the c.v. for a χ^2 with 6 *df* is 12.59, we can just barely reject the null hypothesis. The results suggest that the cities in the Far West have the highest crime rates and those in the Northeast the lowest.

30. $G = 0.747$, $t_b = 0.553$, and $\hat{d}_{yx} = 0.557$. Clearly, as per capita income increases in cities, so does the percentage of white-collar workers in the labor force.

31. $G = 0.068$, $t_b = 0.047$, and $\hat{d}_{yx} = 0.047$. There is no systematic relationship between per capita income and crime rate in the 63-cities data set.

32. $\chi_1^2 = 4.53$, $G = 1.000$, $\phi = 0.312$. Since the c.v. for χ^2 with 1 *df* is 3.84, we can reject the null hypothesis. Income is substantially higher in the West than in the other regions of the country.

33. $\chi_1^2 = 3.45$, $G = .596$, and $\phi = 0.272$. Even though the chi square is not significant, G and ϕ are significant, suggesting that the Far West has gained more in population than the other regions of the country.

Answers to Chapter 10 Problems

1. These hypotheses suggest a causal sequence in which economic instability (in the form of periodic high unemployment) interprets the relationship between a single-product export economy and regime repression. Diagram not shown.

2. Older persons, from earlier generations, probably have had less formal education than younger persons. But older persons may also tend to be more conservative than younger persons, both because of an aging process that can produce greater caution and a higher stake in the status quo, and because of a generational difference in political orientations that could dispose older persons to be more conservative in their outlook. If these assumptions are correct, holding age constant would cause the correlation between education and conservatism to fall to zero.

3. Examples of such relationships would be: (*a*) as more women enter the labor force, they become less financially dependent on their husbands and thus less willing to remain in a troubled marriage; (*b*) as more women enter the labor force, they have fewer children and are thus not so strongly tied to their spouses; (*c*) as women enter the labor force, they may perceive their sex role as demanding more autonomy and equality, thus putting strains on a marriage in which traditional husband-wife relations cannot be maintained. Many other intervening variables are plausible candidates to interpret the causal relation.

4. Diagrams not shown.

5. An interaction of age and type of work on response to most important job characteristic. Diagram not shown.

6. The correlation in the zero-order table is $r_{XY} = -0.23$, with women having lower wages. Controlling for work status provides a partial explanation for this relationship, since it is reduced to $r_{XY} = -0.21$ for full-time work and $r_{XY} = -0.12$ for part-time work. These results also suggest an interaction effect; the relationship between gender and wages is stronger for full-time than for part-time work.

7. The hypothesis is correct, the zero-order $r_{XY} = -0.21$ (women attend less than men do), but controlling for belief that films lead to moral breakdown, the gender differences in attendance vanish (partial $r_{XY} = +0.02$ and $r_{XY} = -0.03$, for "no" and "yes" levels of the control variable, respectively). Thus, holding constant the third variable provides a complete explanation of the original covariation.

8. Age and opinion zero order $r_{XY} = -0.07$; highly religious subtable $r_{XY} = -0.01$; not highly religious subtable (not shown) $r_{XY} = -0.33$. An interaction effect.

9. The correlation between religiosity and opinion is 0.18, and that between religiosity and age is -0.24.

10. The zero-order $r_{XY} = 0.17$ does not disappear in the subtables, as it would if attitude totally explained the correlation of proximity and behavior. Instead, an interaction occurs, with $r_{XY} = 0.11$ among those who say that blacks don't push themselves, and $r_{XY} = 0.27$ among those who say they do.

11. The results show an even stronger interaction, with proximity having a greater impact on behavior among supporters of increased spending ($r_{XY} = 0.33$) but much less covariation of the two variables among respondents against more spending ($r_{XY} = 0.13$).

12. For a large minority, $r_{XY} = 0.17$, for a small minority, $r_{XY} = 0.38$, consistent with the hypothesis. Note also that the percentage differences are of the same magnitude (16.7% difference in the first partial table, 37.5% difference in the second). As the zero-order table has $r_{XY} = 0.29$, holding miniority group size constant shows a partial interaction effect.

13. The correlation of X with Y must be negative. To completely explain the relationship, the correlation must equal -0.22.

14. (*a*) 0.03; (*b*) -0.41; (*c*) 0.46; (*d*) -0.90.

15. (*a*) -0.18; (*b*) -0.02; (*c*) 0.16; (*d*) 0.18.

16. (*a*) $t_{20} = -1.95$, do not reject H_0; (*b*) $t_{197} = 1.70$, do not reject H_0; (*c*) $t_{35} = 2.07$, do not reject H_0; (*d*) $t_{11} = 2.07$, do not reject H_0.

17. (*a*) .09, 25.0%; (*b*) .09, 56.3%; (*c*) .04, 44.4%; (*d*) .64, 100.0%.

18. (*a*) (1) 0.42, (2) 0.26 and 0.65, (3) 0.40, (4) $t_{252} = 6.93$, reject H_0 that $\rho_{YX \cdot W} \leq 0$; (*b*) interaction effect; the relationship is stronger in rural than in urban areas.

19. (*a*) (1) 0.42, (2) 0.12, 0.28, 0.52, and 0.76, (3) 0.00, (4) 0.00, (5) 0.42, (6) $t_{197} = 6.50$; (*b*) The covariation of social and economic attitudes depends on income level, with the magnitude increasing as income rises. However, because income is uncorrelated with either attitude, holding income constant has no effect on reducing the partial correlation below the zero-order correlation.

20. (*a*) −0.28; (*b*) −0.43 for teens, −0.13 for later; (*c*) −0.23; (*d*) $t_{302} = 4.16$, reject H_0.

21. The more educated are slightly more liberal (38.7%) than the less educated (22.95%), $r_{XY} = 0.13$. Holding age constant, among younger respondents the correlation is unchanged ($r_{XY} = 0.15$), but among older respondents it drops to insignificance ($r_{XY} = 0.05$), an interaction effect.

22. Zero order $r_{XY} = -0.19$; Protestant $r_{XY} = -0.20$; Catholic $r_{XY} = -0.18$. There are virtually no differences in correlation of religious intensity and attitude toward abortion for a poor woman.

23. Zero-order $r_{XY} = 0.22$; low prestige $r_{XY} = 0.16$; high prestige $r_{XY} = 0.13$. Holding occupational prestige constant reduces somewhat the covariation between education and income, in accordance with the status attainment model.

24. Zero-order $r_{XY} = -0.04$; partial, for education $r_{XY} = -0.02$; partial, for prestige $r_{XY} = -0.03$; partial, for income $r_{XY} = -.03$.

25.

Community Size	r_{XY}	*Minority Size*
Large, medium-size city	−0.19	16.9%
Suburb	−0.22	5.8%
Small city, town, village	−0.26	10.2%
Rural area	−0.05	5.7%

The pattern of correlations does not support the hypothesis that the larger the minority group size, the greater the difference in rates of intergroup contact.

26. Zero-order $r = 0.43$; Northeast $r = 0.61$; South $r = 0.36$; Midwest $r = 0.17$; Far West $r = 0.65$. Older cities have larger per capita long-term debt, but the magnitude of the covariation varies widely across regions.

27. Zero-order $r = -0.23$, $p = 0.03$; partial $r = -0.20$, $p = 0.07$.

There is very little reduction in the zero-order correlation of FUNCTION and REFORMGV by holding POPULAT constant.

28. Zero-order $r = -0.23, p = 0.03$; partial $r = 0.04, p = .38$. Holding CITYAGE constant reduces the zero-order correlation to insignificance.

29. Zero-order $r = 0.09$. A substantial interaction occurs when crime rate is held constant: $r = +0.39$ for the low-crime cities and $r = -0.20$ for high-crime cities.

30. Cities that lost population were more likely to have larger black populations (61.9% had 14% or more black) than were cities that gained or did not lose population (45.2% had 14% or more black). An interaction effect occurs upon controlling for level of white-collar labor force. The zero-order $r = -0.08$, but in cities less than half white collar, $r = -0.48$, while in cities with a white-collar majority, $r = +0.16$.

Answers to Chapter 11 Problems

1. $a = 65.524$, $b_1 = 4.258$, and $b_2 = -1.511$.

2. $a = 0.088$, $b_1 = 0.097$, and $b_2 = 0.099$.

3. (*a*) $b_1 = 0.100, b_2 = 0.100$; (*b*) controlling for a third variable makes little difference in both cases; (*c*) the fact that $r_{X_1X_2} = 0.02$, i.e., is nearly zero.

4. (*a*) $a = 2.20, b_1 = .013$, and $b_2 = .093$; (*b*) the zero-order regression coefficients change as the correlation between $r_{X_1X_2}$ increases in value.

5. *a*. $\beta_1^* = 0.681$, $\beta_2^* = -.604$;
   b. $Z_Y = 0.681\ Z_1 - 0.604\ Z_2$.

6. (*a*) $\beta_1^* = 0.291$, $\beta_2^* = .396$; (*b*) $Z_Y = 0.291\ Z_1 + 0.396\ Z_2$.

7. $R^2_{Y \cdot X_1X_2} = 0.582$.

8. $R^2_{Y \cdot X_1X_2} = 0.246$.

9. $Z_Y = -0.51\ Z_1 + 0.72\ Z_2 + 0.67\ Z_3$; X_1 has the smallest effect, X_2 the largest.

10. (*a*) 0.604; (*b*) 0.283; (*c*) 0.599; (*d*) 0.294.

11. $F_{5,221} = 5.63$, c.v. $= 3.02$ for $\alpha = .01$. Therefore, reject the null hypothesis.

12. $F_{2,97} = 67.53$, c.v. $= 3.09$ for $\alpha = .05$. Therefore, reject the null hypothesis.

13. $F_{2,7} = 1.14$, c.v. $= 4.74$ for $\alpha = .05$. Therefore you cannot reject the null hypothesis.

14. For $N = 83$ and $\alpha = .05$, the c.v. $= \pm 1.66$ for a one-tailed test. Since the t ratios equal 1.77 and -1.50, respectively, only the first coefficient is significant.

15. For $\alpha = .05$, the c.v. $= \pm 1.65$ for a one-tailed test. Since the t ratios equal 0.28 and 0.03, respectively, neither coefficient is statistically significant.

16. $t_{97} = 2.18$. The c.v. $= 2.37$ for a one-tailed test, given $\alpha = .05$. You cannot reject the null hypothesis.

17. $t_{47} = 3.45$. The c.v. $= 1.68$ for a one-tailed test, given $\alpha = .05$. Reject the null hypothesis.

18. 1.61 ± 1.46 or LCL $= 0.147$ and UCL $= 3.073$.

19. -2.70 ± 2.10 or LCL $= -4.80$ and UCL $= -0.60$.

20. $r_{YX_1 \cdot X_2} = 0.71$, $r_{YX_2 \cdot X_1} = -0.67$.

21. $r_{YX_1 \cdot X_2} = 0.32$, $r_{YX_2 \cdot X_1} = 0.41$

22. Each year of age decreases the civil liberties index by 0.031 points, each year of education raises the index by 0.123, and living in a rural area decreases the predicted score by 0.546 points relative to living in an urban place.

23.

Category	*D1*	*D2*	*D3*	*D4*	*D5*	*D6*
Married	1	0	0	0	0	0
Widowed	0	1	0	0	0	0
Divorced	0	0	1	0	0	0
Separated	0	0	0	1	0	0
Never married	0	0	0	0	1	0
No answer	0	0	0	0	0	1

24. If D_4 is omitted:

$$\hat{Y} = a + b_1X + b_2D_1 + b_3D_2 + b_4D_3$$

$$\hat{Y} = a + b_1X + b_2D_1 + b_3D_2 + b_4D_3 + b_5XD_1 + b_6XD_2 + b_7XD_3$$

Test if R^2 for the first equation is significantly larger than that for the second one.

25. *a.* $F_{2,304} = 10.64$; c.v. $= 4.61$ at $\alpha = .01$. Reject the null hypothesis.

b. $\hat{Y}_{city} = 2.53$; $\hat{Y}_{town} = 1.31$; $\hat{Y}_{country} = 7.91$.

c. Education decreases liking of country music; persons living in towns like it more than those in cities, and country people like it less than those in towns. The more educated townfolk like it less, while the more educated cityfolk like it more.

26. *a.*

Variable	b_j	t_j	β_j^*
Income	−0.049	6.19*	−0.170
Education	−0.065	8.48*	−0.233
Constant	3.229		

*Significant beyond .01 level, given c.v. = 1.96 for two-tailed test.

b. $R^2 = 0.115$; $F_{2,1389} = 90.30$. Since c.v. = 4.61, reject the null hypothesis.

c. The higher one's income (net of education) and the more education one has (net of income), the less likely one is to report poor health. The net effect of education is slightly greater than that for income.

27. Since the beta weight for education is over five times the size of that for father's occupational prestige (0.5103, 0.09), it is much more important in predicting the respondent's own occupational prestige.

28. *a.* No, since its t ratio is only 1.50 and c.v. = 1.96 for a two-tailed test.

b. $R_3^2 - R_2^2 = 0.0002$; $F_{1,1348} = 0.29$, which is far from statistical significance, since c.v. = 6.63. Therefore we conclude that sex of the respondent does not interact with income in predicting self-reported health status.

29. *a.* Yes, $F_{1,1130} = 8.21$ for race, whereas c.v. = 6.63. The occupational prestige of nonwhites is lower on average than for whites controlling for years of education and father's occupational prestige.

b. No, R_3^2 and R_2^2 are equal to each other to the 5th decimal place; ($F_{1,1129} = 0.0003$).

30. *a.* Alpha = 0.615.

b. $Y = 7.42 - 1.00$ MARRIED $+ 0.89$ SEPARATED $- 0.09$ DIVORCED $- 0.34$ WIDOWED; $R^2 = 0.029$.

c.

Source	SS	df	MS	F
Between	374.11	4	93.53	10.87*
Within	12,455.00	1,448	8.60	
Total	12,829.11	1,452		

*Significant at $\alpha = .01$, since c.v. = 3.32.

$\overline{Y}_{MAR} = 6.42$, $\overline{Y}_{SEP} = 8.31$, $\overline{Y}_{DIV} = 7.33$, $\overline{Y}_{WID} = 7.08$, $\overline{Y}_{NEVMAR} = 7.42$ *d.* $\hat{\eta}^2 = 0.029 = R^2$.

31. For married, $F_{1,1448} = 21.51$; for separated, $F_{1,1448} = 3.66$; for divorced, $F_{1,1448} = .08$; for widowed, $F_{1,1448} = 1.31$. Since c.v. = 3.84, only the mean of the married group differs significantly from that of the never married group. Those who never married report being less happy.

32. *a.* Alpha = .905.
b. $\hat{Y} = 3.62 + .94$ Protestant $- .24$ Jewish $+ 1.21$ Catholic; $R^2 = 0.067$.

c.

Source	SS	df	MS	F
Between	170.98	3	56.99	32.24*
Within	2,382.72	1,348	1.77	
Total	2,553.70	1,351		

*Significant at $\alpha = .01$ level, since c.v. = 3.32.

$\overline{Y}_{PROT} = 4.56$, $\overline{Y}_{CATH} = 4.83$, $\overline{Y}_{JEW} = 3.38$, $\overline{Y}_{NONE} = 3.62$
d. $\hat{\eta}^2 = 0.067 = R^2$.

33. For Protestants, $F_{1,1348} = 53.82$; for Catholics, $F_{1,1348} = 74.64$; for Jews, $F_{1,1348} = 0.76$. Since c.v. = 3.84, the means of the Protestants and Catholics are significantly different from that of the "none" group.

34. Table is same as answer to 30*c*.

35. Table is same as answer to 32*c*.

36. *a.* $\hat{Y} = 34.84 - 0.036$ EXPENDIT $+ 3.649$ FUNCTION; $R_1^2 = 0.2810$; $F_{2,59} = 11.53$. Reject null hypothesis, since c.v. = 3.15 for $\alpha = .05$, $F_{1,59} = 1.99$ for EXPENDIT and $F_{1,59} = 3.19$ for FUNCTION. Since c.v. = 3.84, neither is significantly different from zero.
b. $R_2^2 = 3.122$; $R_2^2 - R_1^2 = .0312$; and $F_{2,57} = 1.29$. Since c.v. = 3.16, the null hypothesis cannot be rejected; that is, POPULAT and BLACKPCT do not add significantly to variance explained in HOUSEPCT.

37. $Z_Y = 0.089$ FUNCTION $+ 0.235$ CITYAGE $- 0.282$ WCPCT. Only WCPCT has a significant effect. That is, as percent of white-collar workers in the labor force increases, centralized decision making decreases.

38. *a.* $\hat{Y} = 93.60 + 10.12$ FUNCTION $+ 2.04$ CITYAGE. $F_{1,54} =$

14.82 for city age and $F_{1,59} = 0.06$ for number of functions performed. Therefore city age is significantly related to fiscal status. For each additional year of age the stress index increases 2.04 points, on average. $R_1^2 = 0.2643$, with $F_{2,59} = 10.60$. Since c.v. = 3.15, the null hypothesis can be rejected.

b. $R_2^2 = 0.2771$ and $R_2^2 - R_1^2 = 0.0128$, with $F_{1,58} = 1.03$. Since c.v. = 3.84, we conclude, in explaining fiscal stress in 1970, that there is not an interaction between age of the city and the number of functions it performs.

39. *a.* $\hat{Y} = 60.14 - 9.81$ N.E. $- 7.89$ SOUTH $- 12.53$ MIDWEST.

b.

Source	SS	df	MS	F
Between	1,314.56	3	438.19	9.11*
Within	2,836.33	59	48.07	
Total	4,150.89	62		

*Significant at $\alpha = .05$, since c.v. = 2.76.

$\overline{Y}_{N.E.} = 50.33$, $\overline{Y}_{SOUTH} = 52.25$, $\overline{Y}_{MIDWEST} = 47.61$, $\overline{Y}_{FARWEST} = 60.14$

c. $\hat{\eta}^2 = 0.317 = R^2$.

d. Yes, more than all of them. The F tests (14.49, 9.68, and 25.7) for the Northeast, the Midwest, and the South are all significant, since c.v. = 3.84.

40. *a.* R^2 without city age is 0.3167; with it, 0.3315. $R_2^2 - R_1^2 = 0.0148$ and $F_{1,58} = 1.28$, which is not significant since c.v. = 3.84.

b. R^2 with the interaction terms is 0.3610. $R_3^2 - R_2^2 = 0.0295$ and $F_{3,55} = 0.85$ is not significant, since c.v. = 2.78.

41. Table is same as that for Problem 39*b*.

Answers to Chapter 12 Problems

1. The time order and hence causal ordering between the two attitudes is unclear.
2. An example: "A greater societal discrepancy between expected and actual economic well-being causes increased social turmoil."
3. Effect of X on Q cannot be determined without knowing actual magnitudes, since X indirectly increases W and directly decreases it, in turn affecting the direction of change in Q.

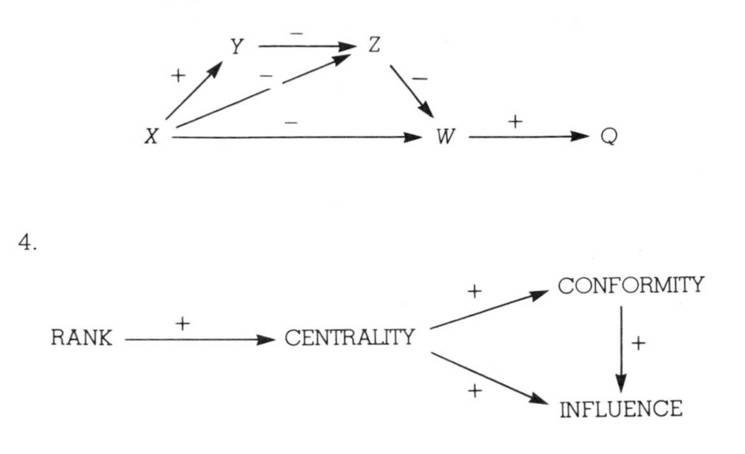

5. $r'_{HJ} = p_{HJ} - p_{HK}\, r_{JK}$

6. (*a*) $r'_{BD} = p_{DC}\, p_{CB} + p_{DC}\, p_{CA}\, r_{BA} + p_{DA}\, r_{BA}$; (*b*) there are none.

7. (*a*) L has a larger indirect effect on H (0.15) than does K (0.10); (*b*) 38% of variance in J is explained.

8. (*a*) B has a larger indirect effect on D (.36) than does A (.30); (*b*) 36% of the variance in D is explained.

9. $r' = 0.50$.

10. $r'_{t,t+2} = 0.56$

11. $r'_{DB} = p_{DC}\, p_{CB} + p_{DC}\, p_{CA}\, r_{AB} + p_{DA}\, r_{AB}$; first term is causal, last two terms are correlated effects.

12.

	y_1	y_2	y_3	y_4	y_5
y_1	1.00	0.25	0.13	0.06	0.03
y_2		1.00	0.25	0.13	0.06
y_3			1.00	0.25	0.13
y_4				1.00	0.25
y_5					1.00

Note: Since the matrix is symmetric, only the upper half is tabled.

13. (*a*) $r'_{BD} = 0.41$; (*b*) $r'_{AD} = 0.56$.

14. (*a*) proof not shown; (*b*) $r'_{UV} = 0.03$.

15.

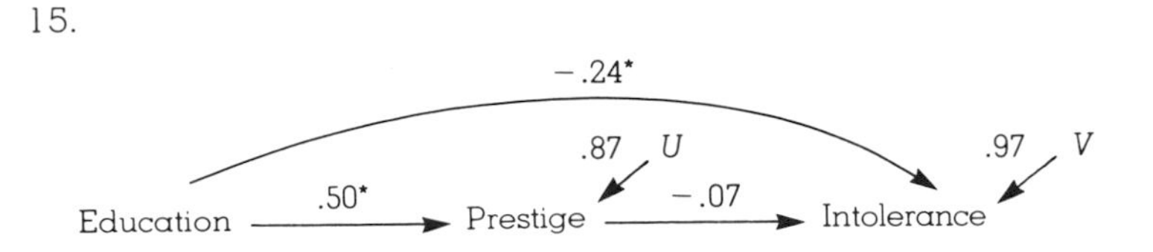

*Significant at or beyond α = .001 level.
The results suggest strongly that education and not occupational prestige accounts for such intolerance.

16.

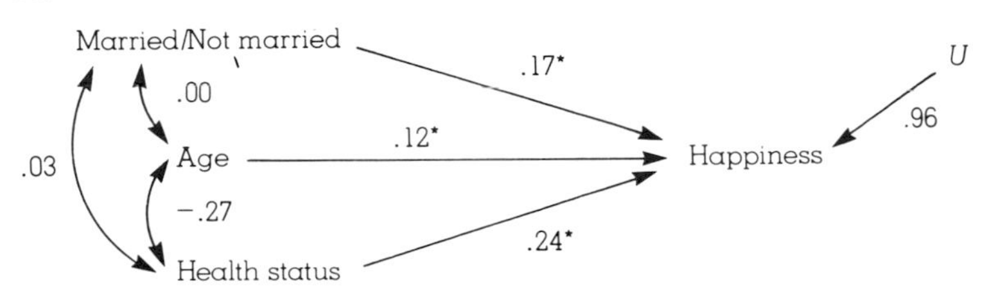

*Statistically significant beyond the α = .001 level.

17.

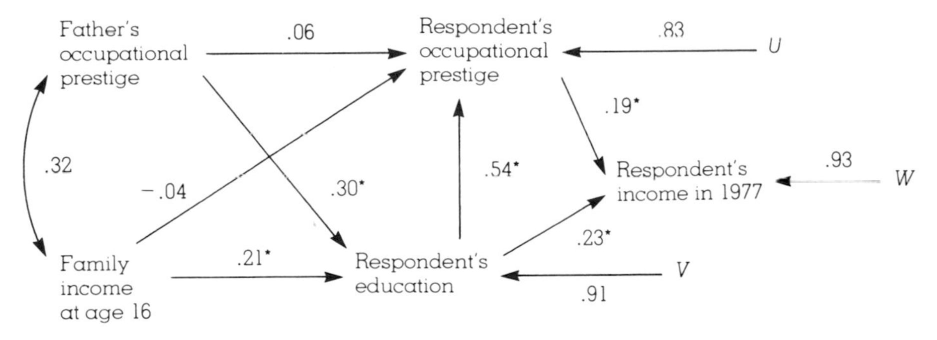

*Significant at or beyond the α = .001 level.
Respondent's education and occupational prestige both determine income, although education is the slightly more important variable. Occupational prestige is heavily influenced by educational attainment and not affected by father's occupational prestige or family income when the respondent was growing up. Educational attainment is determined both by father's occupational prestige and family income when the respondent

was growing up, although the former is more important. More variance can be explained in the respondent's occupational prestige level than income level. You might test an alternative model to see if family income at age 16 has a direct effect on respondent's income.

18. Path coefficient from BLACKPCT to POLICEXP is +0.34, from CRIMRATE to POLICEXP is 0.40.

19.

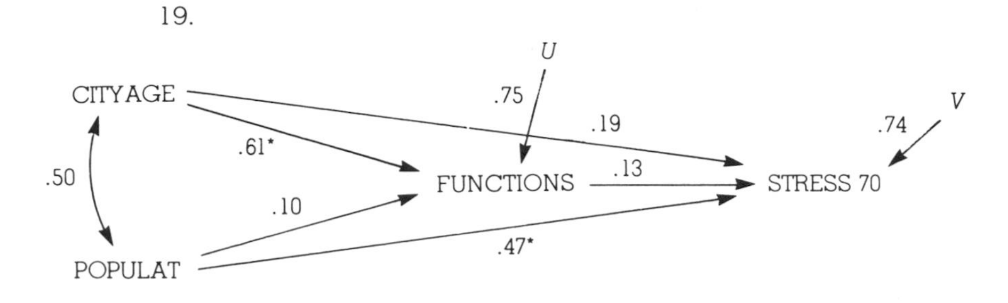

*Significant at or beyond the $\alpha = .001$ level.

20.

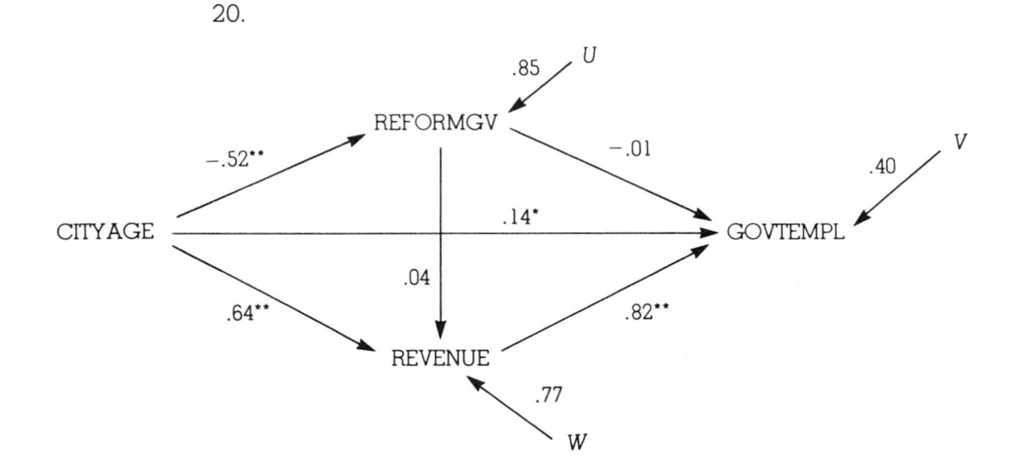

*Significant at $\alpha = .05$ level.
**Significant at or beyond the $\alpha = .001$ level.

Index

THE BOOK MANUFACTURE

Statistics for Social Data Analysis was typeset at Com Com, Allentown, Pennsylvania. Printing and binding was at Kingsport Press, Inc., Kingsport, Tennessee. Cover design was by Willis Proudfoot. The typeface is Stymie Light.